ACTIVITIES TO ACCOMPANY

MATHEMATICS

FOR ELEMENTARY TEACHERS

ACTIVITIES TO ACCOMPANY

MATHEMATICS

FOR ELEMENTARY TEACHERS

SECOND EDITION

SYBILLA BECKMANN

UNIVERSITY OF GEORGIA

Boston San Francisco New York
London Toronto Sydney Tokyo Singapore Madrid
Mexico City Munich Paris Cape Town Hong Kong Montreal

Photo credits: p. 279 U.S. Air Force; pp. 280–281, 304, 344, 348, 405, 408 photos courtesy of the author

ISBN-10: 0-321-44976-2 ISBN-13: 978-0-321-44976-4
5 6 7 8 9 10—CRW—11 10 09

Contents

Preface

The activities in this book are designed to help prospective elementary school teachers develop a deep understanding of elementary school mathematics.

I wrote these activities because I wanted my students to be actively engaged in mathematics in class. Few students seem to get much out of long lectures, and every teacher of math knows that math is not a spectator sport: In order to learn math, you have to *do* math, and you have to *think deeply* about math. Every teacher also knows that when you explain something to someone else, you deepen your own understanding—or you uncover a gap in your understanding. So when students explain their mathematical ideas to each other in class, it helps develop their own thinking.

I have tried to strike a good balance between structure and guidance on the one hand, and leaving room for students to put their own thinking into the activities on the other hand.

Although the activities are designed for college-educated adults, many can be adapted easily for use in elementary school. Indeed, I have used a number of them with children myself, and my students have too. Therefore, I hope that prospective teachers will keep these activities in mind for their own teaching.

Sybilla Beckmann
Athens, Georgia

ACTIVITIES TO ACCOMPANY

MATHEMATICS

FOR ELEMENTARY TEACHERS

1

Problem Solving

1.1 Solving Problems

Class Activity 1A: A *Clinking Glasses* Problem

Use Polya's four problem-solving steps to solve the following problem:

At a party, someone proposes a toast. Each of the 20 people in the room wants to "clink" glasses with everyone else. How many "clinks" will there be? (There is exactly one "clink" for each pair of people.)

1. Understand the problem.
2. Devise a plan.
3. Carry out the plan. If you get stuck, return to step 2 and think up a new plan.
4. Look back. Does your answer seem sensible? What plan worked best? What if there were 30 people at the party? Would your technique work in that case? Did someone in your class or group have a different plan or idea? Are your results compatible? Think of more ways to "look back" and learn from what you've just done.

Class Activity 1B: Problems about Triangular Numbers

The number of dots in a triangular array is called a **triangular number**.

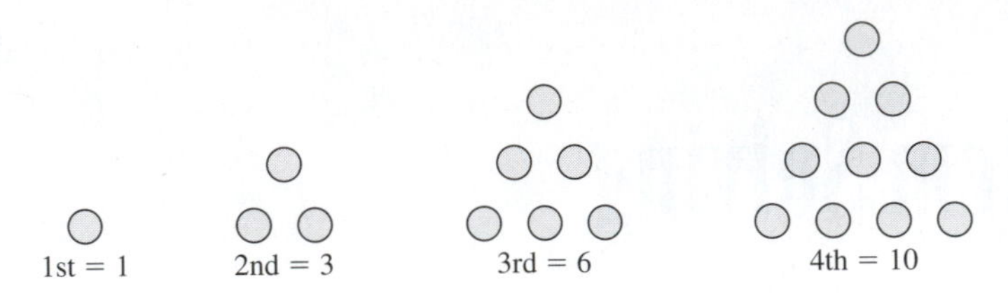

The first triangular number is 1, the second is 3, the third is 6, the fourth is 10, and so on. In general, the Nth triangular number is the number of dots in a triangular array of dots that has N dots on each side.

Use Polya's four problem-solving steps to solve the following problems about triangular numbers:

1. What is the 25th triangular number? What is the 1000th triangular number?

If you have solved Class Activity 1A, can you identify a relationship between this problem and the "clinking glasses" problem?

2. Is there a triangular number that has 91 dots in its shape? If so, which one is it? In other words, where does it appear in the sequence of triangular numbers? Is it the 5th, the 12th, etc.? Is there a triangular number that has 150 dots in its shape? What about 2000 dots? If so, where are these numbers in the sequence of triangular numbers? If there are no numbers with 150 or 2000 dots, how do you know for sure?

Class Activity 1C: What Is a Fair Way to Split the Cost?

Two friends, Eliza and Margo, go shopping and find a *buy-two-pairs-get-one-pair-free* sale at a shoe store. (The least expensive pair is the free one.) Eliza finds two pairs she'd like to buy: one for $83.95 and one for $67.95. Margo is interested in a pair that costs $59.95. Eliza and Margo decide to buy the three pairs together, so the pair that costs $59.95 will be free.

1. How should Eliza and Margo divide fairly the total cost of the three pairs of shoes?

First, solve this problem on your own, then discuss your solution with the group. Explain your solution method and explain why you think it is a fair way to split the cost of the shoes.

Note that different people may have different views of what is "fair."

2. What if Margo's pair of shoes only cost $24.95, and Eliza's shoes still cost $83.95 and $67.95? Now, how should Margo and Eliza divide the cost?

As before, solve this problem on your own first, then discuss your solution with the group. Explain your method of solution and explain why you think it is a fair way to split the cost of the shoes.

Did you recommend a different method for dividing the cost than you did in Problem 1 of this activity? If so, why?

1.2 Explaining Solutions

Class Activity 1D: Who Says You Can't Do Rocket Science?

In this class activity, use Polya's problem-solving steps to *make progress* on solving the next problem. Then explain why the conclusions you have reached are true, whether or not your conclusions fully solve the problem.

Problem: A rocket can use two different types of fuel. Rocket scientists know that it takes 30 minutes to fill the fuel tank with type A fuel from a type A hose, whereas it takes 45 minutes to fill the same fuel tank with type B fuel from a type B hose. (The type B hose is smaller than the type A hose.) The rocket's fuel tank will now be filled simultaneously with type A and type B fuels, each flowing out of their own hoses at a constant rate. How long will it take to fill the fuel tank?

Before you attempt to find an exact answer, determine *approximately* how long it should take for the fuel tank to fill. Answer this either with a specific number of minutes, or by saying, "It will take less than . . . minutes, but more than . . . minutes, to fill the tank." In either case, explain your reasoning.

2

Numbers and the Decimal System

2.1 Overview of the Number Systems

2.2 The Decimal System and Place Value

Class Activity 2A: How Many Are There?

Each person participating in this activity needs a bunch of toothpicks or other small objects, such as coffee stirrers. Each person should have less than 100 toothpicks but enough so that they can't see right away how many toothpicks there are. Each person also needs either several small plastic snack bags, or some rubber bands, or both.

The purpose of this activity is to help you understand the development of our way of writing numbers.

1. Determine how many toothpicks you have *without* counting them all one by one. Do this by arranging your toothpicks so that you can *visually see* how many toothpicks you have. Use your plastic snack bags or rubber bands to help you organize your toothpicks. Describe how you arranged your toothpicks.

2. Does the way you arranged your toothpicks in, above, problem 1 correspond to the way you write the number that represents how many toothpicks you have? If so, explain how. If not, try to arrange your toothpicks so that you can visually see how many toothpicks you have and so that this way of arranging the toothpicks corresponds to the way we write the number that stands for how many toothpicks you have.

3. Put your toothpicks together with the toothpicks of several other people. Once again, arrange the toothpicks so that you can visually see how many toothpicks there are and so that your way of arranging the toothpicks corresponds to the way we write the number that stands for how many toothpicks there are. Use plastic snack bags or rubber bands to help organize the toothpicks. Describe how you arranged the toothpicks.

4. Repeat, the above steps, found in Problem 3, but now with the toothpicks from everyone in the class. How many toothpicks are there in all?

5. Draw rough pictures showing how to bundle 137 toothpicks so that the way the toothpicks are organized corresponds to the way we write the number 137.

Class Activity 2B: Showing Powers of Ten

Mrs. Kubrick wants her students to get a feel for the powers of ten so that they can better understand place value. Mrs. Kubrick decides to use small stars printed on paper for this purpose, as shown in the next set of figures. One star, 10 stars, 100 stars, and 1000 stars all fit on one piece of paper. Even 10,000 stars fit on a piece of ordinary $8\frac{1}{2}$-by-11-inch paper, as on the next page.

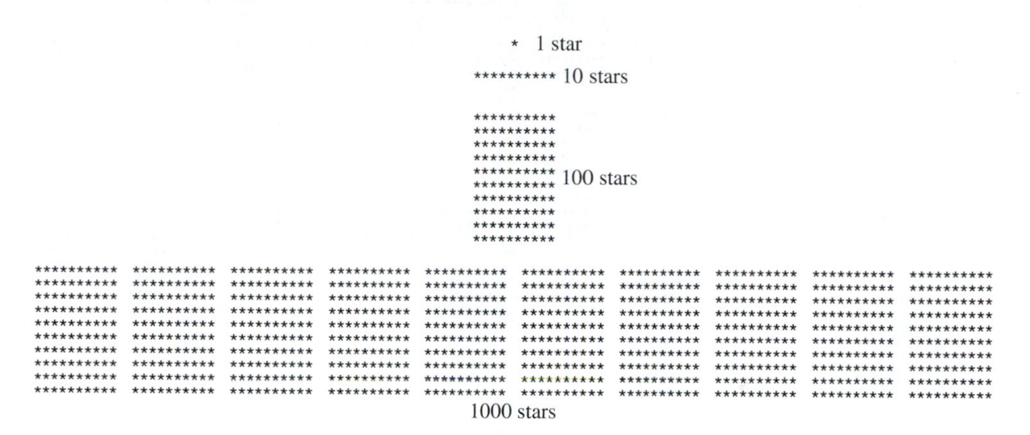

1. How many pieces of paper will Mrs. Kubrick need to show 100,000 stars? How many pieces of paper will Mrs. Kubrick need to show 1,000,000 stars?

2. Mrs. Kubrick's students really want to see one billion stars. How many pieces of paper would Mrs. Kubrick need to show one billion stars?

   A standard package of paper contains 500 sheets. How many packages of paper would Mrs. Kubrick need to show one billion stars? Realistically, can Mrs. Kubrick show one billion stars?

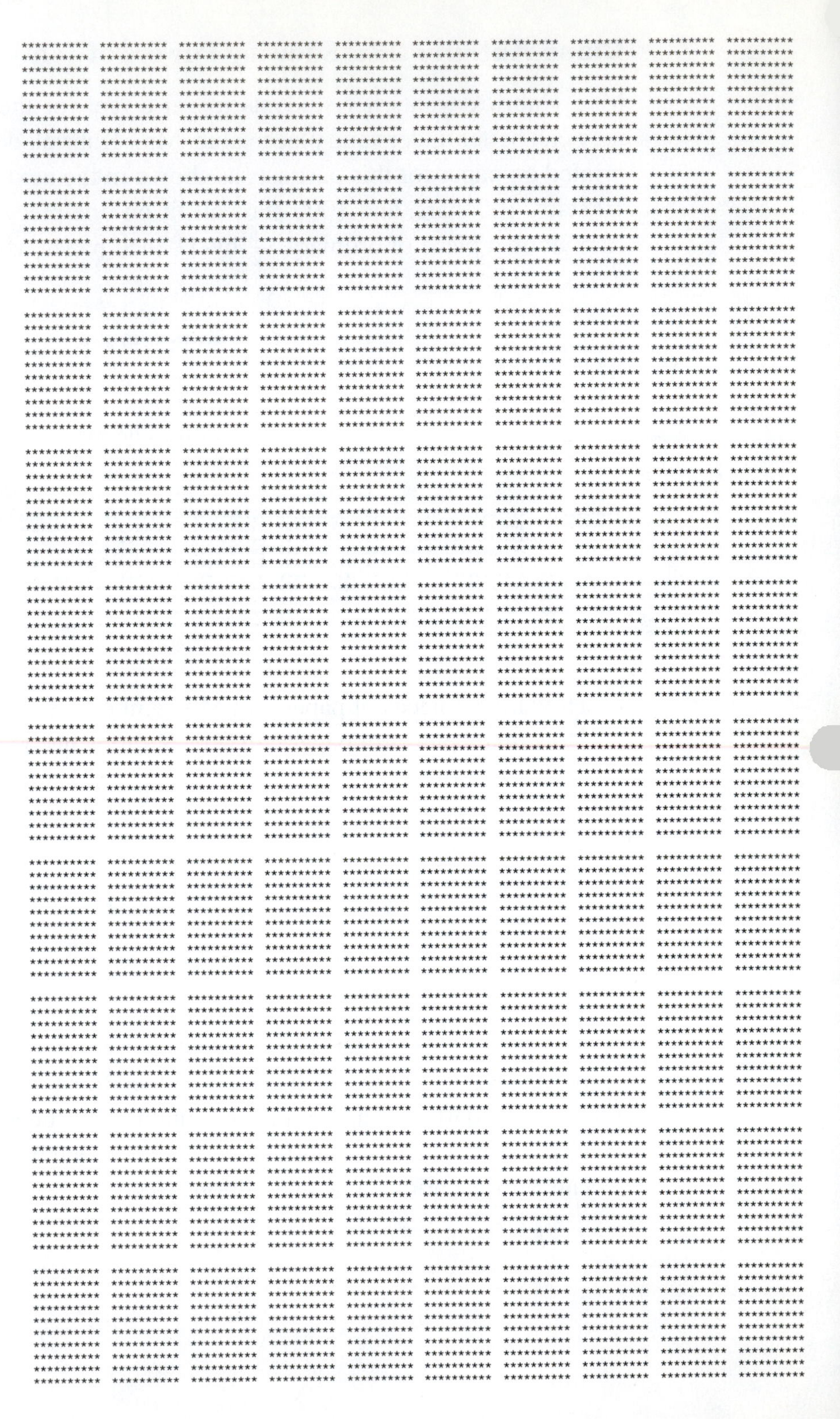

Figure 2B.1
10000 Stars

2.3 Representing Decimal Numbers

Class Activity 2C: Representing Decimal Numbers with Bundled Objects

1. It is not a strange idea to use 1 object to represent an amount less than 1. After all, 1 penny stands for \$0.01, or $\frac{1}{100}$ of a dollar.

   Let's let 1 paperclip stand for 0.001, or $\frac{1}{1000}$. Show simple drawings of bundled paperclips so that the way of organizing the paperclips corresponds to the way we write the following decimal numbers:

   0.034

   0.134

   0.13

2. Let's let 1 small bead stand for 0.0001, or $\frac{1}{10,000}$. Show simple drawings of bundled beads so that the way of organizing the beads corresponds to the way we write the following decimal numbers:

   0.0028

   0.012

3. List at least three different decimal numbers that the toothpicks pictured here could represent. In each case, state the value of the single toothpick.

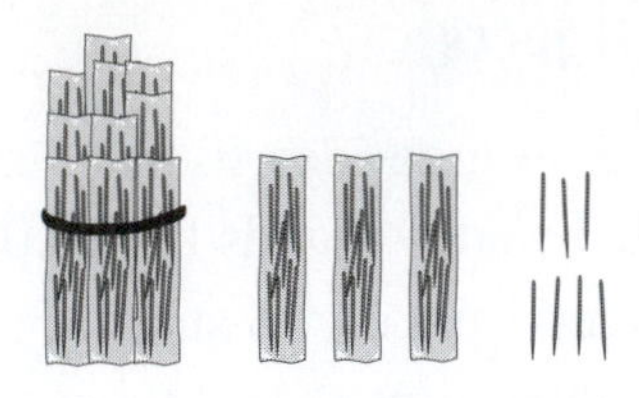

4. Explain why the way the bagged and loose toothpicks pictured here are organized does not correspond to the decimal representation of that number of toothpicks. Show how to reorganize these bagged and loose toothpicks so as to correspond to the decimal representation for the total number of toothpicks.

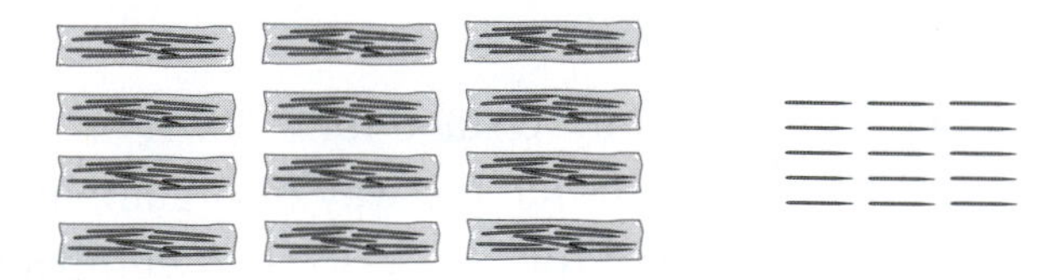

Class Activity 2D: Zooming In and Zooming Out on Number Lines

1. Label the tick marks on the number lines that follow with appropriate decimal numbers. The second, third, and fourth number lines should be labeled as if they are "zoomed in" on the indicated portion of the previous number line.

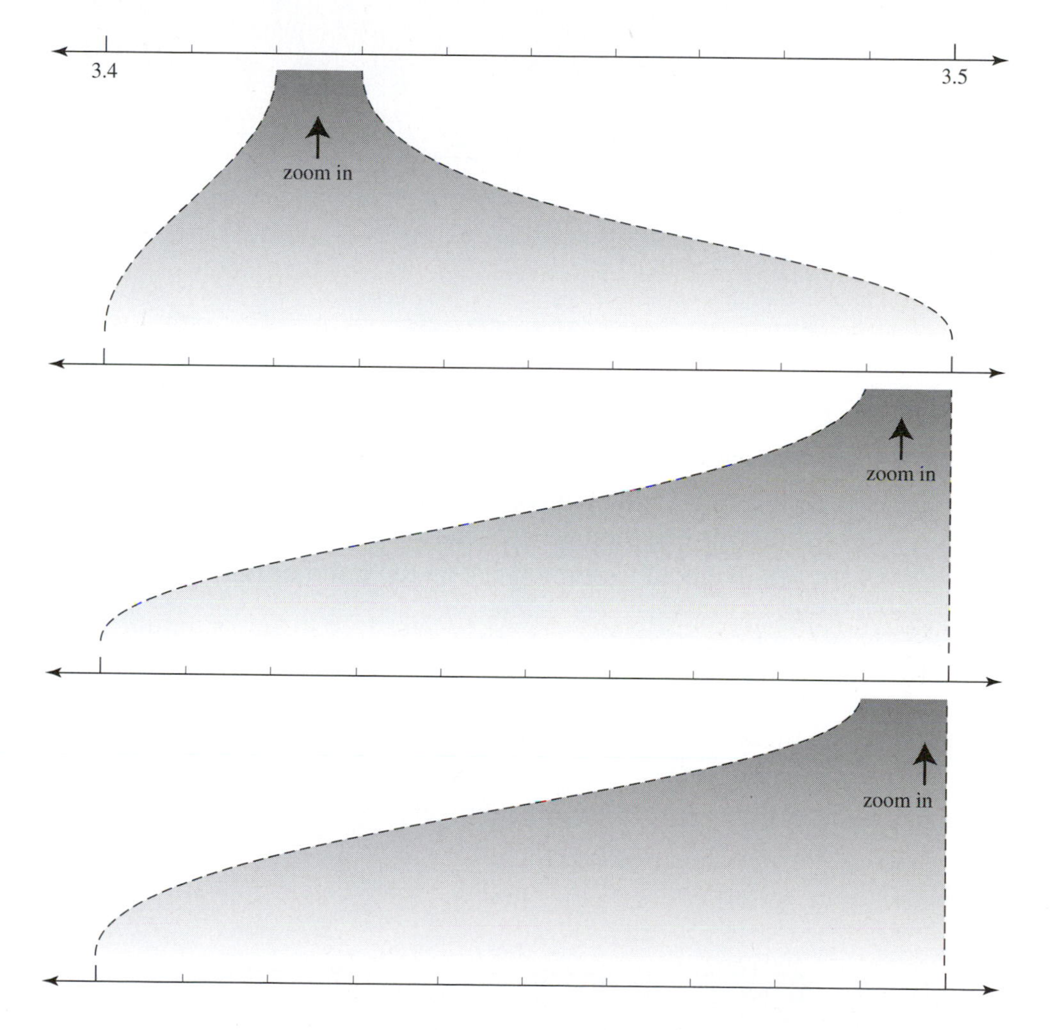

2. Label the tick marks on the next three number lines in three different ways. In each case, your labeling should fit with the fact that the tick marks at the ends of the number lines are longer than the other tick marks. You may further lengthen the tick marks at either end as needed. It may help you to think about "zooming in" on the number line.

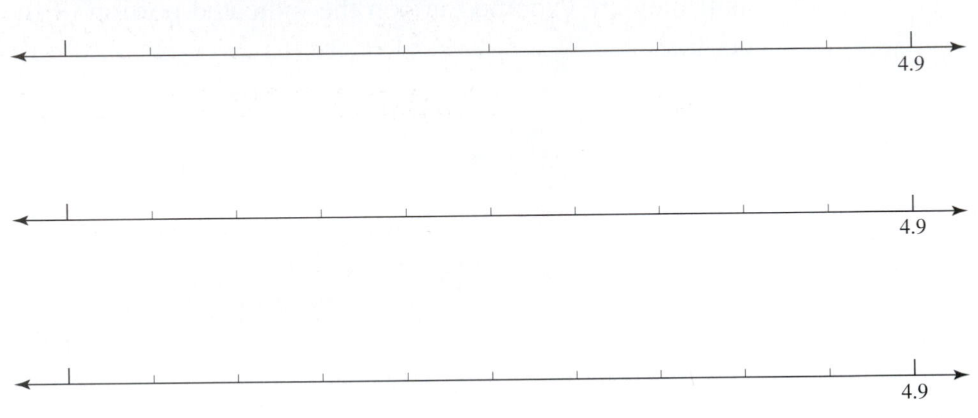

3. Label the tick marks on the next three number lines in three different ways. In each case, your labeling should fit with the fact that the tick marks at the ends of the number lines are longer than the other tick marks. You may further lengthen the tick marks at either end as needed. It may help you to think about "zooming in" on the number line.

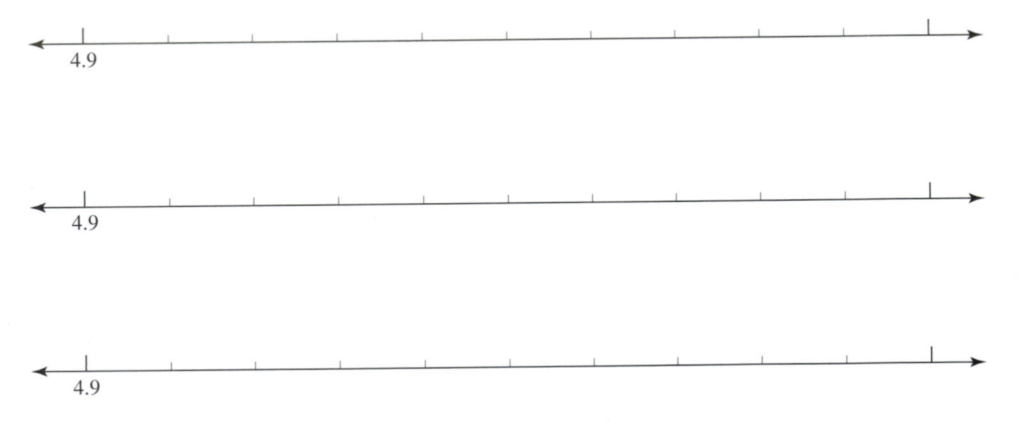

4. Plot 6.728 on each of the following number lines so that adjacent tick marks are the specified distance apart. (The number 6.728 may or may not land on a tick mark.) Your labeling should fit with the fact that the tick marks at the left and right ends are longer than the others. Indicate how each number line is a "zoomed out" version of the number line above it.

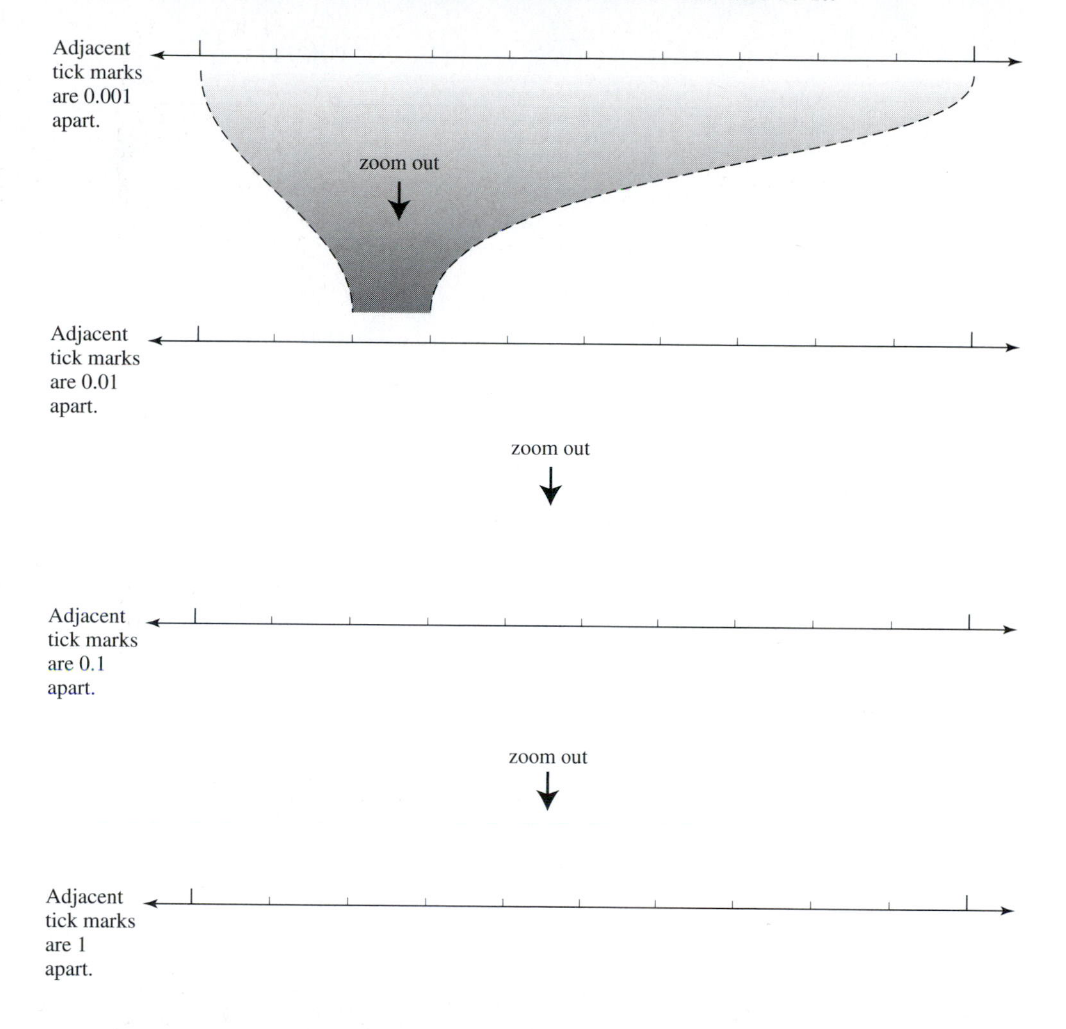

Adjacent tick marks are 0.001 apart.
zoom out
Adjacent tick marks are 0.01 apart.
zoom out
Adjacent tick marks are 0.1 apart.
zoom out
Adjacent tick marks are 1 apart.

5. Label the tick marks on the following number lines so that the tick marks are the specified distance apart and so that the given number can be plotted on the number line. The number need not land on a tick mark. Your labeling should fit with the fact that the tick marks at the left and right ends are longer than the others.

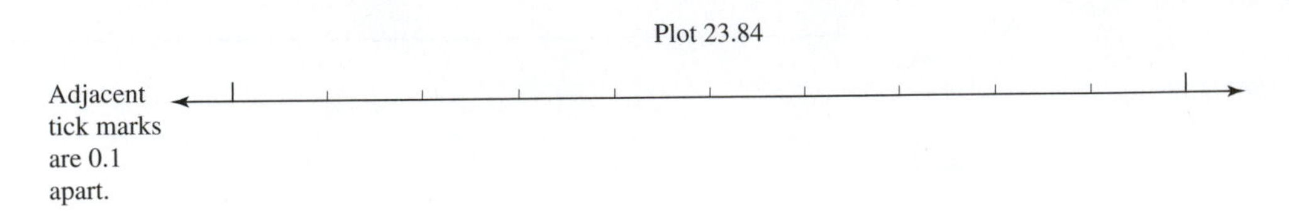
Plot 23.84
Adjacent tick marks are 0.1 apart.

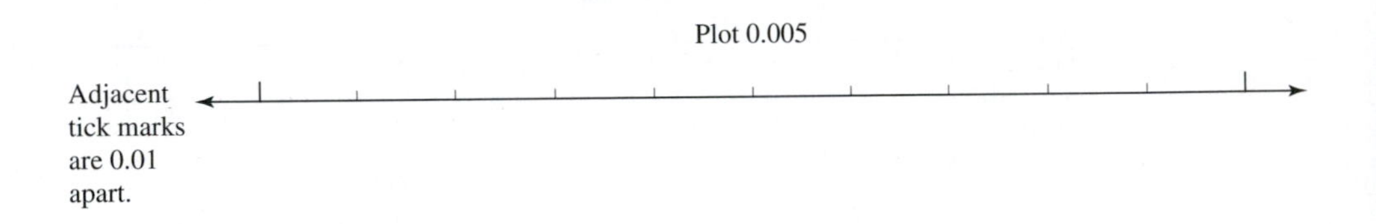
Plot 0.005
Adjacent tick marks are 0.01 apart.

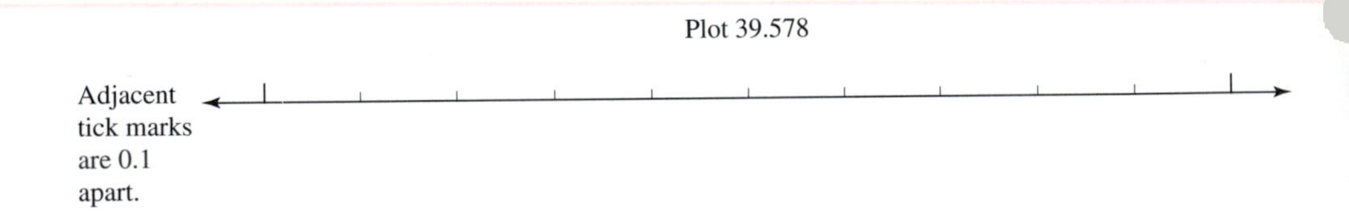
Plot 39.578
Adjacent tick marks are 0.1 apart.

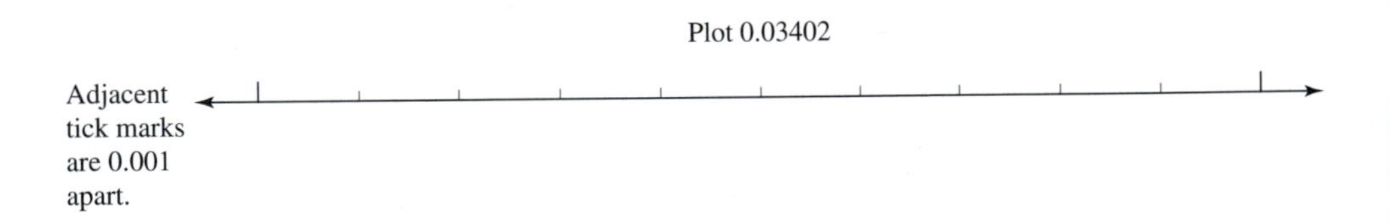
Plot 0.03402
Adjacent tick marks are 0.001 apart.

6. Label the tick marks on the next number line so that the decimal numbers -12.3 and -17 can both be plotted visibly and distinctly. The distance between adjacent tick marks should be a power of ten, such as 0.1, 0.01, or 0.001. The numbers do not necessarily have to fall on tick marks. Plot the two decimal numbers.

7. Label the tick marks on the next number line so that the decimal numbers 4.3572 and 4.3586 can both be plotted visibly and distinctly. The distance between adjacent tick marks should be a power of ten, such as 0.1, 0.01, 0.001, and so on. The numbers do not necessarily have to fall on tick marks. Plot the two decimal numbers.

8. Label the tick marks on the next number line so that the decimal numbers 18.977145 and 18.977638 can both be plotted visibly and distinctly. The distance between adjacent tick marks should be a power of ten, such as 0.1, 0.01, 0.001, and so on. The numbers do not necessarily have to fall on tick marks. Plot the two decimal numbers.

Class Activity 2E: Representing Decimals as Lengths

A good way to represent (positive) decimal numbers is as lengths. Cut out the 4 long strips in Figure A.1 on page 595 and tape them end-to-end without overlaps to make one long strip. The length of this long strip is 1 unit. Cut out the ten 0.1 unit long strips.

1. By placing strips end-to-end without gaps or overlaps, verify the following:

   The 1 unit long strip is as long as 10 of the 0.1 unit long strips.

   A 0.1 unit long strip is as long as 10 of the 0.01 unit long strips.

   A 0.01 unit long strip is as long as 10 of the 0.001 unit long strips.

   Now cut apart the 0.01 and 0.001 unit long strips.

   Represent the following decimals as lengths by placing appropriate strips end-to-end without gaps or overlaps (as best you can). In each case, draw a rough sketch (which need not be to scale) to show how you represented the decimal as a length.

2. 1.234

3. 0.605

4. 1.07

5. 1.007

6. 0.089

7. Use the strips to represent some other decimals as lengths.

2.4 Comparing Decimal Numbers

Class Activity 2F: Places of Larger Value Count More than Lower Places Combined

1. What is the largest decimal number you can make that has nonzero digits only in the tens and ones places?

   make the largest $\longrightarrow$ ___ ___

   What is the smallest decimal number you can make that has a nonzero digit in the hundreds place?

   make the smallest $\longrightarrow$ ___ ___ ___

   Draw rough pictures of bundled objects to compare these two numbers. Which is larger? How can we see this?

2. What is the largest decimal number you can make that has nonzero digits only in the hundreds, tens, and ones places?

make the largest $\longrightarrow$ ___ ___ ___

What is the smallest decimal number you can make that has a nonzero digit in the thousands place?

make the smallest $\longrightarrow$ ___ ___ ___ ___

Draw *very rough* pictures of bundled objects to compare these two numbers. Which is larger? How can we see this?

3. In order to compare two decimal numbers, we first look at the place of largest value in which at least one of the numbers has a nonzero entry. Why does this make sense?

Class Activity 2G: Misconceptions in Comparing Decimal Numbers

The list that follows describes some of the misconceptions children can develop about comparing decimal numbers. The list is based on the work of mathematics education researchers Kaye Stacey and Vicki Steinle, who have gathered data on thousands of children in Australia. (See [13] and [14] for further information, including additional misconceptions and advice on instruction.)

Whole number thinking: Children with this misconception treat the portion of the number to the right of the decimal point as a whole number, thus thinking that $2.352 > 2.4$ because $352 > 4$. These children therefore think that longer decimals are always larger than shorter ones.

Column overflow thinking: Children with this misconception name decimal numbers incorrectly by focusing on the first nonzero digit to the right of the decimal point. For example, they say that 2.34 is "two and thirty-four tenths." These children think that $2.34 > 2.5$ because 34 tenths is more than 5 tenths. These children usually identify longer decimal numbers as larger; they will, however, correctly identify 2.5 as greater than 2.06 because 5 tenths is more than 6 hundredths.

Denominator focused thinking: Children with this misconception think that any number of tenths is greater than any number of hundredths and that any number of hundredths is greater than any number of thousandths, and so on. These children identify 5.47 as greater than 5.632, reasoning 47 hundredths is greater than 632 thousandths because hundredths are greater than thousandths. Children with this misconception identify shorter decimal numbers as larger.

Reciprocal thinking: Children with this misconception view the portion of a decimal number to the right of the decimal point as something like the fraction formed by taking the reciprocal. For example, they view 0.3 as something like $\frac{1}{3}$ and thus identify 2.3 as greater than 2.4 because $\frac{1}{3} > \frac{1}{4}$. These children usually identify shorter decimal numbers as larger, except in cases of intervening zeros. For example, they may say that $0.03 > 0.4$ because $\frac{1}{3} > \frac{1}{4}$.

Money thinking: Children with this difficulty truncate decimal numbers after the hundredths place and view decimal numbers in terms of money. If two decimal numbers agree to the hundredths place, these children simply guess which one is greater, sometimes guessing correctly, sometimes guessing incorrectly. Most of these students recognize that 1.8 is like \$1.80, although some view 1.8 incorrectly as \$1.08.

1. Put the set of decimal numbers that follows in order from least to greatest. Then show how children with the misconceptions just described would probably put the numbers in order.

3.3 3.4 3.05 3.25 3.251

Correct order (least to greatest):	
Whole number thinking:	
Column overflow:	
Denominator focused:	
Reciprocal thinking:	
Money thinking:	

2. Make up a decimal comparison quiz that provides ten pairs of decimal numbers and asks children to circle the larger decimal number in each pair. Try to pick the ten pairs so that children with the different misconceptions previously described will not give exactly the same answers for all ten pairs. For each misconception, show how children with that misconception would probably answer the quiz.

Class Activity 2H: Finding Smaller and Smaller Decimal Numbers

You can play the following game alone, or with two or more people or teams:

Start by picking any two decimal numbers. Call the lesser one the *target decimal number*. Put the greater decimal number at the top of a list. Take turns placing new decimal numbers on the list. Each new decimal number placed on the list must be *less than* the previous decimal number on the list, but *greater than* the target decimal number. You score a point when you are the first to catch an opponent's mistake.

For example: starting decimal numbers 2, 2.491

Target decimal number: 2

The first few entries in the list could be as follows:

2.491
2.37
2.085
2.01

1. Play the game several times. Each person should take at least 5 turns each time.
2. Is it possible for the game to be *forced* to end, or is it always possible to continue playing if you want to?
3. Is there a least decimal number that is greater than 2? If yes, show it; if no, explain why not.

Class Activity 2I: Finding Decimals between Decimals

Two people or two teams can play the following game:

Team 1 goes first and picks a decimal number between 1 and 2. Team 2 then picks a different decimal number between 1 and 2. Each team writes its decimal number on a list. For example, the starting lists might look like this:

Team 1: 1.3

Team 2: 1.4

Team 1 takes a turn and writes on its list a new decimal number that is between the decimal numbers on the two lists. Continuing the previous example, Team 1 could write 1.39 on its list because 1.39 is between 1.3 and 1.4. Thus, the two teams' lists would appear as follows:

Team 1: 1.3, 1.39

Team 2: 1.4

Then Team 2 takes a turn and writes on its list a new decimal number that is between the last decimal numbers on the two lists. Continuing the example, Team 2 could write 1.391 on its list because 1.391 is between 1.39 and 1.4:

Team 1: 1.3, 1.39

Team 2: 1.4, 1.391

The teams continue to take turns writing decimal numbers on their lists. The decimal number written on the list must be between the last decimal numbers on the two lists.

A team scores a point when it is the first to catch a mistake made by its opponent.

The game ends when each team has 6 numbers on its list.

1. Play the game several times.
2. Is it possible for the game to be *forced* to end early, or is it always possible for each team to write 6 numbers on its list? Could teams keep playing and write more than 6 numbers on their lists? If so, is there a limit to how many numbers the teams could write on their lists?
3. Is there a decimal number that is "right next to" 1.1? Explain your answer, and explain how you interpret the meaning of "right next to."

Class Activity 2J: Decimals between Decimals on Number Lines

For each of the pairs of numbers that follow, find a decimal number in between the two numbers. Label the longer tick marks on the number line so that all three numbers can be plotted visibly and distinctly. The distance between adjacent tick marks should be a power of ten, such as 0.1, 0.01, 0.001, and so on. Plot all three numbers. The numbers need not land on tick marks.

1. The numbers 1.6 and 1.7.

   Now, describe how to use money to find a decimal number between 1.6 and 1.7.

2. The numbers 12.54 and 12.541.

3. The numbers 2.781 and 2.7342.

4. The numbers 23.99 and 24.

Class Activity 2K: "Greater Than" and "Less Than" with Negative Decimal Numbers

1. Explain in two different ways why a negative decimal number is always less than a positive decimal number.

2. Johnny says that $-5 > -2$. Describe two different ways to explain to Johnny why this is not correct.

2.5 Rounding Decimal Numbers

Class Activity 2L: Why Do We Round?

Why do we round numbers? Write down several reasons.

Class Activity 2M: Explaining Rounding

1. Label the number lines in Figure 2M.1 so that each number line is "zoomed out" from the previous one and so that 34, 617 can be plotted on each number line. Plot 34, 617 on each number line.
2. Use the same number lines in Figure 2M.1 to help you explain how to round 34, 617 to the nearest ten, the nearest hundred, the nearest thousand, and the nearest ten-thousand.

3. Label the number lines in Figure 2M.2 so that each number line is "zoomed out" from the previous one and so that 99.253 can be plotted on each number line. Plot 99.253 on each number line.
4. Use the number lines in Figure 2M.2 to help you explain how to round 99.253 to the nearest hundredth, the nearest tenth, the nearest one (i.e., to the nearest whole number), and the nearest ten.

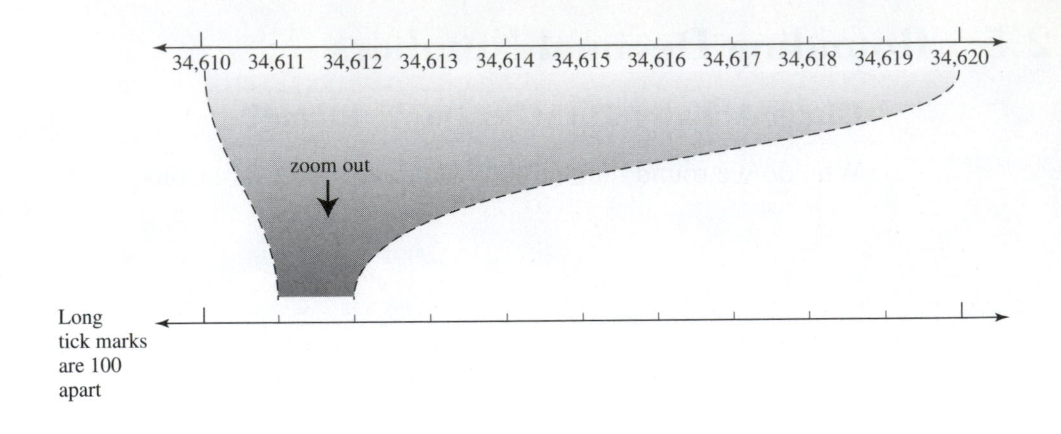

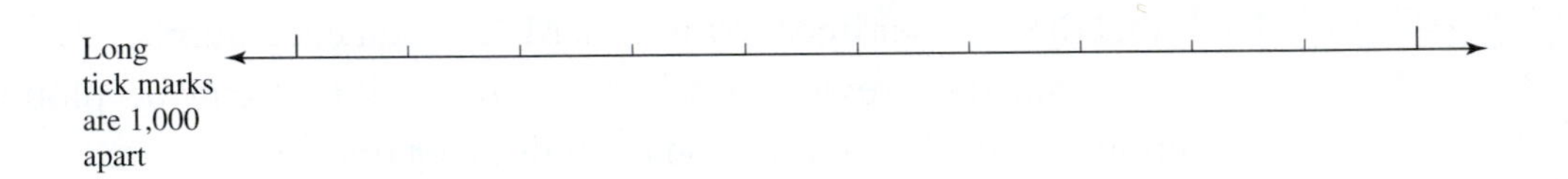

Figure 2M.1
Rounding 34, 617

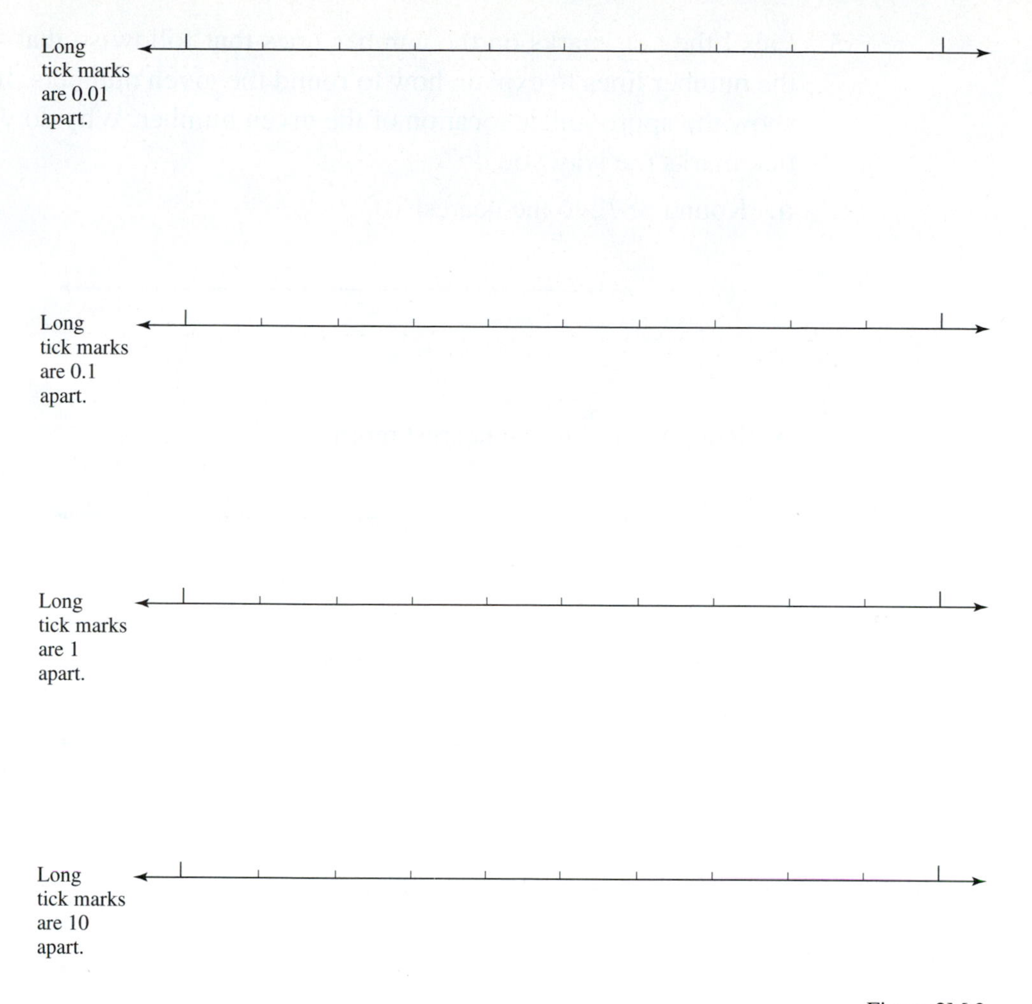
Long tick marks are 0.01 apart.
Long tick marks are 0.1 apart.
Long tick marks are 1 apart.
Long tick marks are 10 apart.

Figure 2M.2

Rounding 99.253

5. Label the tick marks on the number lines that follow so that you can use the number lines to explain how to round the given numbers. In each case, show the approximate location of the given number. Why do you label the tick marks the way you do?

a. Round 3872 to the nearest 10.

b. Round 2.349 to the nearest tenth.

c. Round 6.995 to the nearest hundredth.

d. Round 6.995 to the nearest tenth.

e. Round 2006 to the nearest 100.

f. Round 4.001 to the nearest tenth.

Class Activity 2N: Can We Round This Way?

Maureen has made up her own method of rounding. Starting at the rightmost place in a decimal number, she keeps rounding to the value of the next place to the left until she reaches the place to which the decimal number was to be rounded.

For example, Maureen would use the following steps to round 3.2716 to the nearest tenth:

$$3.2716 \to 3.272 \to 3.27 \to 3.3$$

Try Maureen's method on several examples. Is Maureen's method valid? That is, does it always round decimal numbers correctly? Or are there examples of decimal numbers where Maureen's method does not give the correct rounding?

3

Fractions

3.1 The Meaning of Fractions

Class Activity 3A: Fractions of Objects

1. Take a blank piece of paper and imagine that it is $\frac{4}{5}$ of some larger piece of paper. Fold your piece of paper to show $\frac{3}{5}$ of the larger (imagined) piece of paper. Do this as carefully and precisely as possible without using a ruler or doing any measuring. Explain why your answer is correct. Could two people have different-looking solutions that are both correct?
2. Benton used $\frac{3}{4}$ cup of butter to make a batch of cookie dough. Benton rolled his cookie dough out into a rectangle, as shown in the next figure. Now Benton wants to cut off a piece of the dough so that the portion he cuts off contains $\frac{1}{4}$ cup of butter. How could Benton cut the dough? Explain your answer.

3. The next picture shows two pies of different sizes. Make up a story or situation where you would want to determine $\frac{1}{6}$ of the combined amount of the big pie and the little pie. Using your story or situation, determine $\frac{1}{6}$ of the combined amount of the big pie and the little pie, and shade this amount on the pies in the picture.

4. Shade $\frac{5}{6}$ of the combined amount of the big pie and the little pie shown in the next picture. Use the meaning of fractions to explain why your answer is correct.

5. Mireya has a recipe that calls for 5 cups of flour. Mireya wants to make $\frac{3}{4}$ of the recipe. Instead of figuring out what number $\frac{3}{4}$ of 5 is, Mireya measures $\frac{3}{4}$ of a cup of flour 5 times, and uses this amount of flour for $\frac{3}{4}$ of the recipe. Use the meaning of fractions and the following picture to help you explain why Mireya's strategy is valid:

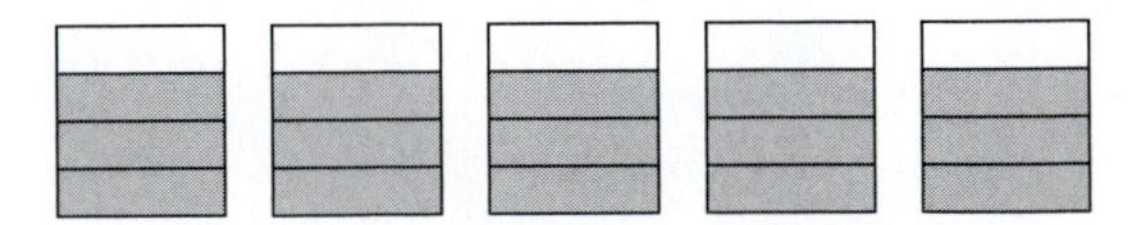

Class Activity 3B: The Whole Associated with a Fraction

1. Maurice says that the next picture shows that $\frac{3}{6}$ is bigger than $\frac{2}{3}$. The shaded portion representing $\frac{3}{6}$ *is* larger than the shaded portion representing $\frac{2}{3}$, so why is Maurice not correct?

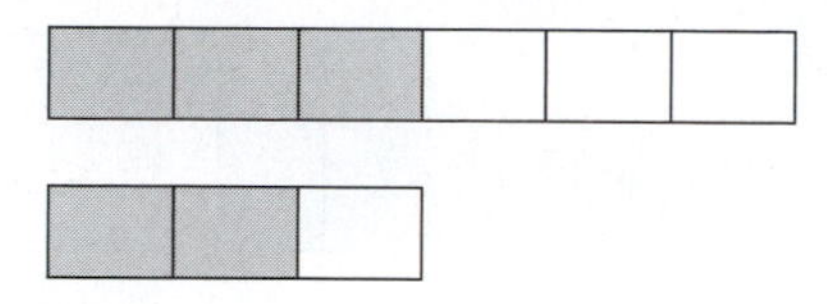

2. Kayla says that the shaded part of the next picture can't represent $\frac{1}{4}$ because there are 3 shaded circles, and 3 is more than 1, but $\frac{1}{4}$ is supposed to be less than 1. What can you tell Kayla about fractions that might help her?

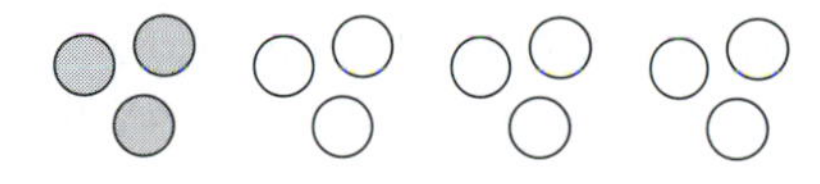

3. When Ted was asked what the 4 in the fraction $\frac{3}{4}$ means, Ted said that the 4 is the whole. Explain why it is not completely correct to say, "4 is the whole." What is a better way to say what the 4 in the fraction $\frac{3}{4}$ means?

4. Explain why you can use the fractions in (a), (b), and (c) to describe the shaded region in the next figure. How can three different numbers describe the same shaded region? How must you interpret the shaded region in each case?

a. $\frac{3}{6}$ **b.** 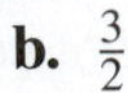 $\frac{3}{2}$ **c.** $\frac{3}{1}$

5. Suppose a textbook problem asks students to "name the fraction shown in the shaded region of the next figure." Assuming that there is no other information given in the problem, explain why the problem is ambiguous. What should the textbook problem clarify?

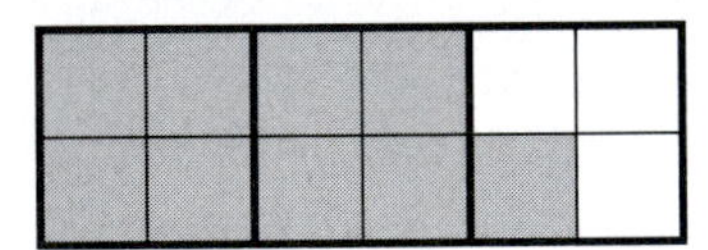

6. At a neighborhood park, $\frac{1}{3}$ of the area of the park is to be used for a new playground. Swings will be placed on $\frac{1}{4}$ of the area of the playground. What fraction of the neighborhood park will the swing area be? Draw a picture to help you solve the problem, and explain your answer. For each fraction in this problem, and in your solution, describe the whole associated with this fraction. In other words, describe what each fraction is *of*. Are all the wholes the same or not?

7. Ben is making a recipe that calls for $\frac{1}{3}$ cup of oil. Ben has a bottle that contains $\frac{2}{3}$ cup of oil. Ben does not have any measuring cups. What fraction of the oil in the bottle should Ben use for his recipe? Draw a picture to help you solve the problem, and explain your answer. For each fraction in this problem, and in your solution, describe the whole associated with this fraction. In other words, describe what each fraction is *of*. Are all the wholes the same or not?

Class Activity 3C: Is the Meaning of Equal Parts Always Clear?

A set of pattern tiles would be useful for Problem 4.

1. Matteo says that he can't show $\frac{1}{3}$ of these marbles because some of the marbles are big and some are little. What do you think?

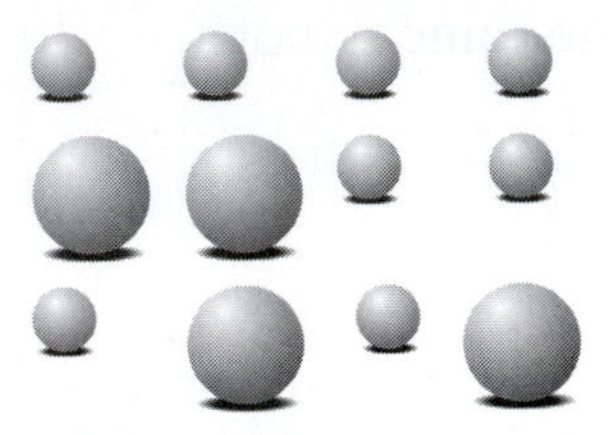

2. A first grade worksheet [7, p. 241] asked children to "show 4 equal parts" and showed a picture like the one on the left in the next figure. Arianna's response is on the right. Although Arianna's work is probably not what the authors had in mind, can it still be considered correct?

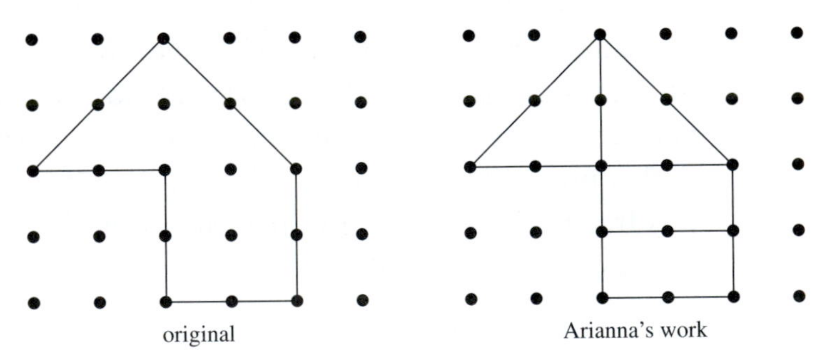

3. The definition of fractions of objects refers to dividing objects into equal parts. Discuss whether the meaning of equal parts is clearly defined in every situation.

4. The design that follows is made with 1 hexagon in the middle and surrounded by 6 rhombuses (diamond shapes) and 6 triangles. If pattern tiles are available, you might want to use them to copy this design.

a. What fraction of the shapes in the design is made of triangles?

b. What fraction of the area of the design is made of triangles? Is this the same fraction as in Part (a)?

c. In order to find the fractions in Part (a) and Part (b), you divided the design into equal parts. Did you interpret "equal parts" the same way in (a) and (b)?

Class Activity 3D: Improper Fractions

1. Show a plot of land that is $\frac{5}{3}$ the area shown on the map. Is more than one correct answer possible?

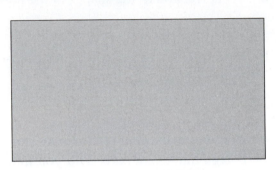

2. On day 1, Denise runs a distance shown on the next diagram. On day 2, Denise runs $\frac{5}{4}$ as far. Show how far Denise runs on day 2. On day 3, Denise runs $\frac{4}{5}$ as far as she ran on day 2. Show how far Denise runs on day 3. Compare the distance Denise ran on day 3 with the distance she ran on day 1. For each fraction in this problem, and in your solution, describe the associated whole. In other words, describe what each fraction is *of*.

Denise's running distance

3. Suppose you use a picture like the next one to talk about the fraction $\frac{5}{4}$. What kind of confusion could arise about this picture? What must we do in order to interpret the shaded region as $\frac{5}{4}$?

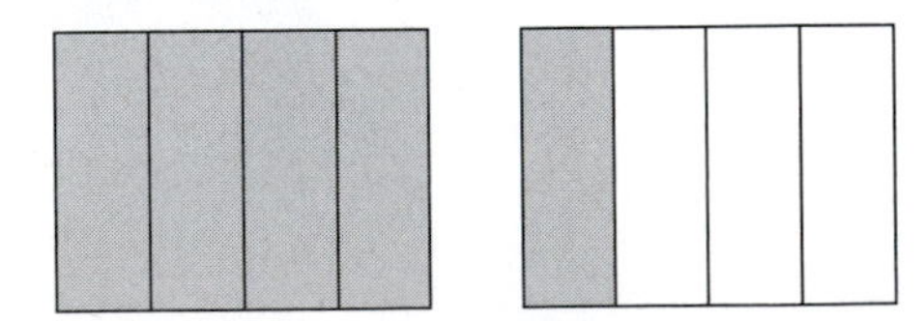

4. Enrico says that it doesn't make sense to talk about $\frac{5}{4}$ of a piece of paper because if you divide a piece of paper into 4 equal pieces, then you only have 4 pieces and can't show 5 pieces. What can you tell Enrico that might help him?

3.2 Fractions as Numbers

Class Activity 3E: Counting along Number Lines

Children first learn to count by ones. Later, children learn to count by 2s, by 5s, by 10s and by other numbers. We can also count by decimal amounts, such as tenths or hundredths, or by other fractional amounts, such as thirds or fifths. Viewed on a number line, we can think of counting by ones, by twos, by tenths, or by thirds as jumping along the number line, where each jump is a fixed distance, namely either one unit, two units, one tenth of a unit, or one third of a unit.

Compare and contrast counting by ones, by tenths, and by thirds. What are similarities and differences in the way we write the numbers as we count? What are similarities and differences in the way we say the numbers as we count?

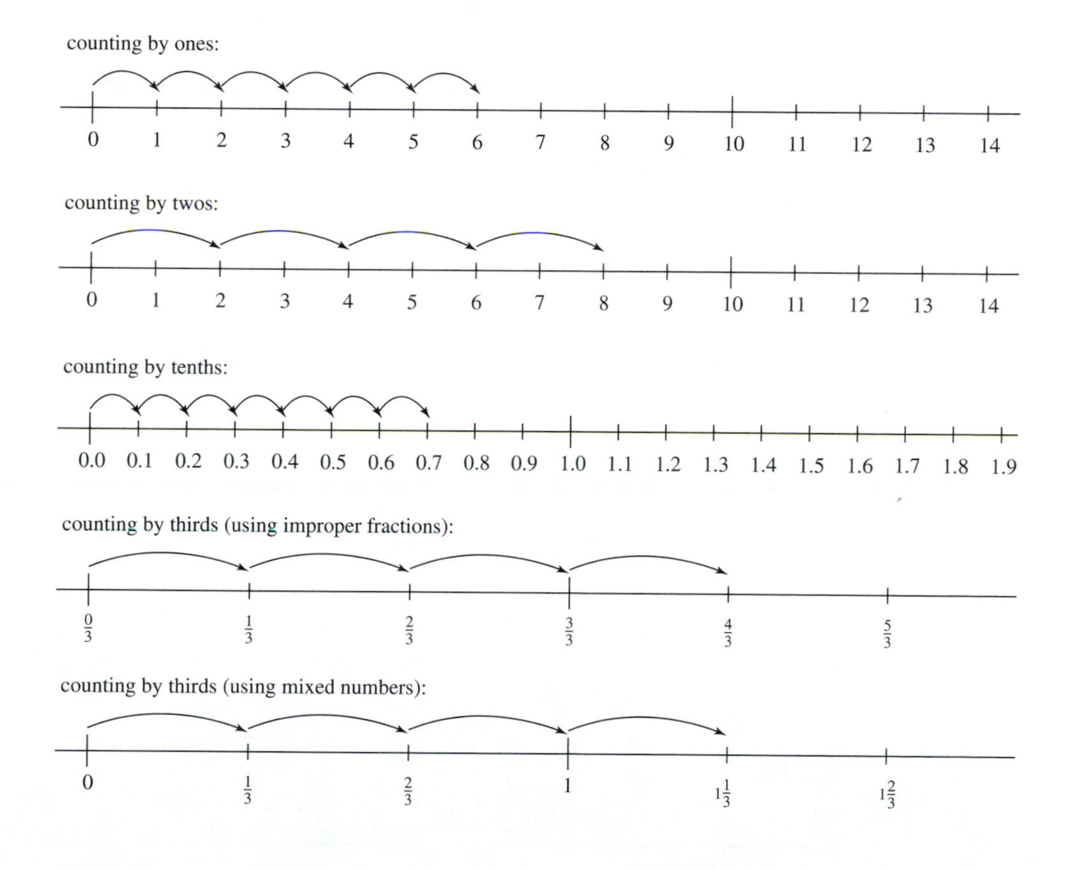

Class Activity 3F: Fractions on Number Lines, Part 1

1. When Tyler was asked to plot $\frac{3}{4}$ on a number line showing 0 and 2, he plotted it as shown on the next number line. What might be the source of Tyler's confusion?

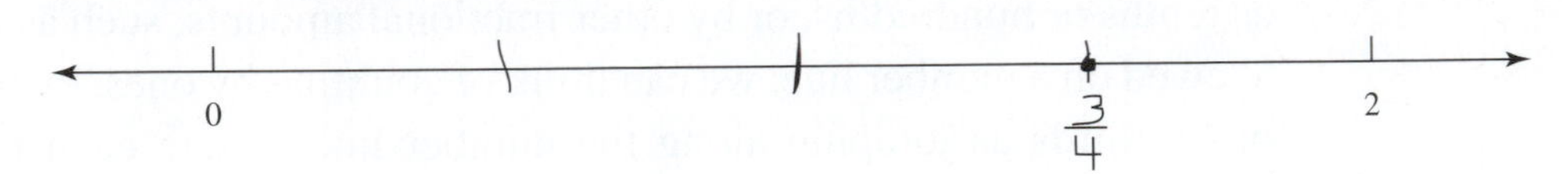

2. Plot 0, 1, $\frac{1}{4}$, $\frac{7}{4}$, and $\frac{11}{4}$ on the number line for this problem in such a way that each number falls on a tick mark. Lengthen the tick marks of whole numbers.

For each of the following problems, place equally spaced tick marks on the number line so that you can plot the requested fraction on a tick mark. You may place the tick marks "by eye"; precision is not needed. Explain your reasoning.

3. Plot $\frac{5}{4}$.

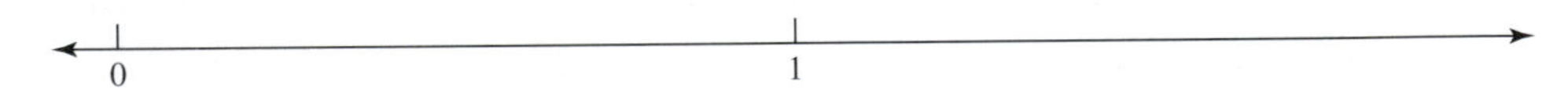

4. Plot 1.

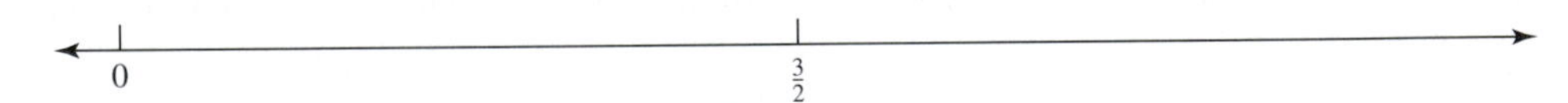

5. Plot 1.

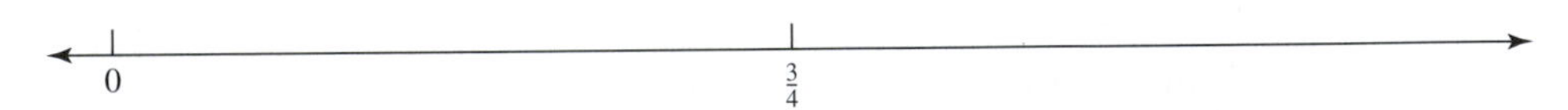

6. Plot $\frac{3}{5}$.

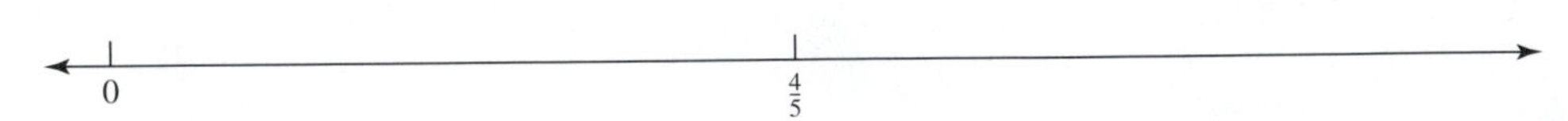

3.3 Equivalent Fractions

Class Activity 3G: Equivalent Fractions

1. Subdivide and label the second, third, and fourth strips and number lines in order to show that

$$\frac{1}{3} = \frac{2}{6} = \frac{3}{9} = \frac{4}{12}$$

$$\frac{2}{3} = \frac{4}{6} = \frac{6}{9} = \frac{8}{12}$$

$$\frac{3}{3} = \frac{6}{6} = \frac{9}{9} = \frac{12}{12}$$

$$\frac{4}{3} = \frac{8}{6} = \frac{12}{9} = \frac{16}{12}$$

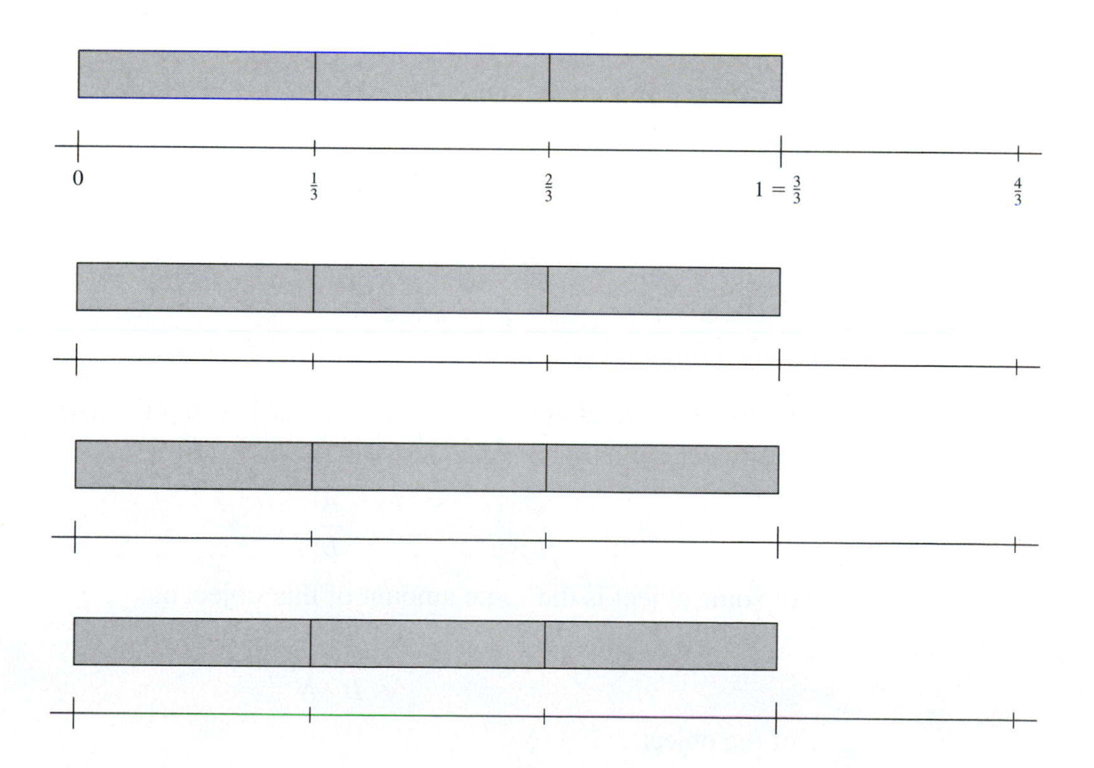

2. Use the meaning of fractions of objects to give a detailed conceptual explanation for why

$$\frac{2}{3}$$

of a cake is the same amount of cake as

$$\frac{2 \cdot 4}{3 \cdot 4}$$

of the cake.

Draw pictures to support your explanation. Discuss how your pictures show the process of multiplying both the numerator and denominator of $\frac{2}{3}$ by 4.

3. Discuss how to modify your explanation for Problem 2 to show that

$$\frac{2}{3}$$

of a cake is the same amount of cake as

$$\frac{2 \cdot 5}{3 \cdot 5}$$

of the cake.

4. Suppose A and B are whole numbers and B is not 0. Also suppose that N is any natural number. Explain why

$$\frac{A}{B}$$

of some object is the same amount of that object as

$$\frac{A \cdot N}{B \cdot N}$$

of the object.

Class Activity 3H: Misconceptions about Fraction Equivalence

1. Anna says,

$$\frac{2}{3} = \frac{6}{7}$$

because, starting with $\frac{2}{3}$, you get $\frac{6}{7}$ by adding 4 to the top and the bottom. If you do the same thing to the top and the bottom, the fractions must be equal.

Is Anna right? If not, why not? What should we be careful about when talking about equivalent fractions?

2. Don says that

$$\frac{11}{12} = \frac{16}{17}$$

because both fractions are one part away from a whole. Is Don correct? If not, what is wrong with Don's reasoning?

Class Activity 3I: Common Denominators

1. Write $\frac{1}{3}$ and $\frac{3}{4}$ with two different common denominators. Show each on the rectangles in the next figure. In terms of the pictures, what are we doing when we give the two fractions common denominators?

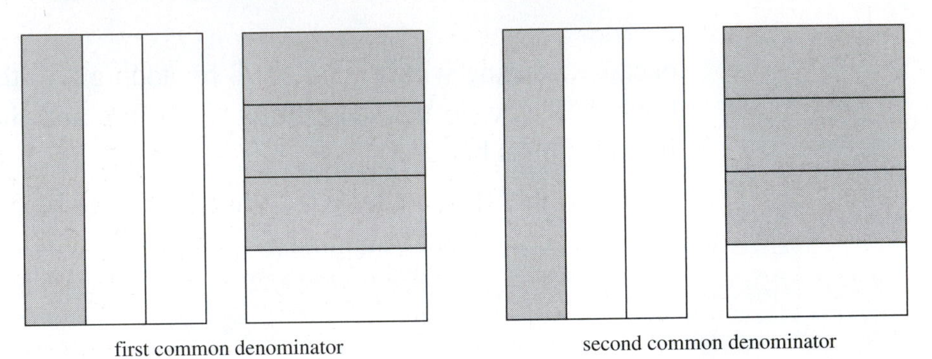

2. Write $\frac{3}{4}$ and $\frac{5}{6}$ with two different common denominators. In terms of the strips and the number lines, what are we doing when we give the two fractions common denominators?

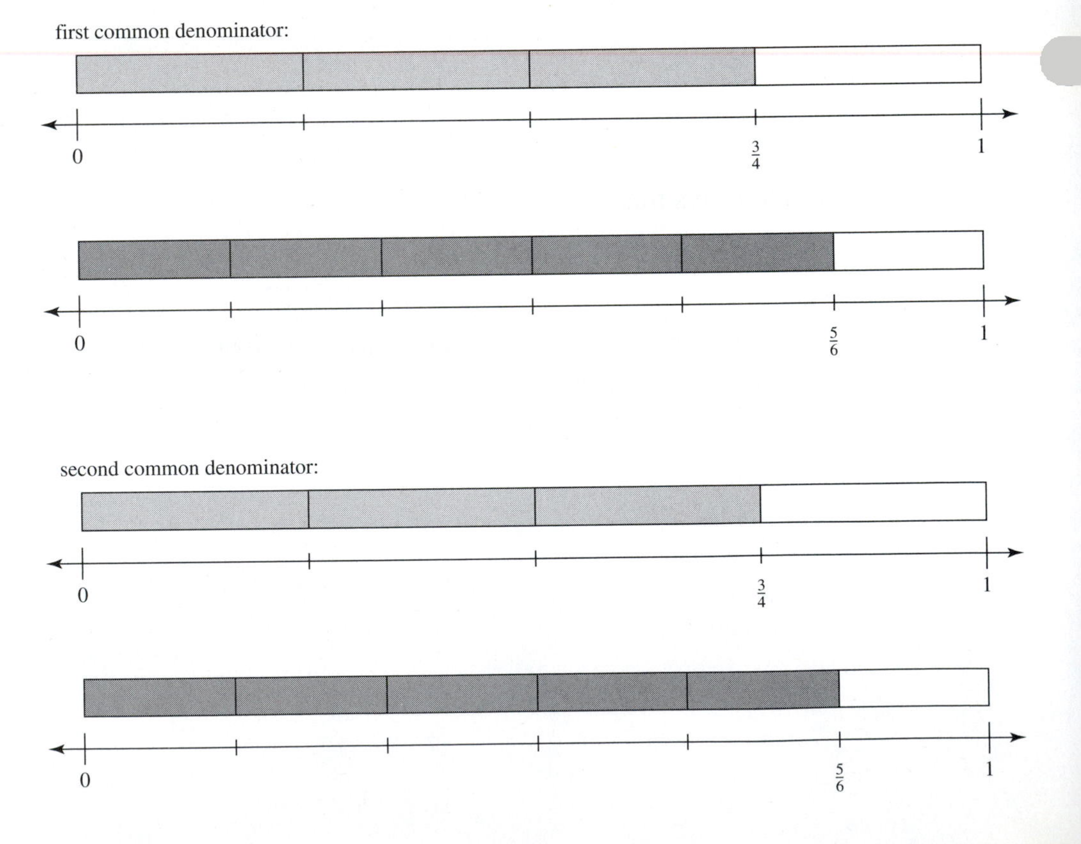

Class Activity 3J: Solving Problems by Changing Denominators

1. Take a blank piece of paper and imagine that it is $\frac{2}{3}$ of some larger piece of paper. Fold your piece of paper to show $\frac{1}{6}$ of the larger (imagined) piece of paper. Do this as carefully and precisely as possible without using a ruler or doing any measuring. Explain why your answer is correct.

   In solving this problem, how does $\frac{2}{3}$ appear in a different form?

   Could two people have different solutions that are both correct?

2. Jeremy first put $\frac{3}{4}$ cup of butter in the sauce he is making. Then Jeremy added another $\frac{1}{3}$ of a cup of butter. How much butter is in Jeremy's sauce? Draw pictures to help you solve this problem. Explain why your answer is correct.

   In solving the problem, how do $\frac{3}{4}$ and $\frac{1}{3}$ appear in different forms?

   For each fraction in this problem, and in your solution, describe the whole associated with this fraction. In other words, describe what each fraction is *of*.

3. Jean has a casserole recipe that calls for $\frac{1}{2}$ cup of butter. Jean only has $\frac{1}{3}$ cup of butter. Assuming that Jean has enough of the other ingredients, what fraction of the casserole recipe can Jean make? Draw pictures to help you solve this problem. Explain why your answer is correct.

In solving this problem, how do $\frac{1}{2}$ and $\frac{1}{3}$ appear in different forms?

For each fraction in this problem, and in your solution, describe the whole associated with this fraction. In other words, describe what each fraction is *of*.

4. One serving of SugarBombs cereal is $\frac{3}{4}$ cup. Joey wants to eat $\frac{1}{2}$ of a serving of SugarBombs cereal. How much of a cup of cereal should Joey eat? Draw pictures to help you solve this problem. Explain why your answer is correct.

In solving the problem, how does $\frac{3}{4}$ appear in a different form?

For each fraction in this problem, and in your solution, describe the whole associated with this fraction. In other words, describe what each fraction is *of*.

Class Activity 3K: Fractions on Number Lines, Part 2

1. Plot 1, $\frac{2}{3}$, and $\frac{5}{2}$ on the number line for this problem in such a way that each number falls on a tick mark. Lengthen the tick marks of whole numbers.

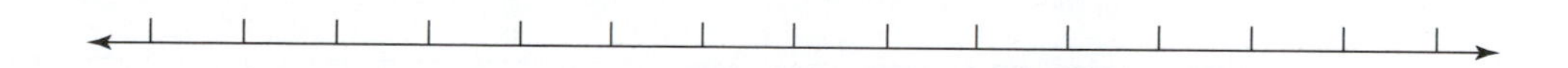

2. Plot 9, $\frac{55}{6}$, and $\frac{33}{4}$ on the number line for this problem in such a way that each number falls on a tick mark. Lengthen the tick marks of whole numbers.

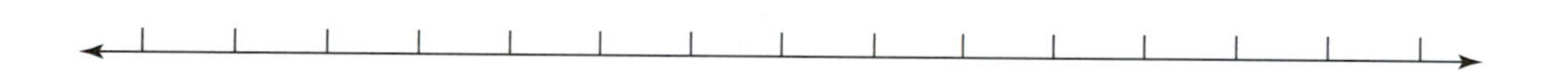

3. Plot 1, 0.7, and $\frac{3}{4}$ on the number line for this problem in such a way that each number falls on a tick mark. Lengthen the tick marks of whole numbers.

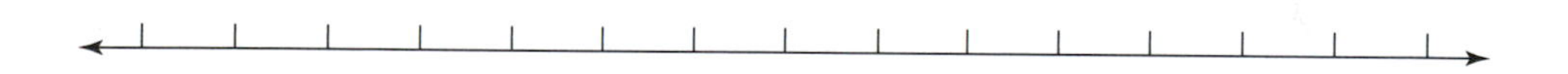

For the next two problems, place equally spaced tick marks on the number line so that you can plot the requested fraction on a tick mark. You may place the tick marks "by eye"; precision is not needed.

4. Plot $\frac{3}{4}$.

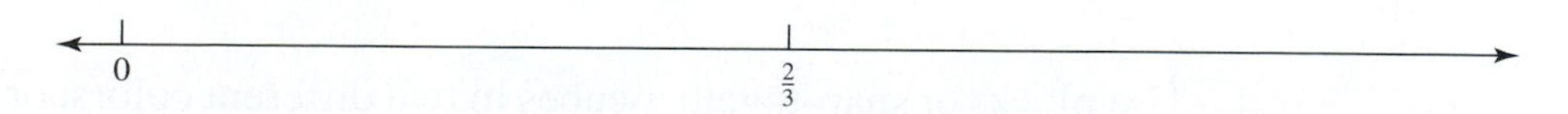

5. Plot $\frac{3}{5}$.

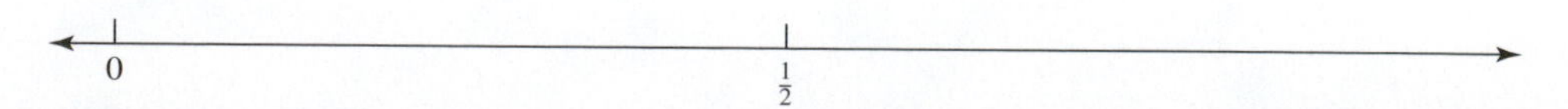

Class Activity 3L: Simplifying Fractions

1. Use the next diagram to help you explain why the following equations that put $\frac{6}{15}$ in simplest form make sense:

$$\frac{6}{15} = \frac{2 \cdot 3}{5 \cdot 3} = \frac{2}{5}$$

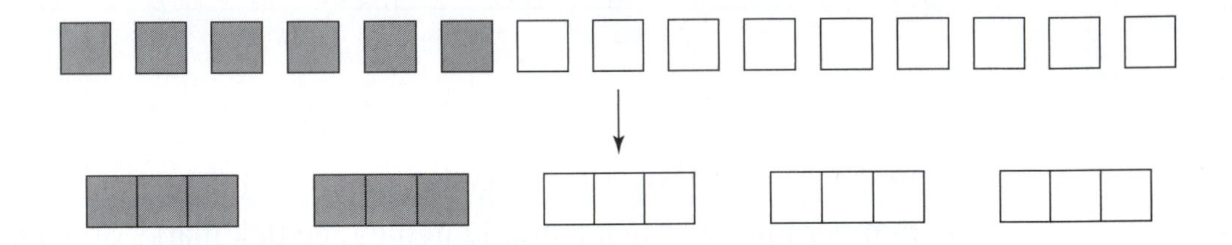

2. Use blocks or snap-together cubes in two different colors, or draw pictures, to help you demonstrate why the following equations that put $\frac{8}{12}$ in simplest form make sense:

$$\frac{8}{12} = \frac{2 \cdot 4}{3 \cdot 4} = \frac{2}{3}$$

3. Show two different ways to write equations that put $\frac{12}{18}$ in simplest form. (One way can take several steps to simplify the fraction.)

4. Use blocks or snap-together cubes in two different colors, or draw pictures, to help you explain why your equations, above, in Problem 3, make sense.

Class Activity 3M: When Can We "Cancel" to Get an Equivalent Fraction?

Some people say that

$$\frac{20}{30} = \frac{2}{3}$$

because you can "cancel the zeros" in $\frac{20}{30}$:

$$\frac{2\not{0}}{3\not{0}}$$

But if you can "cancel the zeros" in $\frac{20}{30}$, then why can't you "cancel the 5s" in

$$\frac{52}{53}$$

and say that

$$\frac{52}{53} = \frac{2}{3}?$$

Can you "cancel the 5s" in

$$\frac{25}{35}$$

and say that

$$\frac{25}{35} = \frac{2}{3}?$$

What is a better way to explain why

$$\frac{20}{30} = \frac{2}{3}$$

than saying that we "cancel the zeros"? What are you really doing when you "cancel the zeros"?

3.4 Comparing Fractions

Class Activity 3N: Can We Compare Fractions this Way?

You will need a calculator for Problem 1 of this activity.

One method for comparing fractions is to determine the decimal representation of each fraction and to compare the decimal numbers.

When you are asked to compare two fractions to see whether they are equal, your first inclination might be to use a calculator to divide, converting each fraction to a decimal number, so that you can compare the decimal numbers. This usually works well, but watch out, because calculators can display only so many digits.

Compare

$$\frac{111{,}111{,}111{,}111}{1{,}000{,}000{,}000{,}000} \quad \text{and} \quad \frac{111{,}111{,}111{,}111}{999{,}999{,}999{,}999}$$

by dividing, thereby converting both fractions to decimal numbers. Use a calculator for the second fraction. (If your calculator holds only 8 digits, then work with the fractions $\frac{11{,}111{,}111}{100{,}000{,}000}$ and $\frac{11{,}111{,}111}{99{,}999{,}999}$ instead.)

1. According to your calculator's display, do the two fractions appear to be equal?

2. In fact, *are* the two fractions equal? If not, which is bigger, and why? Explain this by reasoning about the fractions.

Class Activity 3O: What Is Another Way to Compare these Fractions?

For each of the pairs of fractions shown, determine which fraction is greater in a way other than finding common denominators or converting to decimals. Explain your reasoning.

$$\frac{1}{49} \qquad \frac{1}{39}$$

$$\frac{7}{37} \qquad \frac{7}{45}$$

Class Activity 3P: Comparing Fractions by Reasoning

Use reasoning other than finding common denominators, cross-multiplying, or converting to decimal numbers to compare the sizes (=, <, or >) of the following pairs of fractions:

$$\frac{27}{43} \qquad \frac{26}{45}$$

$$\frac{13}{25} \qquad \frac{34}{70}$$

$$\frac{17}{18} \qquad \frac{19}{20}$$

$$\frac{9}{40} \qquad \frac{12}{44}$$

$$\frac{51}{53} \qquad \frac{65}{67}$$

$$\frac{13}{25} \qquad \frac{5}{8}$$

Class Activity 3Q: Can We Reason this Way?

1. Claire says that

$$\frac{4}{9} > \frac{3}{8}$$

because

$$4 > 3 \quad \text{and} \quad 9 > 8$$

Discuss whether or not Claire's reasoning is correct.

2. Conrad says that

$$\frac{4}{11} > \frac{3}{8}$$

because

$$4 > 3 \quad \text{and} \quad 11 > 8$$

Discuss whether or not Conrad's reasoning is correct.

3.5 Percent

Class Activity 3R: Pictures, Percentages, and Fractions

1. For each of diagrams 1–5, determine the percent of the diagram that is shaded, explaining your reasoning. Write each percent as a fraction in simplest form, and explain how to see that this fraction of the diagram is shaded. You may assume that portions of each diagram which appear to be the same size really are the same size.

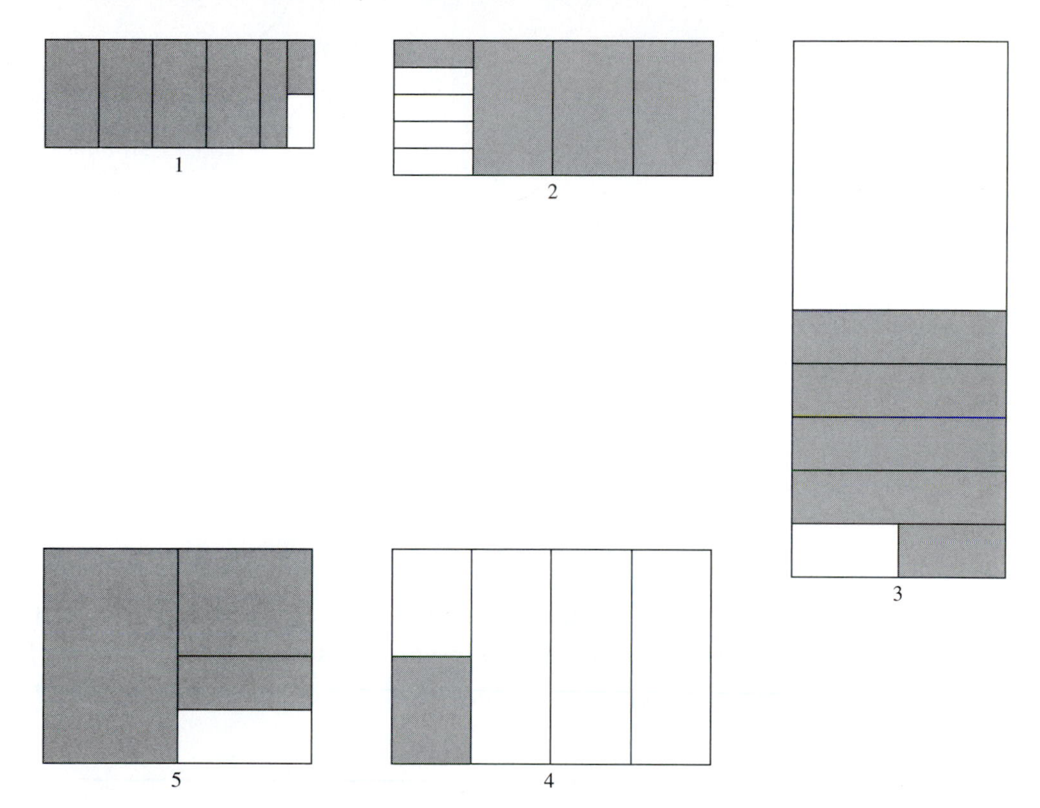

2. For each of diagrams 6 and 7, determine approximately what percent of the diagram is shaded. Give your answer rounded to the nearest multiple of 5 (i.e., 5, 10, 15, 20, 25, …), and explain your reasoning.

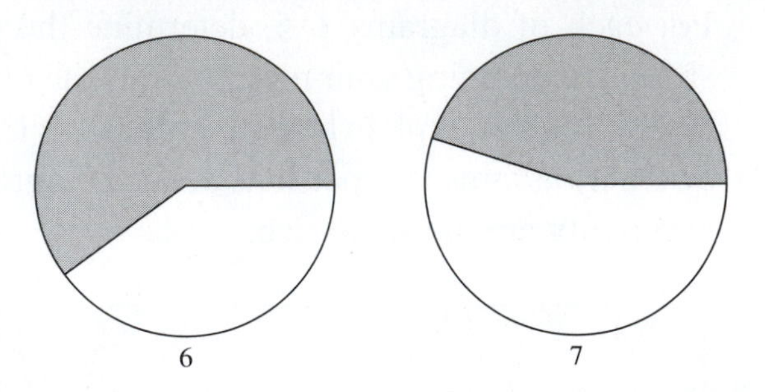

3. Show 125% of the line segment that follows. Then show 80% of your new line segment. It may help you to use common fractions. How does your final line segment compare with the original line segment?

Class Activity 3S: Calculating Percents of Quantities by Using Benchmark Fractions

We can make some percentage problems easy to solve mentally by working with a common fraction that is easy to calculate with. Pictures can help you understand this technique. Simple "percent diagrams" can help you record and clarify your thinking.

1. Mentally calculate 95% of 80,000 by calculating $\frac{1}{10}$ of 80,000, calculating half of that result, and then taking this last amount away from 80,000. Use the following picture and the "percent diagram" to help you explain why this method makes sense and to record your thinking:

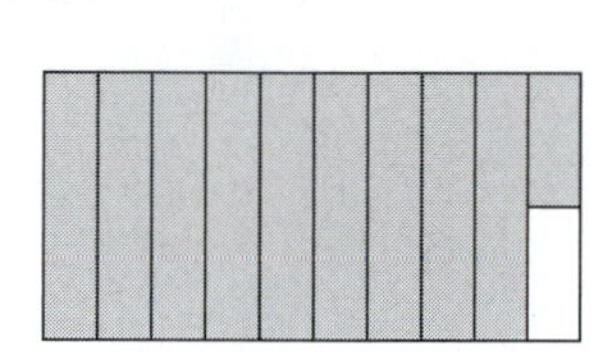

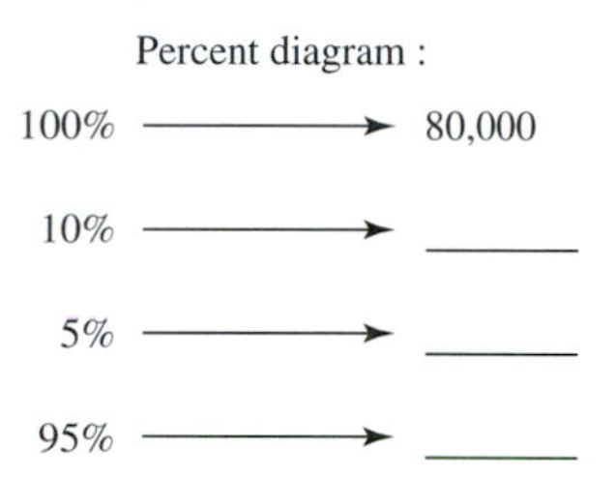

2. Mentally calculate 15% of 6500 by first calculating $\frac{1}{10}$ of 6500. Use the following picture and a "percent diagram" to help you explain why this method makes sense and to record your thinking.

Percent diagram :

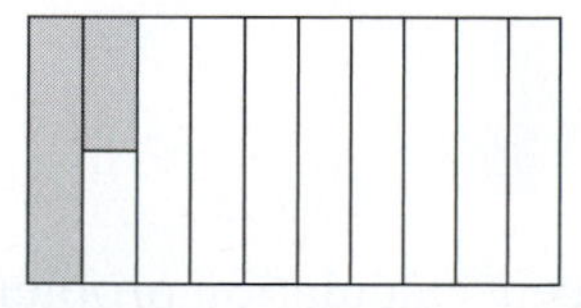

3. Mentally calculate 7% tax on a $25 purchase by first finding 1%. Use a percent diagram to record the strategy.

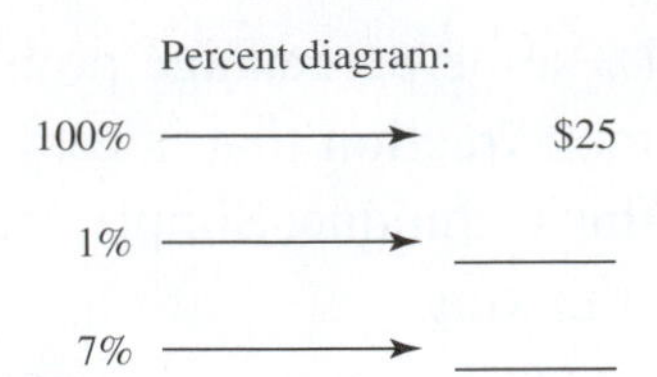

4. Mentally calculate 60% of 810 by first calculating $\frac{1}{2}$ of 810. Use the following picture and a "percent diagram" to help you explain why this method makes sense.

Percent diagram :

5. Find another way to calculate 60% of 810 mentally. Use a percent diagram to show your method.

6. Mentally calculate 55% of 180. Use a percent diagram to show your method.

7. Make up a mental percent calculation problem, and explain how to solve it.

Class Activity 3T: Calculating Percentages

1. Mentally determine what percent 660 is of 1200 by first observing that $\frac{1}{2}$ of 1200 is 600. Use a percent diagram to help you explain why this method makes sense.

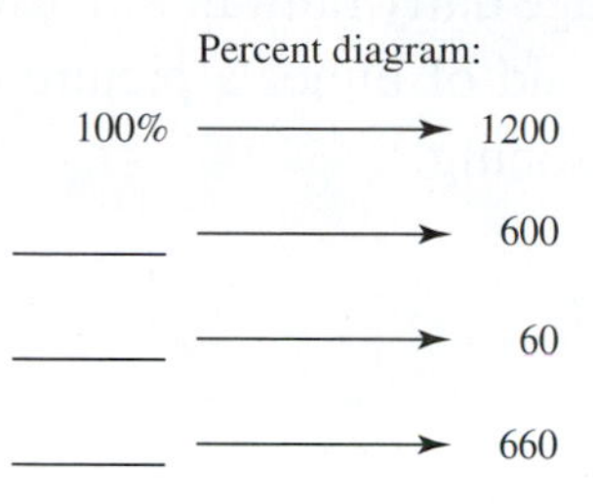

2. Show how to determine what percent 660 is of 1200 by finding equivalent fractions. Explain why you should look for a fraction that has denominator 100.

$$\frac{660}{1200} =$$

3. Mentally determine what percent 60 is of 400. Use a percent diagram to help you explain why your method makes sense.

4. Show how to determine what percent 60 is of 400 by finding equivalent fractions.

5. Mentally determine what percent 225 is of 500. Use a percent diagram to help you explain why your method makes sense.

6. Show how to determine what percent 225 is of 500 by finding equivalent fractions.

Class Activity 3U: Calculating Percentages with Pictures and Percent Diagrams

1. In Green Valley, the average daily rainfall is $\frac{5}{8}$ of an inch. Last year, the average daily rainfall in Green Valley was only $\frac{3}{8}$ of an inch. What percent of the average daily rainfall fell last year in Green Valley? Solve this problem with the aid of either a picture or a percent diagram, or both, explaining your reasoning.

2. If a $\frac{1}{4}$ cup serving of cheese provides your full daily value of calcium, then what percentage of your daily value of calcium is provided by $\frac{1}{3}$ cup of the cheese? Solve this problem with the aid of either a picture or a percent diagram, or both, explaining your reasoning.

Class Activity 3V: Calculating Percentages by Going through 1

We have seen how to calculate a percentage of a quantity by first calculating 1% of the quantity. This was the method of "going through 1%." In this activity, we'll see a different way to "go through 1."

1. Use the next percent diagram to help you determine what percent 21 is of 500. Explain your reasoning.

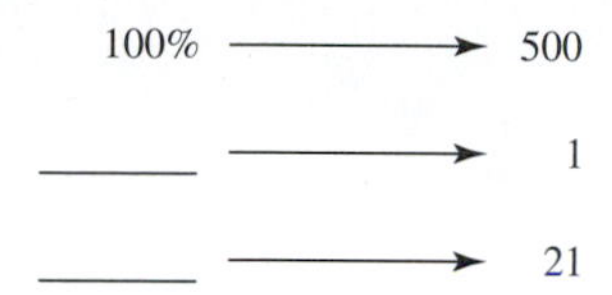

2. Observe that the previous problem asked you to "go through 1" to determine a percentage. Use a percent diagram and the method of "going through 1" to determine what percent 9 is of 40.

3. The method of "going through 1" can also be shown with equivalent fractions, although the numerators of the fractions may not always be whole numbers. Show how to determine what percent 3 is of 7 by "going through 1" with equivalent fractions.

$$\frac{3}{7} = \frac{\quad}{1} = \frac{\quad}{100}$$

4. Now show how to determine what percent 3 is of 7 by going through 1 with a percent diagram. Compare the calculations you did in Problem 3 with the ones you do here.

Class Activity 3W: Calculating a Quantity from a Percentage of It

1. Lenny has received 6 boxes of paper, which is 30% of the paper he ordered. How many boxes of paper did Lenny order? Draw a picture to help you solve this problem. Explain your reasoning.

2. Solve the problem about Lenny's paper in Problem 1 by completing the next percent diagram. Explain your reasoning.

$$30\% \longrightarrow 6$$

3. Now solve the problem about Lenny's paper in Problem 1 by making equivalent fractions (without "cross-multiplying"). Explain your reasoning.

$$\frac{30}{100} = \frac{6}{}$$

4. There are 30 blue marbles in a bag, which is 40% of the marbles in the bag. How many marbles are in the bag? Solve this problem in three ways: 1) with the aid of a picture, 2) with the aid of a percent diagram, and 3) by making equivalent fractions (without "cross-multiplying"). Explain your reasoning in each case.

5. Ms. Jones paid $2.10 in tax on an item she purchased. The tax was 7% of the price of the item. What was the price of the item (not including the tax)? Solve this problem with the aid of a percent diagram. Explain your reasoning.

4

Addition and Subtraction

4.1 Interpretations of Addition and Subtraction

Class Activity 4A: Addition and Subtraction Story Problems

The basic addition and subtraction equations are of the form

$$A + B = C \qquad D - E = F$$

Addition and subtraction problems arise when two out of the three quantities in an addition or subtraction equation are known and the other quantity is to be found.

1. For each of the following equations, write a story problem that is formulated naturally by the equation.

   a. $6 + 9 = ?$

   b. $6 + ? = 15$

   c. $? + 6 = 15$

   d. $15 - 6 = ?$

   e. $15 - ? = 6$

   f. $? - 6 = 9$

For the addition equations (a), (b), (c) in Problem 1, you may have written "add to" (or "join") problems, and for the subtraction equations (d), (e), (f), you may have written "take away" (or "separate") problems. These are the most common type of addition and subtraction problems. However, there are several other types of addition and subtraction story problems. Here is a list of the types of addition and subtraction story problems mathematics education researchers have identified (see [9], page 185, [2], p. 12, or [4], page 70).

Add to or "join" problems. Example for the equation $6 + ? = 15$:

> Asia had 6 stickers. After Asia got some more stickers, she had 15 stickers. How many stickers did Asia get?

Take away or "separate" problems. Example for the equation $? - 6 = 9$:

> Asia had some stickers. After Asia gave 6 of her stickers away, she had 9 stickers left. How many stickers did Asia have at first?

Part-Part-Whole problems. Example for the equation $6 + 9 = ?$:

> Asia has 6 small stickers and 9 large stickers (and no other stickers). How many stickers does Asia have in all?

Part-part-whole problems involve two distinct parts that make a whole. Notice that the equations $15 - ? = 6$ and $? - 6 = 9$ don't fit naturally with part-part-whole problems.

Compare problems. Example for the equation $6 + ? = 15$:

> Asia has 6 stickers. Taryn has 15 stickers. How many more stickers does Taryn have than Asia?

Example for the equation $? - 6 = 9$:

> Asia has some stickers. Taryn has 9 stickers, and that is 6 stickers fewer than Asia has. How many stickers does Asia have?

Compare problems involve comparing two quantities.

2. Write a "part-part-whole" problem for the equation $6 + ? = 15$. (Your problem will probably fit naturally with the equation $? + 6 = 15$ as well.)

3. Write a "compare" problem for the equation $6 + 9 = ?$.

4. Write a "compare" problem for the equation $? + 6 = 15$.

5. Write a "compare" problem for the equation $15 - 6 = ?$.

6. Write a "compare" problem for the equation $15 - ? = 6$.

Class Activity 4B: Solving Addition and Subtraction Story Problems

1. Young children often solve addition and subtraction story problems by modeling the actions in the problems. For each story problem that follows, show how a child could use small objects (such as counters or blocks) to solve the problem by modeling. Which of these problems do you think might be hardest for young children? Why? (See [2] and [4] for a detailed discussion of children's addition and subtraction solution strategies, including levels of development of strategies.)

   a. "Add to" problem for $3 + ? = 9$. Clare had 3 bears. After she got some more bears, Clare had 9 bears. How many bears did Clare get?

   b. "Add to" problem for $? + 3 = 9$. Clare had some bears. After she got 3 more bears, Clare had 9 bears. How many bears did Clare get?

   c. "Part-part-whole" problem for $? + 3 = 9$. Clare has some red bears and 3 blue bears. All together, Clare has 9 bears, and all of them are either red or blue. How many red bears does Clare have?

   d. "Take away" problem for $9 - ? = 3$. Clare had 9 bears. After she gave away some bears, Clare had 3 bears left. How many bears did Clare give away?

   e. "Take away" problem for $? - 3 = 6$. Clare had some bears. After she gave away 3 bears, she had 6 bears left. How many bears did Clare have at first?

   f. "Compare" problem for $3 + 6 = ?$. Clare has 3 red bears. She also has 6 more blue bears than red bears. How many blue bears does Clare have?

   g. "Compare" problem for $3 + ? = 9$ (or for $9 - 3 = ?$). Clare has 9 red bears and 3 blue bears. How many more red bears does Clare have than blue bears?

2. Older children and adults usually just add or subtract appropriate numbers in order to solve the problems in Problem 1. For each problem in Problem 1, what calculation could an adult or child perform to solve the problem?

Some of the problems in Problem 1 are formulated with addition, but can be solved with subtraction, and some are formulated with subtraction, but can be solved with addition. Which problems are these? Children frequently solve these problems incorrectly, adding numbers that should have been subtracted or subtracting numbers that should have been added (See [4]).

3. Simple pictures can be helpful in deciding how to solve a problem. For each of the pictures A–F that follow, determine which of the following equations could fit with the picture. (Each picture fits with *more than one* equation. You may also write other equations that fit with the pictures.)

$$30 + 45 = ? \qquad 30 + ? = 75 \qquad ? + 30 = 75$$

$$75 - 30 = ? \qquad 75 - ? = 30 \qquad ? - 30 = 45 \qquad ? - 45 = 30$$

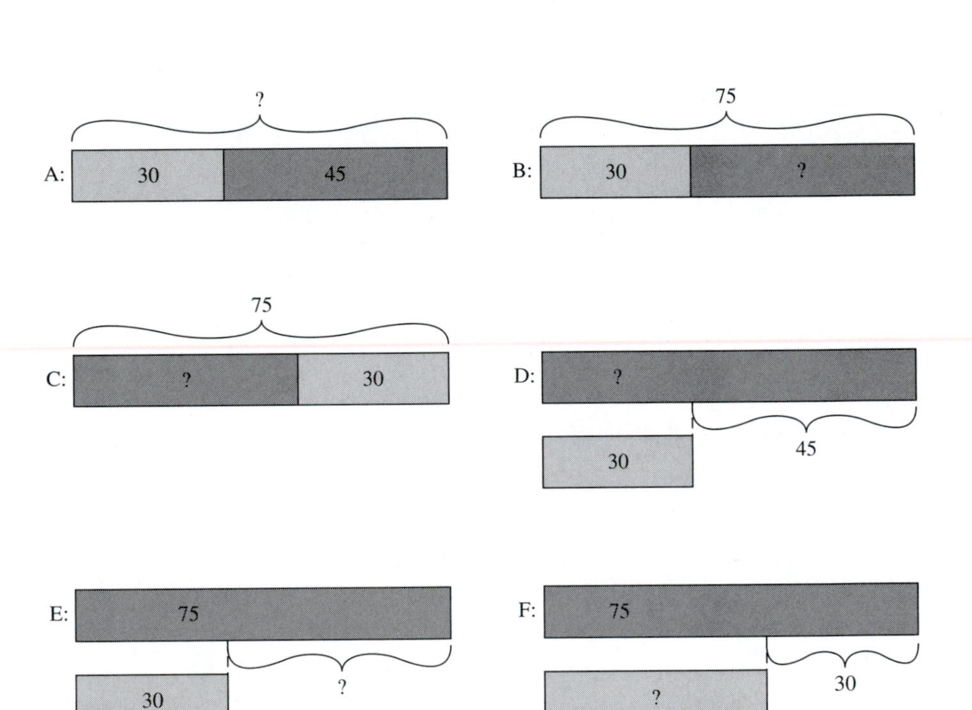

4. For each picture A–F in Problem 3, imagine drawing the picture in order to solve a story problem. Which of the following would you draw the picture for: an "add to" problem, a "take away" problem, a "part-part-whole" problem, or a "compare" problem?

5. For each picture A–F in Problem 3, use the picture to determine whether to add or subtract to solve for the unknown amount.

Class Activity 4C: The Shopkeeper's Method of Making Change

When a patron of a store gives a shopkeeper \$$A$ for a \$$B$ purchase, we can think of the change owed to a patron as what is left from \$$A$ when \$$B$ are taken away.

In contrast, the shopkeeper might make change by starting with \$$B$ and handing the customer money while adding on the amounts until they reach \$$A$.

1. Describe in detail how a shopkeeper could use this method to give a patron change from a \$10 bill on a \$3.26 purchase.

2. Describe in detail how a shopkeeper could use this method to give a patron change from a \$20 bill on a \$7.43 purchase.

3. Reconcile the method used by shopkeepers with the view that change is what is left from \$$A$ when \$$B$ are taken away. Why is the shopkeeper's method for making change valid?

Class Activity 4D: Addition and Subtraction Story Problems with Negative Numbers

Negative numbers are nicely interpreted as amounts owed, temperatures below zero, locations below ground (such as in a mine or a building that has stories below ground), or locations below sea level.

1. Write and solve an "add to" story problem for $(-5) + 5 = ?$.

2. Write and solve a "take away" story problem for $(-2) - 5 = ?$.

3. Write and solve a "compare" story problem for $(-2) - 5 = ?$.

4. Write and solve a "compare" story problem for $(-5) + ? = 2$ (or $2 - (-5) = ?$).

4.2 Why the Common Algorithms for Adding and Subtracting Decimal Numbers Work

Class Activity 4E: Adding and Subtracting with Ten-Structured Pictures

The hypothetical student work below is similar to actual work of urban Latino first graders whose performance on 2-digit addition and subtraction tasks with regrouping was substantially above the performance of first graders of higher socioeconomic status and of older children. (See [5].)

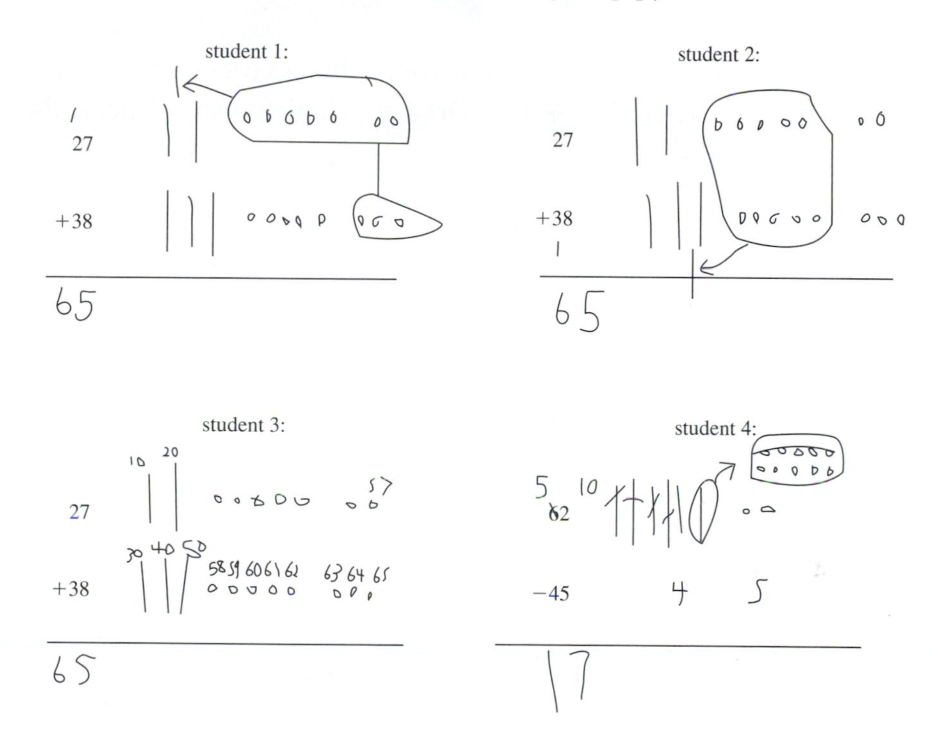

1. Examine and discuss the students' work. Compare the work of students 1, 2, and 3. In particular, compare the methods of student 1 and student 2 for adding $7 + 8$. What mental method for subtracting $12 - 5$ does the work of student 4 suggest?

2. Show how students 1, 2, and 3 might solve the addition problem $36 + 27$ and how student 4 might solve the subtraction problem $43 - 18$.

Class Activity 4F: Understanding the Common Addition Algorithm

Bundled toothpicks (or other small objects) would be useful for this activity.

1. Add the numbers, using the standard paper-and-pencil method. Notice that regrouping is involved.

$$\begin{array}{r} 147 \\ +\,195 \\ \hline \end{array}$$

2. If available, use bundles of toothpicks (or other objects) to solve the addition problem 147 + 195. Draw rough pictures to indicate the process you used.

3. If available, use bundles of toothpicks (or other objects) to solve the addition problem 147 + 195 *in a way that corresponds directly to the addition algorithm you used in Problem 1*. This may be different from what you did in Problem 2. Draw rough pictures to indicate the process you used. Compare with Problem 2.

4. Use expanded forms to solve the addition problem 147 + 195. First add like terms, and then rewrite (regroup) the resulting number so that it is the expanded form of a number. This rewriting is the regrouping process.

$$\begin{array}{r} 1(100) + 4(10) + 7(1) \\ + \ 1(100) + 9(10) + 5(1) \\ \hline \end{array}$$

← First add like terms, remaining in expanded form.

← Then regroup so that you have the expanded form of a decimal number. You might want to take several steps to do so.

5. Compare and contrast your work in Problems 1–4.

Class Activity 4G: Understanding the Common Subtraction Algorithm

Bundled toothpicks (or other small objects) would be useful for this activity.

1. Subtract the following numbers, using the standard paper-and-pencil method. Notice that regrouping is required.

$$\begin{array}{r} 125 \\ -\ \ 68 \\ \hline \end{array}$$

2. If available, use bundles of toothpicks (or other objects) to solve the subtraction problem 125 – 68. Draw rough pictures to indicate the process.

3. If available, use bundles of toothpicks (or other objects) to solve the subtraction problem 125 − 68 *in a way that corresponds directly to the subtraction algorithm you used in Problem 1*. This may be different from what you did in Problem 2. Draw rough pictures to indicate the process. Compare with Problem 2.

4. Solve the subtraction problem $125 - 68$, but now use expanded forms. Start by rewriting the number 125 in expanded form. Rewrite the number in several steps, so that it will be easy to take away 68. This rewriting is the regrouping process.

$$125 = 1(100) + 2(10) + 5(1) =$$

Write your regrouped number here $\rightarrow$

Subtract 68: $\underline{-\ [6(10) + 8(1)]}$

5. Compare and contrast your work in Problems 1–4.

Class Activity 4H: Subtracting across Zeros

1. David solves the subtraction problem 203 − 7 as follows:

```
  1 13
  20̸3̸
−    7
 ─────
   106
```

What mistake is David making? Explain why David's method is incorrect.

2. Some subtraction problems require regrouping "across a 0," as does the problem

```
  203
−  86
 ────
```

Some people like to replace an intermediate 0 with a 9 when regrouping:

```
                1913
  203           2̸0̸3̸
−  86   →   −    86
 ────          ─────
                 117
```

Others prefer to regroup in two steps, first changing an intermediate 0 to a 10 and then changing the 10 to a 9:

```
                               9
                110          11̸0̸13
  203           2̸0̸3           2̸0̸3̸
−  86   →   −    86   →   −    86
 ────          ─────         ─────
                               117
```

a. Explain why the first regrouping method described is just a shortcut of the second method.

b. Using bundled toothpicks, or rough drawings of bundled toothpicks, explain why both of the regrouping methods described make sense.

3. DeShun solves the subtraction problem 1002 – 248 by crossing out 100 and replacing it with 99, as shown. Is this legitimate? To help your thinking, consider the following: The 100 is 100 of what?

$$\begin{array}{r} \scriptstyle 99\,12 \\ \not{1}\not{0}\not{0}\not{2} \\ -248 \\ \hline 754 \end{array}$$

4. Can DeShun's method from the previous part be used in other situations? For example, could you use DeShun's method to subtract 20047 – 321? Explain.

Class Activity 4I: Regrouping with Dozens and Dozens of Dozens

When we regroup decimal numbers we use the base-ten structure of place value: The value of each place is ten times the value of the next place to the right. We can also use the regrouping idea in other situations—for example, when objects are bundled in groups of a dozen instead of in groups of ten. The following problem asks you to regroup with dozens and dozens of dozens:

A store owner buys small, novelty party favors in bags of one dozen and boxes of one dozen bags (for a total of 144 favors in a box). The store owner has 5 boxes, 4 bags, and 3 individual party favors at the start of the month. At the end of the month, the store owner has 2 boxes, 9 bags, and 7 individual party favors left. How many favors did the store owner sell? Give the answer in terms of boxes, bags, and individual favors.

Solve this problem by working with a sort of *expanded form* for these party favors—in other words, working with the following representation:

$$5(\text{boxes}) + 4(\text{bags}) + 3(\text{individual})$$
$$2(\text{boxes}) + 9(\text{bags}) + 7(\text{individual})$$

Solve this problem by regrouping among the boxes, bags, and individual party favors.

Class Activity 4J: Regrouping with Seconds, Minutes, and Hours

When we regroup decimal numbers we use the base-ten structure of place value: The value of each place is ten times the value of the next place to the right. We can also use the regrouping idea in other situations—for example, with time. Just as 1 hundred is 10 tens and 1 ten is 10 ones, one hour is 60 minutes and one minute is 60 seconds. The following problem asks you to regroup among hours, minutes, and seconds:

Ruth runs around a lake two times. The first time takes 1 hour, 43 minutes, and 38 seconds. The second time takes 1 hour, 48 minutes, and 29 seconds. What is Ruth's total time for the two laps? Give the answer in hours, minutes, and seconds.

Solve this problem by working with a sort of *expanded form* for time, in other words, by working with

$$1(\text{hour}) + 43(\text{minutes}) + 38(\text{seconds})$$
$$1(\text{hour}) + 48(\text{minutes}) + 29(\text{seconds})$$

Solve this problem by regrouping among hours, minutes, and seconds.

Class Activity 4K: A Third Grader's Method of Subtraction

When asked to compute 423 − 157, Pat (a third grader) wrote the following:

4−
30−
34−
300
266

"You can't take 7 from 3; it's 4 too many, so that's negative 4. You can't take 50 from 20; it's 30 too many, so that's negative 30; and with the other 4, it's negative 34. 400 minus 100 is 300, and then you take the 34 away from the 300, so it's 266."[1]

1. Discuss Pat's idea for calculating 423 − 157. Is her method legitimate? Analyze Pat's method in terms of expanded forms.

2. Could you use Pat's idea to calculate 317 − 289? If so, write what you think Pat might write, and also use expanded forms.

[1] This is taken from [1, p.263]

4.3 Adding and Subtracting Fractions

Class Activity 4L: Fraction Addition and Subtraction

Paper, scissors, and tape would be useful for this activity.

1. Cut (or tear) a strip of paper, and label it "1 whole," indicating that the strip is 1 whole unit long. Cut about 5 more strips of the same length. By folding, cutting, and joining strips, create and label strips of the following lengths:

$$\frac{1}{2}+\frac{1}{3}, \quad \frac{2}{3}-\frac{1}{2}, \quad \frac{2}{3}+\frac{1}{2}$$

2. Use your strips from Problem 1, as well as drawings and number lines, to help you explain why the procedures for adding and subtracting fractions make sense. In particular, explain why we need common denominators.

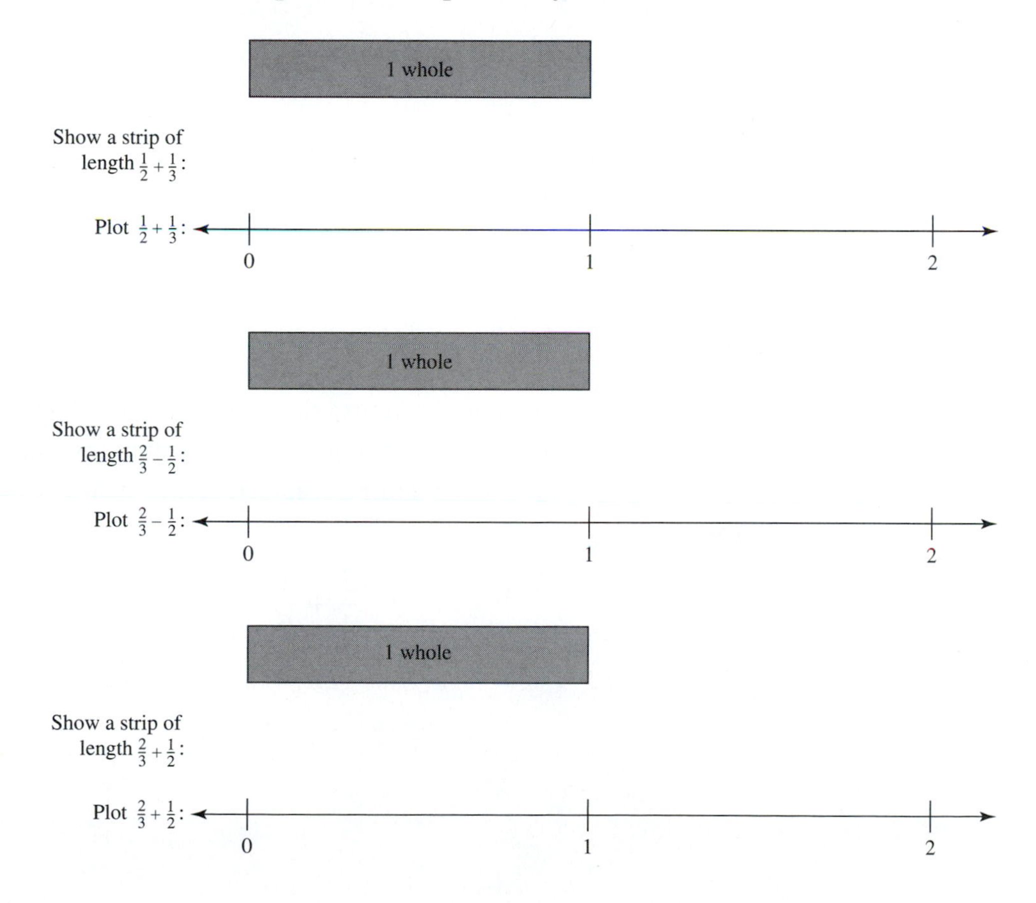

3. Patti thinks that it would be easier to add fractions by "adding the tops and adding the bottoms." For example, Patti wants to add $\frac{2}{3}$ and $\frac{4}{5}$ this way:

$$\frac{2}{3} + \frac{4}{5} = \frac{2+4}{3+5} = \frac{6}{8}$$

Patti uses the picture below to explain why her method makes sense. Why is Patti's method not a valid way to add fractions, and why doesn't Patti's picture prove that fractions can be added in her way? Do not just state the proper way to add fractions; explain what is wrong with Patti's reasoning.

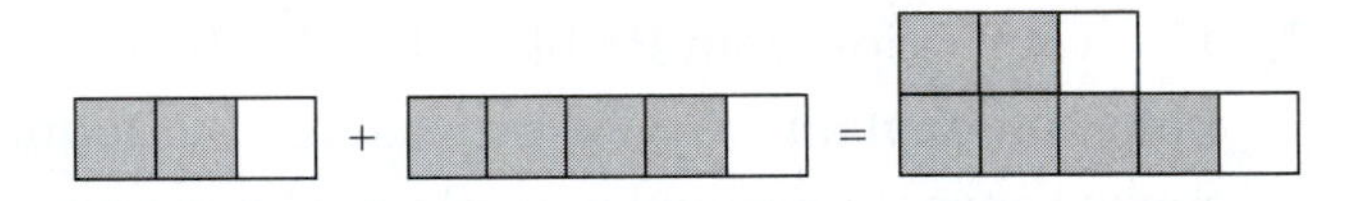

4. When we add fractions, *must* we use the least common denominator? Why or why not? Are there any advantages or disadvantages to using the least common denominator when adding fractions? Illustrate your answers with the example

$$\frac{3}{4} + \frac{1}{6}$$

and with other examples.

Class Activity 4M: Mixed Numbers and Improper Fractions

1. Draw a picture to show how to convert $2\frac{3}{5}$ to an improper fraction. Relate your picture to the following numerical procedure for converting $2\frac{3}{5}$ to an improper fraction:

$$2\frac{3}{5} = \frac{2 \cdot 5 + 3}{5} = \frac{13}{5}$$

2. Sue showed Ramin the next diagram to explain why $1\frac{3}{4} = \frac{7}{4}$, but Ramin says that it shows $\frac{7}{8}$, not $\frac{7}{4}$. What must Sue and Ramin clarify?

3. Label the first number line shown with fractions and mixed numbers. Label the second number line with proper and improper fractions. Then use the number lines to help you explain why the procedure for converting mixed numbers to improper fractions is valid.

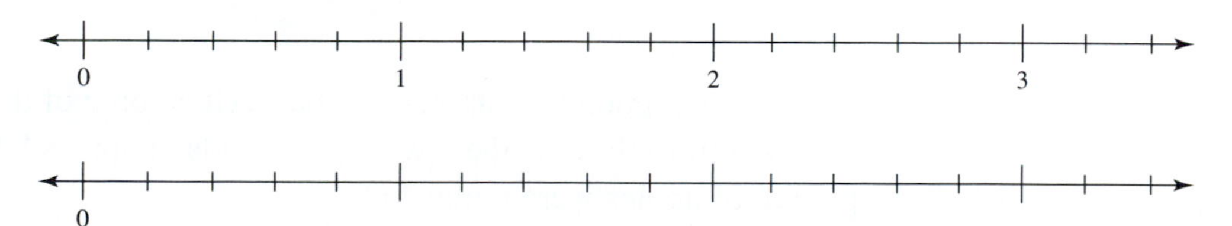

Class Activity 4N: Adding and Subtracting Mixed Numbers

1. Each of the problems that follow shows some student work. Discuss the work: What is correct and what is not correct? In each case, either complete the work or modify it to make it correct. *Use the student's work; do not start from scratch.*

   a. Subtract: $3\frac{1}{4} - 1\frac{3}{4}$.

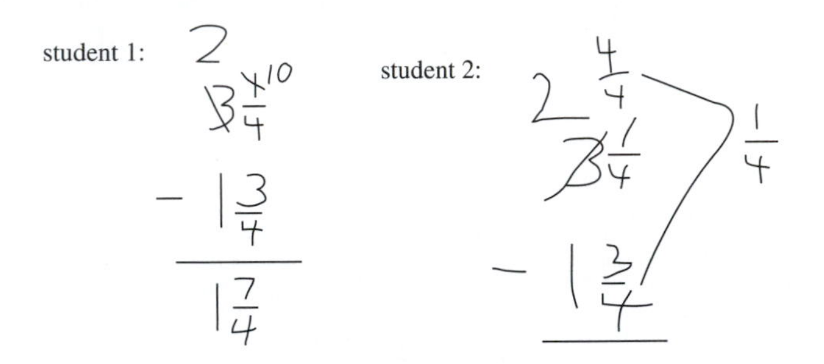

   b. There are $2\frac{1}{3}$ cups of milk in a bowl. How much milk must be added to the bowl so that there will be 3 cups of milk in the bowl?

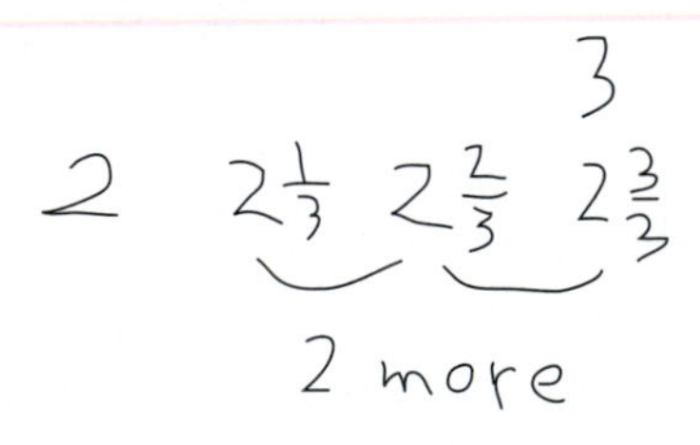

   c. There were 5 pounds of apples in a bag. After some of the apples were removed from the bag, there were $3\frac{1}{4}$ pounds of apples left. How many pounds of apples were removed?

5 $\frac{4}{4}$ $\frac{4}{4}$ $\frac{4}{4}$ $\frac{4}{4}$ $\frac{4}{4}$

$\frac{1}{4}$ $\frac{3}{4}$

d. Add: $2\frac{2}{3} + 1\frac{2}{3}$.

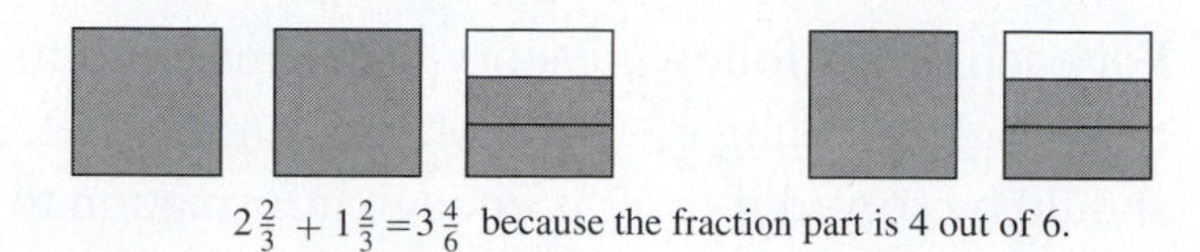

$2\frac{2}{3} + 1\frac{2}{3} = 3\frac{4}{6}$ because the fraction part is 4 out of 6.

2. Find at least two different ways to calculate

$$7\frac{1}{3} - 4\frac{1}{2}$$

and to give the answer as a mixed number. In each case, explain why your method makes sense.

Class Activity 4O: Are These Story Problems for $\frac{1}{2} + \frac{1}{3}$?

For each of the following story problems, determine whether the problem can be solved by adding $\frac{1}{2} + \frac{1}{3}$. If not, explain why not, and explain how the problem should be solved if there is enough information to do so. If there is not enough information to solve the problem, explain why not.

1. Tom pours $\frac{1}{2}$ cup of water into an empty bowl. Then Tom pours $\frac{1}{3}$ cup of water into the bowl. How many cups of water are in the bowl now?

2. Tom pours $\frac{1}{2}$ cup of water into an empty bowl. Then Tom pours in another $\frac{1}{3}$. How many cups of water are in the bowl now?

3. Starting at her apartment, Sally runs $\frac{1}{2}$ mile down the road. Then Sally turns around and runs $\frac{1}{3}$ mile back towards her apartment. How far has Sally run since leaving her apartment?

4. Starting at her apartment, Sally runs $\frac{1}{2}$ mile down the road. Then Sally turns around and runs $\frac{1}{3}$ mile back towards her apartment. How far down the road is Sally from her apartment?

5. $\frac{1}{2}$ of the land in Heeltoe County is covered with forest. $\frac{1}{3}$ of the land in the adjacent Toejoint County is covered with forest. What fraction of the land in the two-county Heeltoe–Toejoint region is covered with forest?

6. $\frac{1}{2}$ of the land in Heeltoe County is covered with forest. $\frac{1}{3}$ of the land in the adjacent Toejoint County is covered with forest. Heeltoe and Toejoint County have the same land area. What fraction of the land in the two-county Heeltoe–Toejoint region is covered with forest?

7. $\frac{1}{2}$ of the children at Martin Luther King Elementary School say they like to have pizza for lunch. $\frac{1}{3}$ of the children at Martin Luther King Elementary School say they like to have a hamburger for lunch. What fraction of the children at Martin Luther King Elementary School would like to have either pizza or a hamburger for lunch?

8. $\frac{1}{2}$ of the children at Timothy Elementary School like to have pizza for lunch, the other half does not like to have pizza for lunch. Of the children who do not like to have pizza for lunch, $\frac{1}{3}$ like to have a hamburger for lunch. What fraction of the children at Timothy Elementary School like to have either pizza or a hamburger for lunch?

Class Activity 4P: Are These Story Problems for $\frac{1}{2} - \frac{1}{3}$?

For each of the following story problems, determine whether the problem can be solved by subtracting $\frac{1}{2} - \frac{1}{3}$. If not, explain why not, and explain how the problem should be solved if there is enough information to do so. If there is not enough information to solve the problem, explain why not.

1. Zelha pours $\frac{1}{2}$ cup of water into an empty bowl. Then Zelha pours out $\frac{1}{3}$. How much water is in the bowl now?

2. Zelha pours $\frac{1}{2}$ cup of water into an empty bowl. Then Zelha pours out $\frac{1}{3}$ cup of water. How much water is in the bowl now?

3. Zelha pours $\frac{1}{2}$ cup of water into an empty bowl. Then Zelha pours out $\frac{1}{3}$ of the water that is in the bowl. How much water is in the bowl now?

4. Yesterday James ate $\frac{1}{2}$ of a pizza. Today James ate $\frac{1}{3}$ of a pizza of the same size. How much more pizza did James eat yesterday than today?

5. Yesterday James ate $\frac{1}{2}$ of a pizza. Today James ate $\frac{1}{3}$ of the whole pizza. Nobody else ate any of that pizza. How much pizza is left?

6. Yesterday James ate $\frac{1}{2}$ of a pizza. Today James ate $\frac{1}{3}$ of the pizza that was left over from yesterday. Nobody else ate any of that pizza. How much pizza is left?

Class Activity 4Q: What Fraction Is Shaded?

For each square shown, determine the fraction of the square that is shaded. Explain your reasoning. You may assume that all lengths that appear to be equal really are equal. Do not use any area formulas.

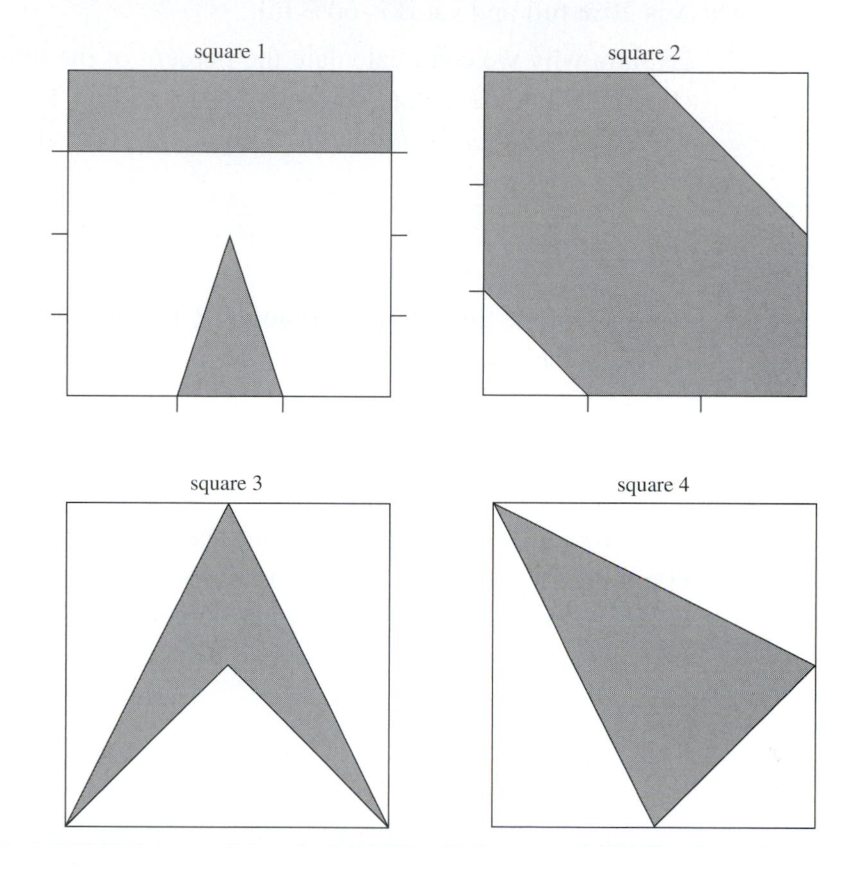

4.4 When Do We Add Percentages? *

Class Activity 4R: * Should We Add These Percentages?

A factory has two vats designed to contain liquid chocolate. At the moment, vat A is 20% full and vat B is 60% full.

1. Explain why we can't calculate the percent of the total capacity for liquid chocolate that the factory currently has by adding $20\% + 60\%$, even though we are combining the amounts of liquid chocolate in the two vats.

2. For each of the three cases in Figure 4R.1, determine the percent of its total capacity for liquid chocolate that the factory currently has.

3. Even though this is not a realistic scenario, suppose that vat A is huge compared with vat B. Assume that vat A is still 20% full and vat B is still 60% full. Approximately what percent of its total capacity for liquid chocolate would the factory have in this case?

4. Now suppose it's the other way around and vat B is much larger than vat A, but vat A is still 20% full and vat B is still 60% full. Approximately what percent of its total capacity for liquid chocolate would the factory have in this case?

5. Based on your work in Problems 1 to 4 above, if you are only given the information that there are two vats, one 20% full and one 60% full, what theoretical range of percentages is possible for the total capacity of liquid chocolate that the factory has?

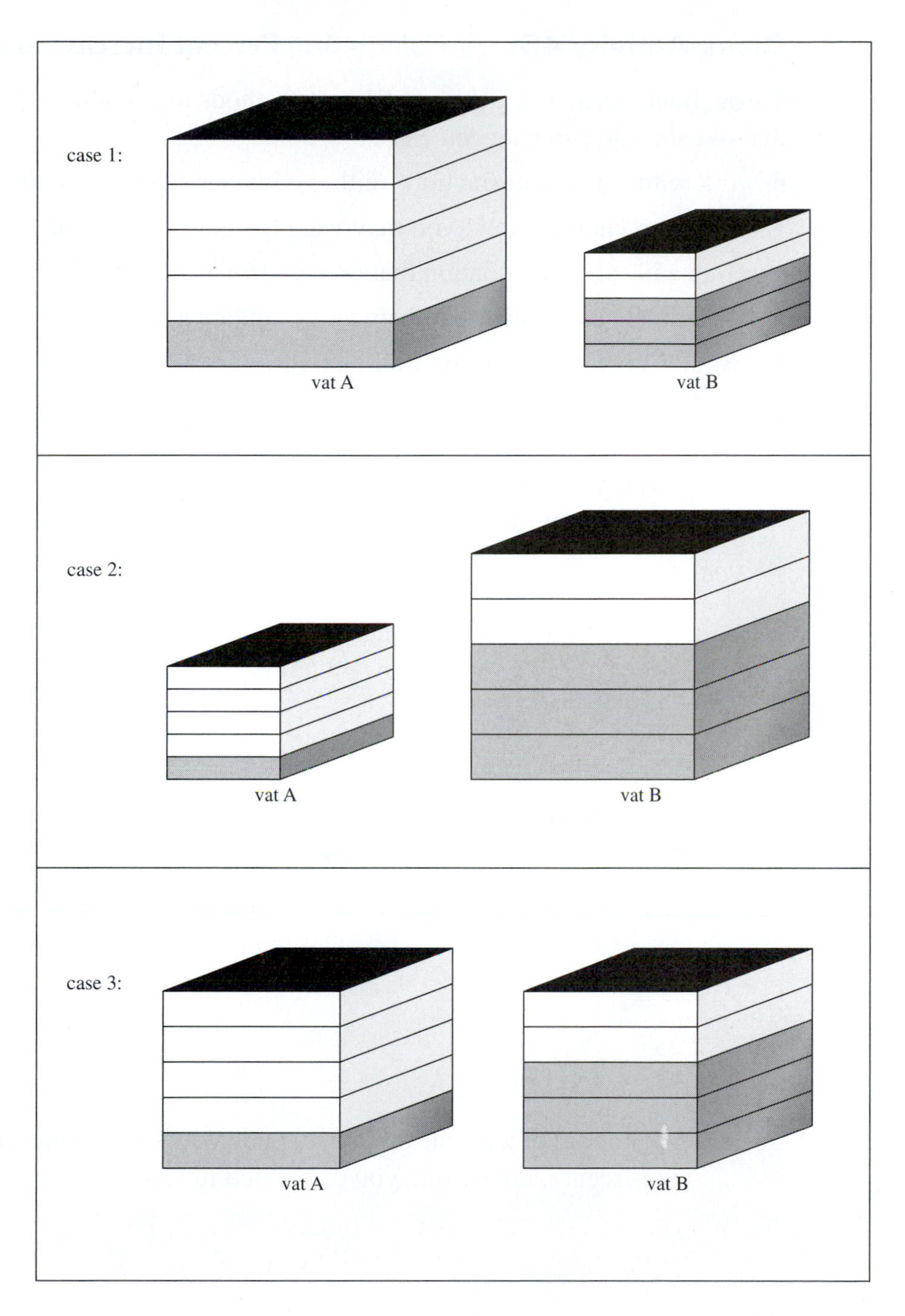

Figure 4R.1
Vats of Chocolate

4.5 Percent Increase and Percent Decrease

Class Activity 4S: Calculating Percent Increase and Decrease

Throughout this activity, use a variety of methods to calculate percents. Recall that to calculate percents you can

- work with equivalent fractions (either with or without cross-multiplying);
- use a "percent diagram" (you may want to "go through 1");
- set up and solve an equation that derives from "$P\%$ of Q is R";
- (sometimes) use a picture to help you calculate a percent.

1. Brand A cereal used to be sold in a 20-ounce box. Now Brand A cereal is sold in a 23-ounce box.

a. Calculate the increase in the weight of cereal in a Brand A box as a percentage of the original weight. This percentage is the percent increase in the weight of a Brand A box of cereal.

b. Now calculate the new weight of a Brand A box of cereal as a percentage of the original weight, and subtract 100%.

c. With the aid of a simple picture, explain why Part (b) must also produce the percent increase that you calculated in (a).

2. There were 80 gallons of gas in a tank. Now there are only 50 gallons left.
   a. Calculate the decrease in the amount of gas in the tank as a percentage of the original. This percentage is the percent decrease in the amount of gas in the tank.

   b. Now calculate the new amount of gas in the tank as a percentage of the original, and subtract it from 100%.

   c. With the aid of a simple picture, explain why Part (b) must also produce the percent decrease that you calculated in (a).

3. Whoopiedoo makeup used to be sold in 4-ounce tubes. Now it's sold in 5-ounce tubes for the same price. Ashlee says the label should read "25% more free," whereas Carolyn thinks it should read "20% more free." Who is right, who is wrong, and why? Help the person with the incorrect answer understand her error and how to correct it.

4. The Film Club increased from 63 members to 214 members. By what percent did membership in the Film Club increase?

Class Activity 4T: Calculating Amounts from a Percent Increase or Decrease

1. A Loungy Chair had cost $400. The price of the Loungy Chair just went up by 20%.

   a. Calculate the new price of the Loungy chair, explaining your reasoning.

   b. Now complete the next percent diagram (fill in steps as needed):

   100% ⟶ $400

   120% ⟶ ____

   c. With the aid of a simple picture, explain why the blank in Part (b) must be equal to your answer in (a).

2. A set of sheets was $60. The sheets are now on sale for 15% off.

   a. Calculate the new price of the sheets, after the reduction.

   b. Now complete the next percent diagram (fill in steps as needed):

   100% ⟶ $60

   85% ⟶ ____

c. With the aid of a simple picture, explain why the blank in Part (b) must be equal to your answer in (a). Where does the 85% in the percent diagram come from?

3. The price of a comforter just went up by 15%. The new price, after the increase, is $60.

a. Explain why you can use the next percent diagram to calculate the price of the comforter before the increase (fill in steps as needed).

115% ⟶ $60

100% ⟶ ____

Where does the 115% come from, and why is it equated with $60?

b. Explain why you *can't* calculate the price of the comforter before the increase by decreasing $60 by 15%.

4. The price of a sofa went down by 20%. The new, reduced price is $400.

a. Explain why you can use the next percent diagram to calculate the price of the sofa before the reduction. (Fill in steps as needed.)

80% ⟶ $400

100% ⟶ ____

Where does the 80% come from, and why is it equated with $400?

b. Explain why you *can't* calculate the price of the sofa before the reduction by increasing $400 by 20%.

Class Activity 4U: Percent *of* versus Percent Increase or Decrease

1. The price of an oven was marked up from $539 to $639. Fill in the blanks, and justify your answers. Simple drawings may be helpful.

   a. The new price is ____% of the old price.

   b. The new price is ____% more than the old price.

   c. The old price is ____% lower than the new price.

   d. The old price is ____% of the new price.

   e. Compare your answers to Part (b) and Part (c). Why does it make sense that they are *not* the same? (If your answers are the same, go back and rethink your methods.)

2. Explain the difference between the following concepts, and give examples to illustrate each one.
   - 200% of an amount
   - a 200% increase in an amount

Class Activity 4V: Percent Problem Solving

Simple pictures may help you solve some of these problems.

1. At first, Prarie had 10% more than the cost of a computer game. After Prarie spent \$7.50, she had 15% less than the cost of the computer game. How much did the computer game cost? How much money did Prarie have at first? Explain your reasoning.

2. One mouse weighs 20% more than another mouse. Together, the two mice weigh 66 grams. How much does each mouse weigh? Explain your reasoning.

3. There are two vats of orange juice. After 10% of the orange juice in the first vat is poured into the second vat, the first vat still has 3 times as much orange juice as the second vat. By what percent did the amount of juice in the second vat increase when the juice from the first vat was poured into it? Explain your reasoning.

4.6 The Commutative and Associative Properties of Addition and Mental Math

Class Activity 4W: Mental Math

Try to find ways to make the problems that follow easy to do *mentally*. In each case, explain your method.

1. $7999 + 857 + 1$

2. $367 + 98 + 2$

3. $153 + 19 + 7$

4. $7.89 + 6.95 + .05$

Class Activity 4X: Using Properties of Addition in Mental Math

For each addition problem in this activity, the equations show how one or more properties of addition have been used to make the problem easier to do mentally. In each case,

- identify the property or properties that have been used;
- describe the strategy that is shown, using groups of marbles, parts of pies, or money.

Example:

$$\begin{aligned} 297 + 35 &= 297 + (3 + 32) \\ &= (297 + 3) + 32 \\ &= 300 + 32 \\ &= 332 \end{aligned}$$

Solution: The associative property was used at the second equal sign to say that

$$297 + (3 + 32) = (297 + 3) + 32$$

If you had 297 marbles and got 35 more marbles, you could first take 3 of the 35 marbles and put them with the 297 marbles to make 300 marbles. Then you could add on the remaining 32 marbles. This shows that you have 332 marbles in all.

1.

$$\begin{aligned} 153 + 19 + 7 &= 153 + 7 + 19 \\ &= 160 + 19 \\ &= 179 \end{aligned}$$

2.

$$\begin{aligned} 687 + 799 &= (686 + 1) + 799 \\ &= 686 + (1 + 799) \\ &= 686 + 800 \\ &= 1486 \end{aligned}$$

3.

$$\begin{aligned} 2\frac{5}{8} + 6\frac{3}{8} &= 2 + \frac{5}{8} + 6 + \frac{3}{8} \\ &= 2 + 6 + \frac{5}{8} + \frac{3}{8} \\ &= 8 + 1 \\ &= 9 \end{aligned}$$

4.

$$\begin{aligned} 3.95 + 2.87 &= 3.95 + (.05 + 2.82) \\ &= (3.95 + .05) + 2.82 \\ &= 4.00 + 2.82 \\ &= 6.82 \end{aligned}$$

Class Activity 4Y: Using Properties of Addition to Aid Learning of Basic Addition Facts

In school, children must learn the basic addition facts from $1 + 1 = 2$ to $10 + 10 = 20$. These are 100 separate facts, but many of these facts are related by properties of addition. By knowing how some facts are related to other facts, children can structure their understanding of the basic facts and can lighten their memorization load.

1. Examine the 4 regions (1 shaded, 3 unshaded) in the addition table in Figure 4Y.1 on page 106. Describe how the unshaded regions are related to the shaded region and to each other. The shaded region is labeled "core facts" because you can derive the other facts from these core facts.

2. In some countries, children systematically learn to turn an addition problem of the form

$$\text{smaller number} + \text{larger number}$$

into a problem of the form

$$\text{larger number} + \text{smaller number}$$

and to solve the second problem. For example, if you asked a child to solve $2 + 7$, the child would say

$$2 + 7 = 7 + 2 = 9$$

a. Which property of arithmetic does this practice use?

b. Relate this practice to the addition table on page 106. If children learn this practice, then what portion of the addition table must the children focus on memorizing?

3. We can derive each of the facts in the rightmost unshaded region from the core facts (or facts related to core facts by the commutative property) by using the "make-a-10" strategy. For example, to solve $8 + 7$ you can break 7 into $2 + 5$, join the 2 with the 8 to make 10, and add the 5 to make 15. The following dot picture illustrates this strategy:

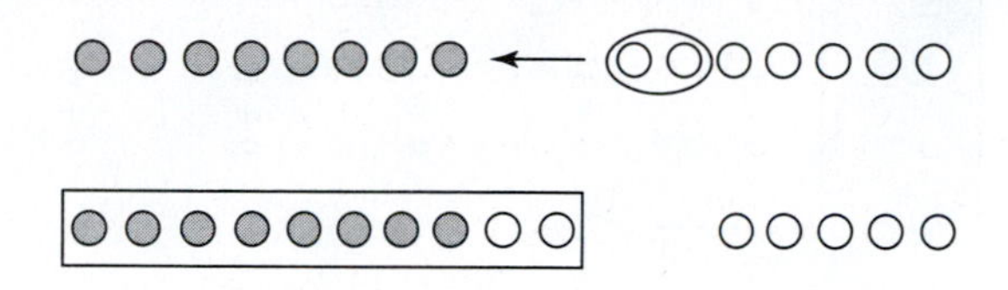

Pick several facts from the rightmost unshaded region of the addition table in Figure 4Y.1. In each case, do the following:

a. Explain how you can derive that fact from core facts (or facts related to core facts by the commutative property) by using a make-a-10 strategy.

b. Draw a dot picture like the previous one to illustrate the make-a-10 strategy you described in Part (a).

c. Write equations that go along with the make-a-10 strategy you described in Part (a). Which property of arithmetic does your equation illustrate?

Observe that the make-a-10 strategy requires children to know the core facts "forwards and backwards." In other words, to use the make-a-10 strategy, a child must know not only that $7 + 2 = 9$ but also that 9 decomposes into 7 plus 2 or 2 plus 7.

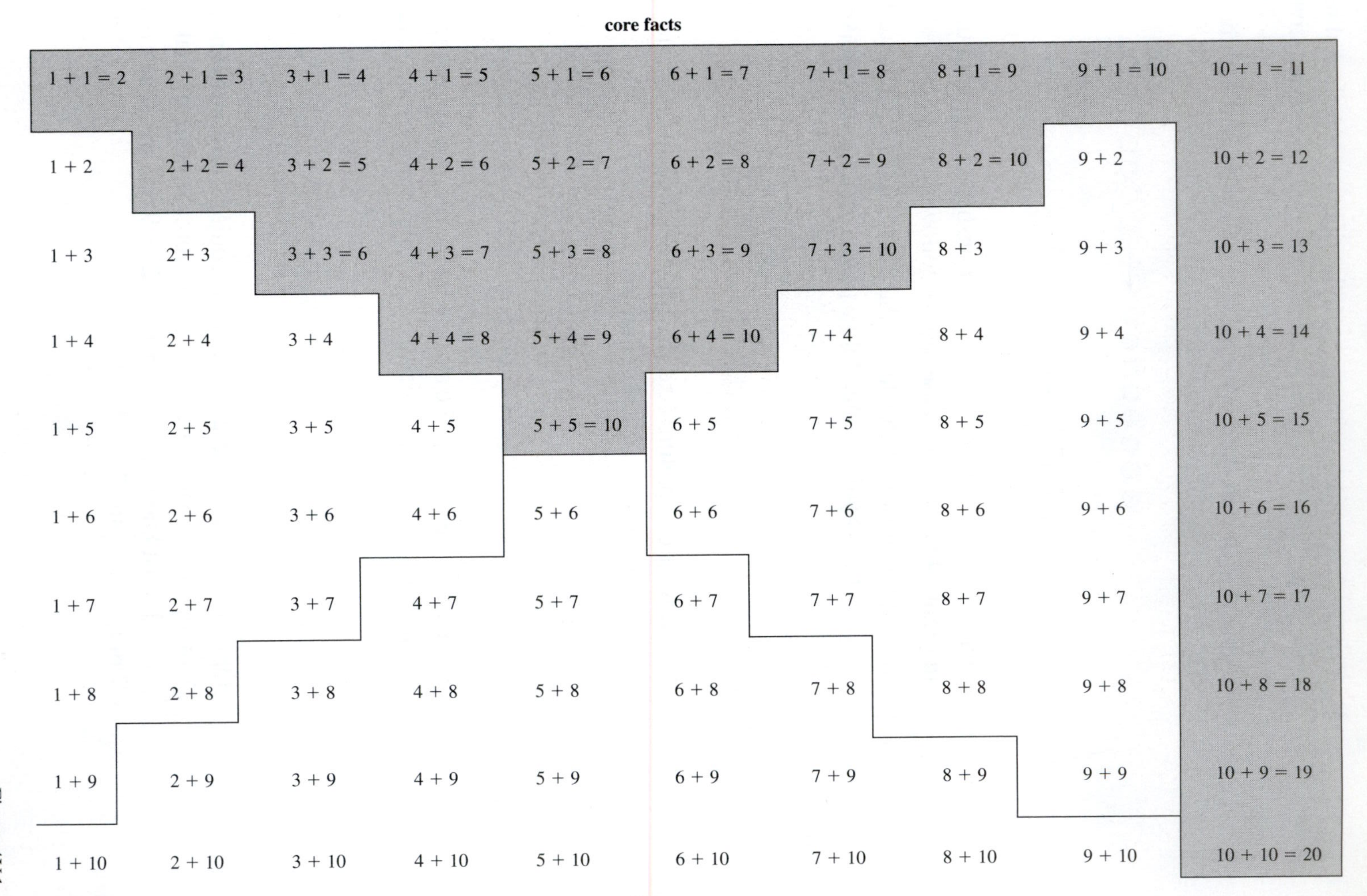

Figure 4Y.1

Basic Addition Facts

Class Activity 4Z: Writing Correct Equations

The problems that follow are hypothetical examples of student work. In each case, the student has a good idea for how to solve the problem, but doesn't use the equal sign correctly. Describe the student's strategy in words, and explain briefly why the strategy makes sense. Then rewrite the student's work, using the same ideas, but using the $=$ sign correctly.

1. The problem: $136 + 57$

Student solution:

$$136 + 50 = 186 + 7$$
$$= 193$$

Your correction:

2. The problem: $378 - 102$

Student solution:

$$378 - 100 = 278 - 2$$
$$= 276$$

Your correction:

3. The problem: $235 - 65$

Student solution:

$$235 - 5 = 230 - 30$$
$$= 200 - 30$$
$$= 170$$

Your correction:

4. The problem: $416 + 99$

Student solution:

$$416 + 100 = 516 - 1$$
$$= 515$$

Your correction:

5. The problem: $114 - 97$

Student solution:

$$97 + 3 = 100 + 14$$
$$= 114$$

So, the answer is:

$$3 + 14 = 17$$

Your correction:

Class Activity 4AA: Writing Equations That Correspond to a Method of Calculation

Write equations which correspond to the solution strategies that follow and that are described in words. In each case, explain briefly why the strategy makes sense. For Problems 1 and 2, write your equations in the following form:

$$\begin{aligned} \text{original problem} &= \text{some expression} \\ &= \text{some expression} \\ &= \vdots \\ &= \text{final answer} \end{aligned}$$

1. *Problem:* $268 + 496$
   Solution: $268 + 500 = 768$, but that's 4 too many, so take away 4. The answer is 764.

2. *Problem:* $123 - 58$
   Solution: $120 - 60 = 60$, plus 3 is 63. I took away 2 too many, so add on 2 to make 65.

3. *Problem:* $153 - 76$
   Solution: 76 and 4 make 80, and 20 make 100, and 53 make 153. So the answer is $4 + 20 + 53$, which is 77.

Class Activity 4BB: Other Ways to Add and Subtract

1. John and Anne want to solve $253 - 99$ by first solving $253 - 100$. They calculate

$$253 - 100 = 153$$

John says that they must now *subtract* 1 from 153, but Anne says that they must *add* 1 to 153.

a. Draw a number line (which need not be perfectly to scale) to help you explain who is right and why. Do not just say which answer is numerically correct; use the number line to help you explain why the answer must be correct.

b. Explain in another way who is right and why.

2. Jamarez says that he can solve $253 - 99$ by adding 1 to both numbers and solving $254 - 100$ instead.

a. Draw a number line (which need not be perfectly to scale) to help you explain why Jamarez's method is valid.

b. Explain in another way why Jamarez's method is valid.

c. Could you adapt Jamarez's method to other subtraction problems, such as to the problem $324 - 298$? Explain, and give several examples.

3. Find ways to solve the addition and subtraction problems that follow *other than* by using the standard addition or subtraction algorithms. In each case, explain your reasoning, and except for Part (g) also write equations that correspond to your line of reasoning.

   a. $183 + 99$

   b. $268 + 52$

   c. $600 - 199$

   d. $164 - 70$

   e. $999 + 9999$

   f. $10.00 − $2.99

   g.
   $$\begin{array}{r} 2.99 \\ 3.99 \\ 1.99 \\ +\,4.99 \\ \hline \end{array}$$
   (No equations are needed.)

5

Multiplication

5.1 The Meaning of Multiplication and Ways to Show Multiplication

Class Activity 5A: Showing Multiplicative Structure

1. Using the meaning of multiplication, explain why you can determine the number of ladybugs in the following picture by multiplying:

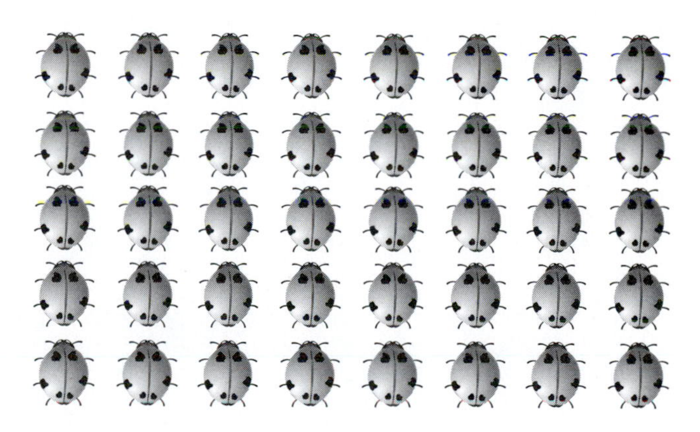

2. Fran has 3 pairs of pants, pants 1, 2, and 3, that coordinate perfectly with 4 different shirts, shirts A, B, C, and D. How many different outfits consisting of a pair of pants and a shirt can Fran make from these clothes?

   Show how to solve this problem by using an organized list and a tree diagram. Use the meaning of multiplication to explain why you can solve the problem by multiplying.

3. You have two bags. The bag on the left contains an orange marble, a red marble, and a purple marble. The bag on the right contains a yellow marble, a green marble, and a blue marble. How many combinations of a marble from the bag on the left and a marble from the bag on the right are there?

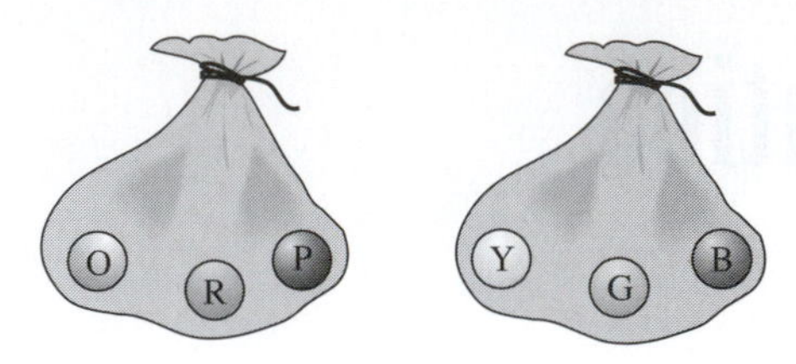

Show how to solve this by using an organized list and a tree diagram. Use the meaning of multiplication to explain why you can solve the problem by multiplying.

Suppose we simply wanted to find out how many marbles were in the two bags altogether. How would this problem be like or unlike finding the different combinations of marbles selected from the two bags?

4. Jenna and Katie work on Problem 3 about picking pairs of marbles from two bags.

Jenna draws the tree diagram in Figure 5A.1 and says that there are 6 ways to pick a pair of marbles from the two bags. What's wrong with Jenna's reasoning?

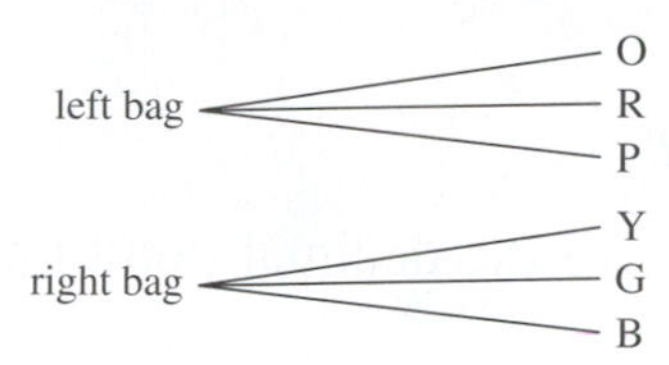

Figure 5A.1
Jenna's Tree Diagram

Katie draws the tree diagram in Figure 5A.2 and counts all the labels (letters) on her tree diagram. Katie says that there are

$$3 + 9 = 12$$

ways to pick a pair of marbles from the two bags. What's wrong with Katie's reasoning?

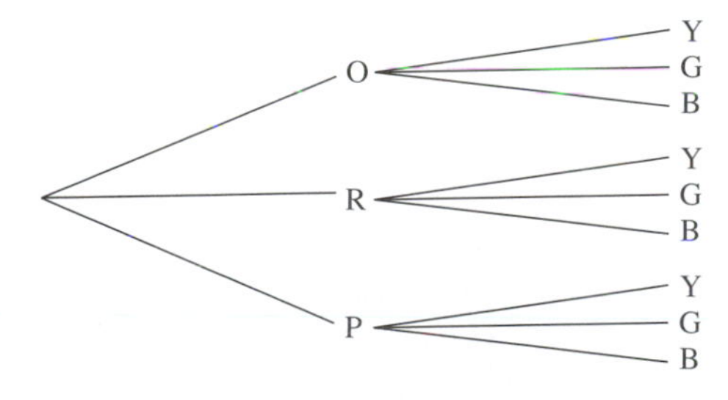

Figure 5A.2
Katie's Tree Diagram

5. Suppose that a certain type of small key is made with two notches and that each notch is either deep, medium, or shallow. How many different possible keys can be made this way?

What if the key is made with four notches and each notch is either deep, medium, or shallow?

Show how to solve the preceding problems by using organized lists and tree diagrams. Use the meaning of multiplication to explain why you can solve the problems by multiplying.

5.2 Why Multiplying Decimal Numbers by 10 Is Easy

Class Activity 5B: Multiplying by 10

1. Is the following statement correct?

To multiply a number by 10, put a 0 at the end of the number.

2. Is the following statement correct?

To multiply a number by 10, move the decimal point one place to the right.

3. A different way to describe how to multiply by 10 is as follows:

To multiply a number by 10, move all the digits one place to the left.

Explain why this statement is true, using a picture of 10×23 to illustrate.

Class Activity 5C: If We Wrote Numbers Differently, Multiplying by 10 Might Not Be So Easy

Multiplying by 10 is easy with decimal numbers because of the special structure of the decimal system. What if we wrote numbers differently?

Let's consider a different way of writing numbers. We'll call our new system for writing numbers **money notation.** To write a whole number in money notation, use the coins and bills of *largest value* to represent that many cents. Then write down from right to left how many cents, nickels, dimes, quarters, fifty-cent pieces, dollar bills, 5-dollar bills, etc., are needed to make the given number of cents. A few examples will illustrate:

Decimal Notation	Money Notation	Explanation
27	1002	1 quarter, 0 dimes, 0 nickels, and 2 pennies make 27 cents.
85	11100	1 fifty-cent piece, 1 quarter, 1 dime, 0 nickels, and 0 pennies make 85 cents.
149	101204	1 dollar, 0 fifty-cent pieces, 1 quarter, 2 dimes, 0 nickels, and 4 pennies make 149 cents.

Notice that, even though you can also make 27 cents with 2 dimes and 7 cents, you shouldn't write 27 as "207" in money notation. This is because you should always use the *largest value* coins and bills possible.

1. Write the following numbers in money notation:

    34

    68

    180

2. In decimal notation, to multiply a number by 10, we shift each digit one place to the left. Does this same procedure work when we write numbers in money notation?

   Write the following numbers in money notation:

   340

   680

   1800

   Compare these results to the results of shifting the digits in your answers to Problem 1 one place to the left.

3. Why is it that, to multiply a number in decimal notation by 10, we have to shift each digit one place to the left, but the same procedure does not work for numbers written in money notation? What creates this difference in the two systems?

Class Activity 5D: * Multiplying by Powers of 10 Explains the Cycling of Decimal Representations of Fractions

When you use a calculator or long division, you will find that

$$\frac{1}{7} = 0.142857142857142857\ldots$$

Notice the repeating pattern—it turns out that it continues forever.

1. Look carefully at the sequence of digits in the decimal representations of the fractions listed. For $\frac{1}{7}$, this sequence is

$$1,\ 4,\ 2,\ 8,\ 5,\ 7,\ 1,\ 4,\ 2,\ 8,\ 5,\ 7,\ldots$$

   How does the sequence of digits in these other decimal representations compare to this sequence?

   $\frac{1}{7} = 0.142857142857\ldots$

   $\frac{2}{7} = 0.285714285714\ldots$

   $\frac{3}{7} = 0.428571428571\ldots$

   $\frac{4}{7} = 0.571428571428\ldots$

   $\frac{5}{7} = 0.714285714285\ldots$

   $\frac{6}{7} = 0.857142857142\ldots$

2. By way of comparison, look at the decimal representations for various fractions with denominator 37. How is this similar to and how is this different from the situation in Problem 1?

$\frac{1}{37} = 0.027027027027\ldots$
$\frac{2}{37} = 0.054054054054\ldots$
$\frac{3}{37} = 0.081081081081\ldots$
$\frac{4}{37} = 0.108108108108\ldots$
$\frac{5}{37} = 0.135135135135\ldots$
$\frac{10}{37} = 0.270270270270\ldots$
$\frac{26}{37} = 0.702702702702\ldots$

The next several parts will be used to explain why the relationships you discovered in Problem 1 exist.

3. Calculate the following without the use of a calculator:

$10 \times 0.142857142857142857\ldots =$

$100 \times 0.142857142857142857\ldots =$

$1{,}000 \times 0.142857142857142857\ldots =$

$10{,}000 \times 0.142857142857142857\ldots =$

$100{,}000 \times 0.142857142857142857\ldots =$

$1{,}000{,}000 \times 0.142857142857142857\ldots =$

4. Write the following numbers as mixed numbers—in other words, with a whole number and a fractional part:

$10 \times \frac{1}{7} = \frac{10}{7} = 1\frac{3}{7}$

$100 \times \frac{1}{7} = \frac{100}{7} =$

$1{,}000 \times \frac{1}{7} = \frac{1000}{7} =$

$10{,}000 \times \frac{1}{7} = \frac{10{,}000}{7} =$

$100{,}000 \times \frac{1}{7} = \frac{100{,}000}{7} =$

$1{,}000{,}000 \times \frac{1}{7} = \frac{1{,}000{,}000}{7} =$

5. What is the relationship between the lists of numbers in Problem 3 and 4? How are the fractional parts in Problem 4 related to your answers in Problem 3? Use this to explain the relationship you discovered in Problem 1.

5.3 The Commutative Property of Multiplication and Areas of Rectangles

Class Activity 5E: Multiplication, Areas of Rectangles, and the Commutative Property

1. The large rectangle shown here is 9 centimeters wide and 5 centimeters tall. Use the meaning of multiplication to explain why you can find the area of the rectangle by multiplying.

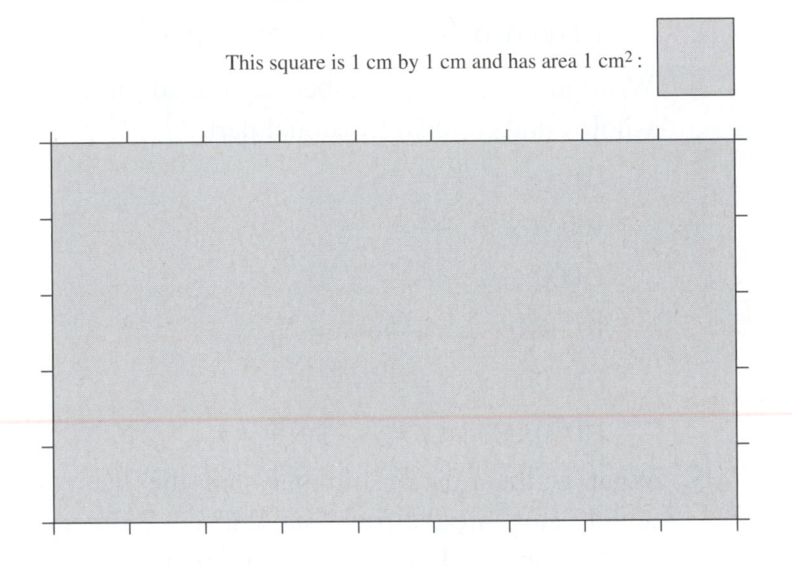

This square is 1 cm by 1 cm and has area 1 cm^2 :

2. Using the rectangle in Problem 1, explain why

$$5 \times 9 = 9 \times 5$$

3. Explain why

$$A \times B = B \times A$$

is true for all counting numbers A and B.

Class Activity 5F: Explaining the Commutative Property of Multiplication

If available, tiles, counters, or other small objects would be useful.

1. Write one story problem for 4×3 and another for 3×4. Before determining the answers to the problems, discuss whether it would necessarily be obvious to a child that the answers are the same.
2. Use a collection of tiles, counters, or other small objects to discuss and explain the commutative property of multiplication.
3. A third-grade math book explained the commutative property of multiplication by using a number line to show that

$$5 + 5 + 5$$

is equal to

$$3 + 3 + 3 + 3 + 3$$

Why is this not a good explanation? Why would it be better to make a rectangle out of tiles, as in Problem 2, or to draw an array like the one shown here?

4. A marching band is arranged in 20 rows of 4. Everybody in the band turns 90° to their left. Now how is the band arranged? How can you use the marching band to illustrate the commutative property of multiplication?

Class Activity 5G: Using the Commutative Property of Multiplication

1. There are 21 bags with 2 marbles in each bag. Ben calculates the number of marbles there are in all by counting by 2s 21 times. Kaia calculates $21 + 21 = 42$ instead. Why is Kaia's approach legitimate?

2. This problem will help you investigate the following question: Which of the following two options will result in the lower price for a pair of pants?

- The price of the pants is marked up by 10% and then marked down by 20% from the increased price.
- The price of the pants is marked down by 20% and then marked up by 10% from the discounted price.

Both options involve marking up by 10% and marking down by 20%. The difference is the order in which the marking up and marking down occurs.

a. Before you do any calculations, make a guess about which of the two ways should result in a lower price.

b. Determine which of the two ways will result in a lower price by using the fact that if a price increases by 10%, then you can find the increased price by multiplying by 1.10 (which is $100\% + 10\% = 110\%$), and if a price decreases by 20%, then you can find the decreased price by multiplying by 0.80 (which is $100\% - 20\% = 80\%$).

c. Explain how the commutative property of multiplication is relevant to the question of which of the two ways results in a lower price of the pair of pants.

Class Activity 5H: Using Multiplication to Estimate How Many

1. Use multiplication to estimate how many curlicues there are in the figure.

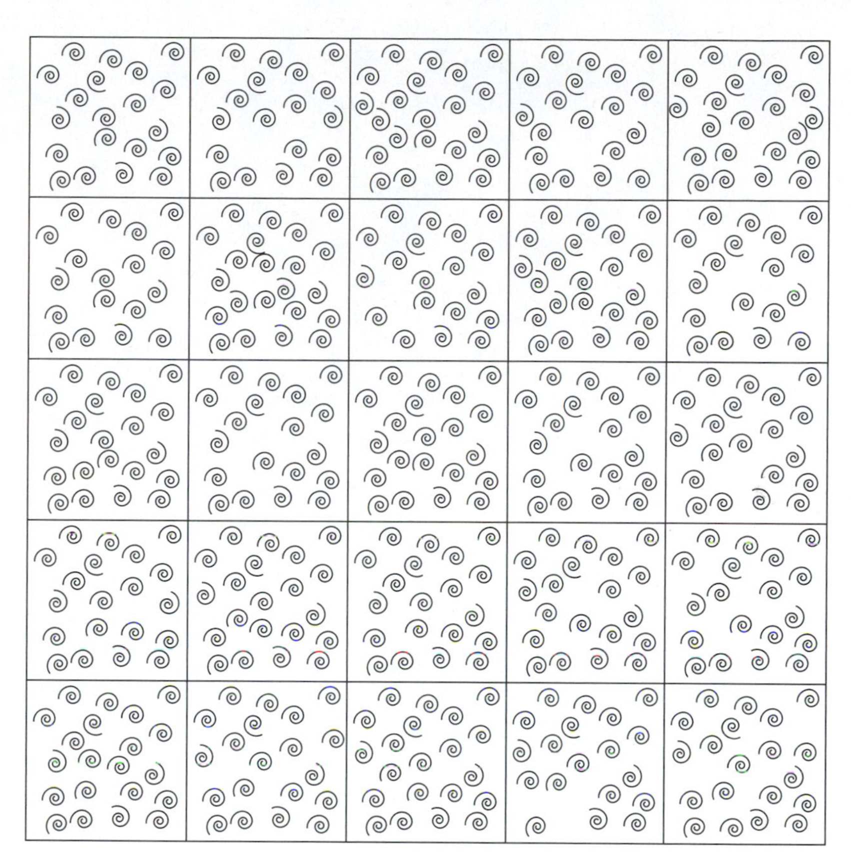

2. Use multiplication to estimate how many dots are shown. Describe your strategy.

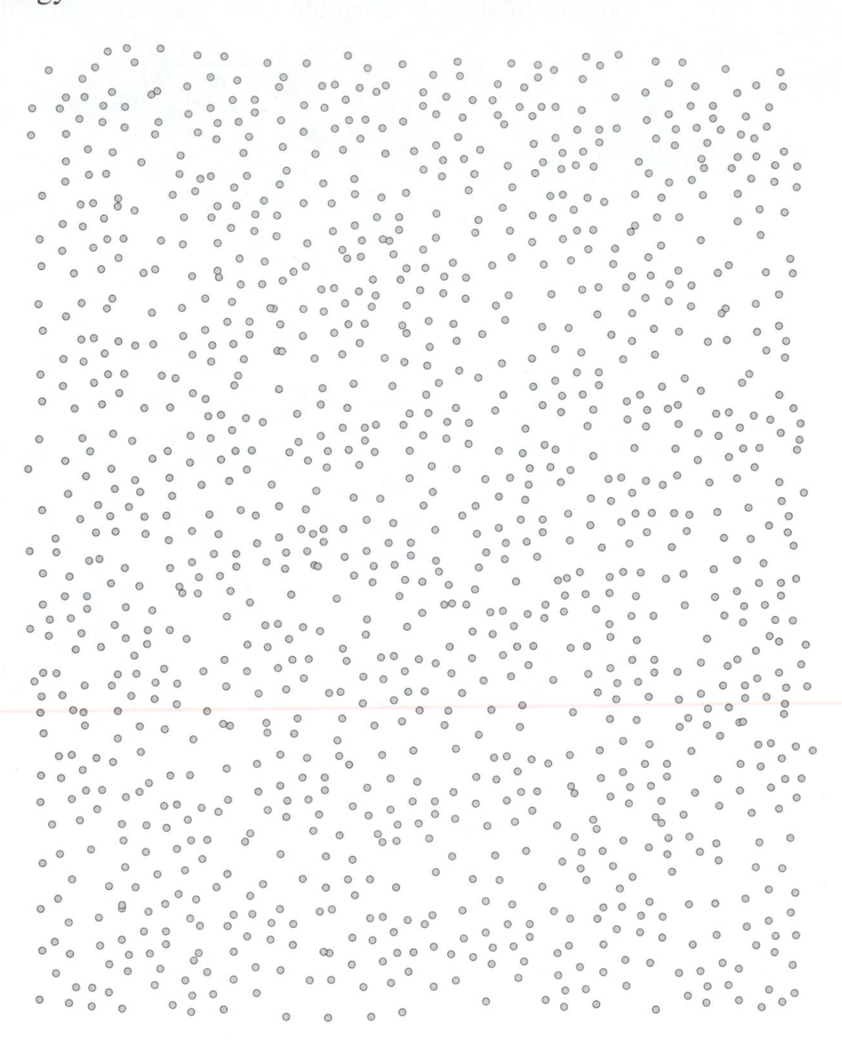

3. Describe a *realistic* strategy for determining approximately how many hairs are on a person's head. If possible, carry out your strategy.

5.4 The Associative Property of Multiplication and Volumes of Boxes

Class Activity 5I: Ways to Describe the Volume of a Box with Multiplication

You will need a set of blocks for Problems 1 and 2 of this activity.

1. If you have cubic-inch blocks available, build a box that is 3 inches wide, 2 inches long, and 4 inches tall. It should look like this:

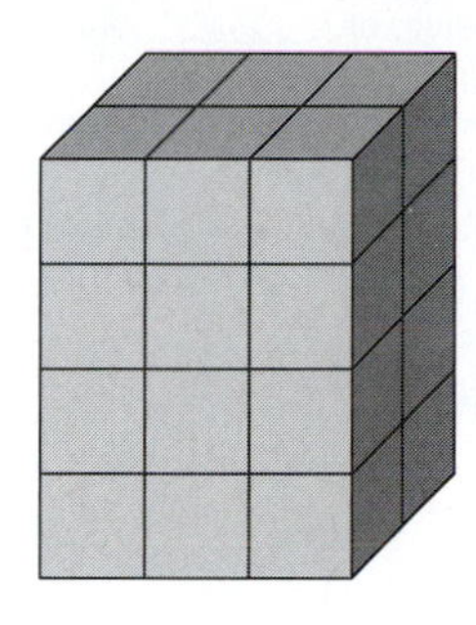

2. Subdivide your box into natural groups of blocks, and describe how you subdivided the box. How many groups were there and how many blocks were in each group? Using multiplication, write the corresponding expressions for the total number of blocks in the box. Now repeat, this time subdividing your box into natural groups in a different way.

3. The next two pages show different ways of subdividing a box into groups. If you have built a box from blocks, subdivide your box in the ways shown in the figures. In each case, describe the number of groups and the number of blocks in each group. Then write an expression for the total number of blocks. Your expression should use multiplication, parentheses, and the numbers 2, 3, and 4.

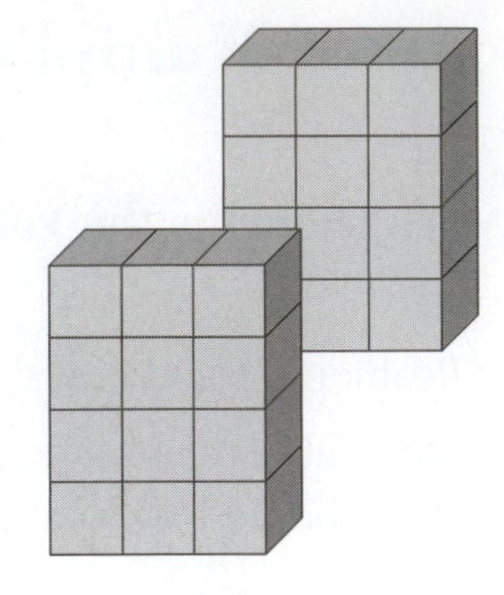

________ groups of ________ blocks

On the following line, write an expression using multiplication, parentheses, and the numbers 2, 3, and 4 for the total number of blocks:

____

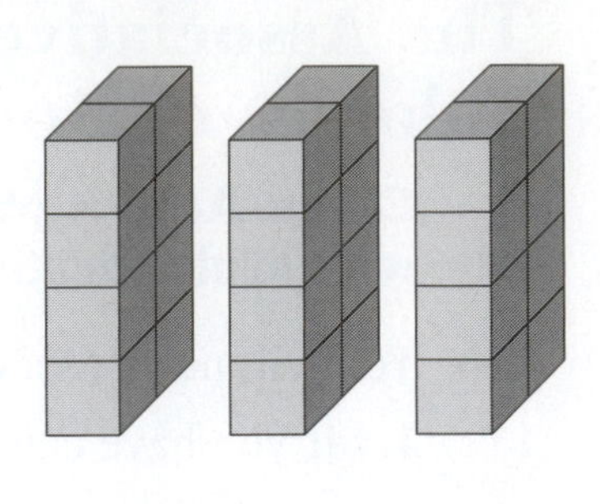

________ groups of ________ blocks

On the following line, write an expression using multiplication, parentheses, and the numbers 2, 3, and 4 for the total number of blocks:

____

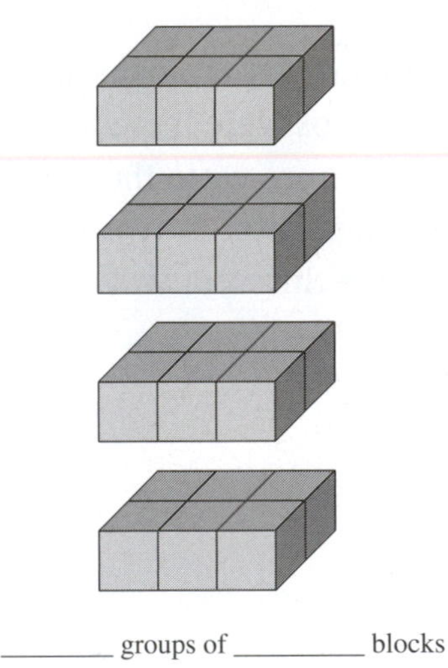

________ groups of ________ blocks

On the following line, write an expression using multiplication, parentheses, and the numbers 2, 3, and 4 for the total number of blocks:

____

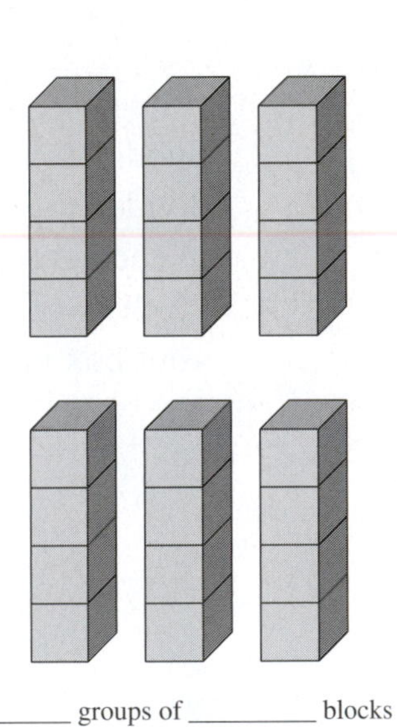

________ groups of ________ blocks

On the following line, write an expression using multiplication, parentheses, and the numbers 2, 3, and 4 for the total number of blocks:

____

________ groups of ________ blocks

On the following line, write an expression using multiplication, parentheses, and the numbers 2, 3, and 4 for the total number of blocks:

____

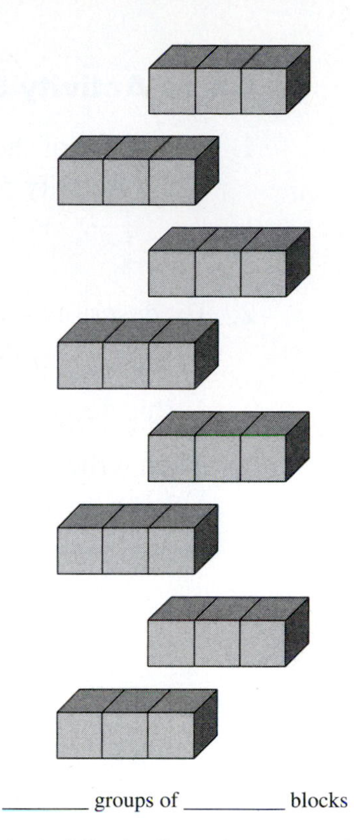

________ groups of ________ blocks

On the following line, write an expression using multiplication, parentheses, and the numbers 2, 3, and 4 for the total number of blocks:

____

Class Activity 5J: Explaining the Associative Property

1. Use some of the pictures and the expressions you wrote for Problem 3 of Class Activity 5I to help you explain why

$$(4 \times 2) \times 3 = 4 \times (2 \times 3)$$

2. By describing the groups, explain how to see the following:

   The next design shows 6 groups of dots with 4×2 dots in each group; this design also shows 6×4 groups of dots with 2 dots in each group.

   Then write an equation to show that you must get the same number of dots, either way you count them.

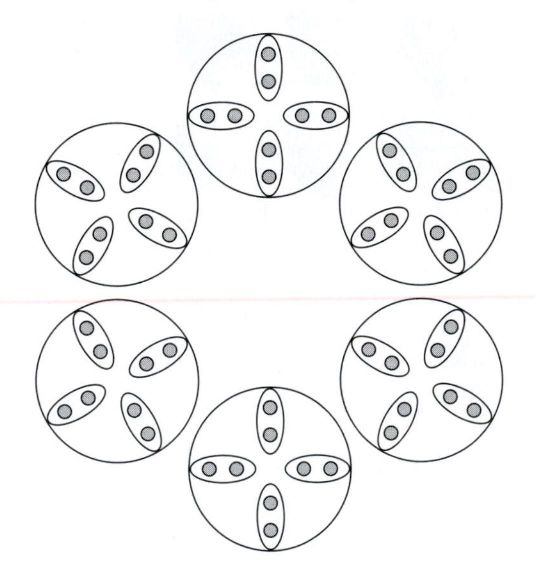

3. Use the next design to help you explain why

$$(5 \times 3) \times 4 = 5 \times (3 \times 4)$$

   Your explanation should be general, in the sense that it explains why the equation is true when you replace the numbers 5, 3, and 4 with other counting numbers, and the design is changed accordingly.

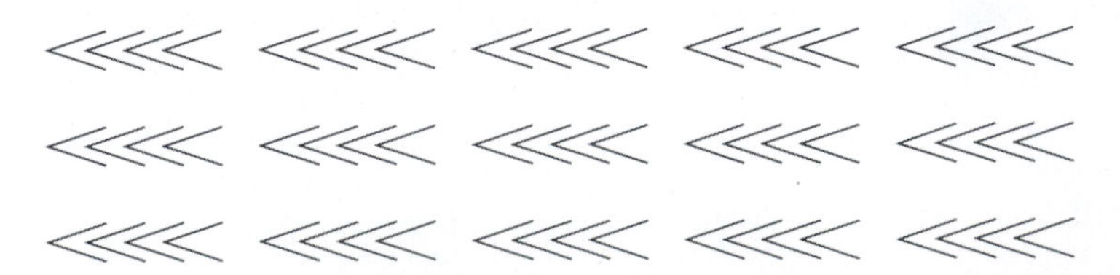

Class Activity 5K: Using the Associative and Commutative Properties of Multiplication

1. Which properties of arithmetic are used in the following calculations? Where are the properties used?

$$\begin{aligned} 8 \times 70 &= (8 \times 7) \times 10 \\ &= 56 \times 10 \\ &= 560 \end{aligned}$$

2. Which properties of arithmetic are used in the following calculations? Where are the properties used?

$$\begin{aligned} 60 \times 4000 &= 6 \times 10 \times 4 \times 1000 \\ &= 6 \times 4 \times 10 \times 1000 \\ &= 24 \times 10 \times 1000 \\ &= 240 \times 1000 \\ &= 240{,}000 \end{aligned}$$

3. To multiply 4×60 mentally, we can multiply 4×6 first, and then put a 0 at the end to get the answer, 240. Write equations to show that this mental technique uses the associative property of multiplication. Use the next picture to explain why the mental technique makes sense.

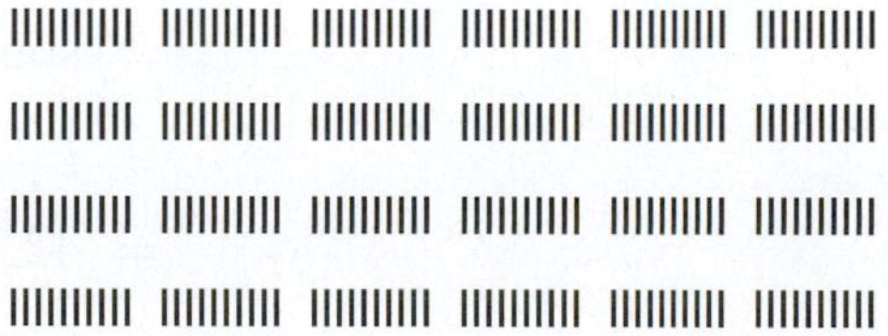

4. To multiply 30×40 mentally, we can multiply 3×4 first, and then put two zeros at the end to get the answer, 1200. Write equations to show that this mental technique uses the associative and commutative properties of multiplication. Use the next picture to explain why the mental technique makes sense.

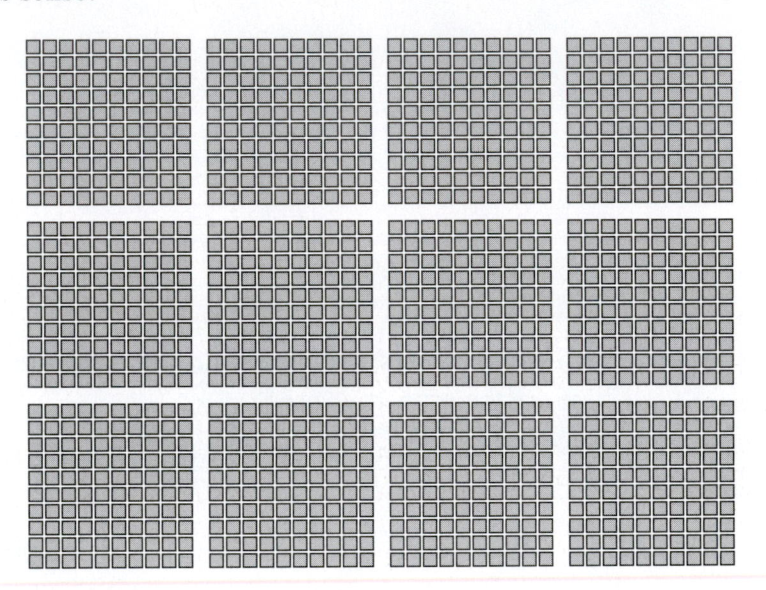

5. Marcie calculated 60×700 mentally as follows:

6 times 7 is 42, 10 times 100 is 1000, 42 times 1000 is 42,000.

Write a sequence of equations that shows why Marcie's method of calculation is valid. Which properties of arithmetic are involved? Write your equations in the following form:

$$\begin{aligned} 60 \times 700 &= \text{some expression} \\ &= \text{some expression} \\ &= \vdots \\ &= 42{,}000 \end{aligned}$$

Class Activity 5L: Different Ways to Calculate the Total Number of Objects

1. Use only the numbers 3, 4, and 5, the multiplication symbol $\times$ (or $\cdot$), and parentheses to write at least two different expressions for the total number of hearts in the next figure. In each case, evaluate your expression to calculate the total number of hearts.

2. Write at least two different expressions for the total number of stars in Figure 5L.1. In each case, evaluate your expression in order to calculate the total number of stars.

Figure 5L.1
How Many Stars?

Class Activity 5M: How Many Gumdrops?

You will need a box of cereal and standard kitchen measuring cups or measuring spoons for Problem 2 of this activity.

1. The container shown has a square base, so that all four vertical sides of the container are identical; the picture shows one of those four sides. Use multiplication to determine approximately how many gumdrops are in the container. Explain your method.

2. Describe several different realistic methods for determining approximately how many pieces of cereal are in a box of cereal. In each case, carry out your method.

5.5 The Distributive Property

Class Activity 5N: Order of Operations

You will need a calculator for Problem 1 Part (b) of this activity.

1. **a.** Evaluate the expression $4 \times 5 - 18 \div 3 + 12 \times 10$ without a calculator.

   b. Using a calculator, enter the numbers and operations in the expression $4 \times 5 - 18 \div 3 + 12 \times 10$ in the order in which they appear, pressing the = sign or "enter" button each time after you enter a number. Do you get the correct answer this way? If not, why not?

2. For each of the following expressions, write a story problem for that expression and explain briefly why your story problem is appropriate for the expression.

   a. $3 + 4 \times 6$

   b. $3 + 4 \qquad \times 6$ Does it matter if there is a big space between the $\times$ and the 4?

   c. $(3 + 4) \times 6$

   d. $6 \times 4 + 3 \times 5$

Class Activity 5O: Explaining the Distributive Property

1. a. There are 6 goodie bags. Each goodie bag contains 3 eraser tops and 4 stickers. Write an expression using the numbers 6, 3, and 4; the symbols $\times$ (or $\cdot$) and $+$; and parentheses, if needed, for the total number of items in the 6 goodie bags. If you use parentheses, explain why you need them; if you do not use parentheses, explain why you do not need them.

   b. Write a different expression for the total number of items in Part (a) and explain why it must be equal to the expression you wrote in Part (a). Draw pictures for the two expressions.

   c. Write an equation which states that the two expressions in Parts (a) and (b) are equal. Which property have you illustrated?

2. Use the shading shown in the rectangle, and use the meaning of multiplication to explain why

$$3 \times (2 + 4) = 3 \times 2 + 3 \times 4$$

Your explanation should be general in the sense that you could use it to explain why

$$A \times (B + C) = A \times B + A \times C$$

for *all* counting numbers A, B, and C.

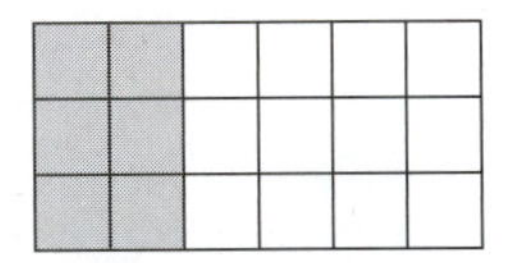

3. On the graph paper shown, draw a rectangle and shade a portion of it to illustrate the equation

$$8 \times (10 + 5) = 8 \times 10 + 8 \times 5$$

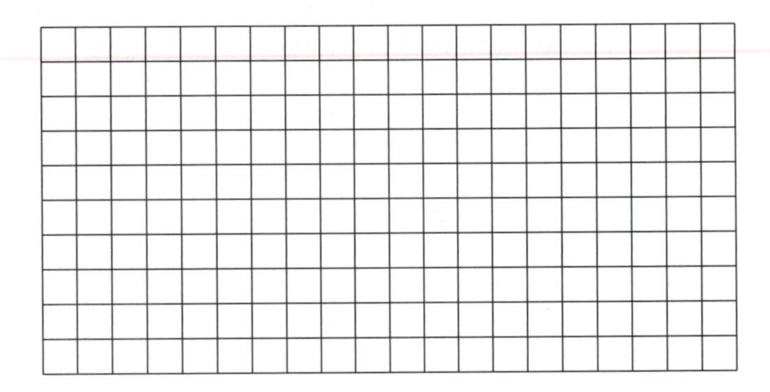

Class Activity 5P: The Distributive Property and FOIL

If you have studied algebra, then you probably learned the FOIL method for multiplying expressions of the form

$$(A + B) \cdot (C + D)$$

FOIL stands for *First, Outer, Inner, Last,* and reminds us that

$$(A + B) \cdot (C + D) = A \cdot C \; + \; A \cdot D \; + \; B \cdot C \; + \; B \cdot D$$

where $A \cdot C$ is *First,* $A \cdot D$ is *Outer,* $B \cdot C$ is *Inner,* and $B \cdot D$ is *Last.*

This class activity will help you explain in several different ways why the FOIL equation is valid.

1. Use the shading shown in the next rectangle and use the meaning of multiplication to explain why

$$(2 + 3) \cdot (7 + 4) = 2 \cdot 7 \; + \; 2 \cdot 4 \; + \; 3 \cdot 7 \; + \; 3 \cdot 4$$

Your explanation should be general in the sense that you could use it to explain why

$$(A + B) \cdot (C + D) = A \cdot C \; + \; A \cdot D \; + \; B \cdot C \; + B \cdot D$$

is true for all counting numbers A, B, C, and D.

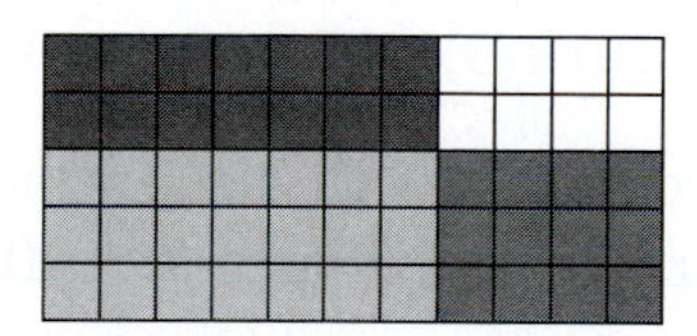

2. Relate the preceding subdivided rectangle to the following sequence of equations:

$$\begin{aligned}(2 + 3) \cdot (7 + 4) &= 2 \cdot (7 + 4) \; + \; 3 \cdot (7 + 4) \\ &= 2 \cdot 7 \; + \; 2 \cdot 4 \; + \; 3 \cdot 7 \; + \; 3 \cdot 4\end{aligned}$$

Which properties of arithmetic do these equations use? Where are the properties used?

3. On the graph paper shown, draw and shade a rectangle to show why the following equation is true:

$$(10 + 3) \cdot (20 + 4) = 10 \cdot 20 \ + \ 10 \cdot 4 \ + \ 3 \cdot 20 \ + \ 3 \cdot 4$$

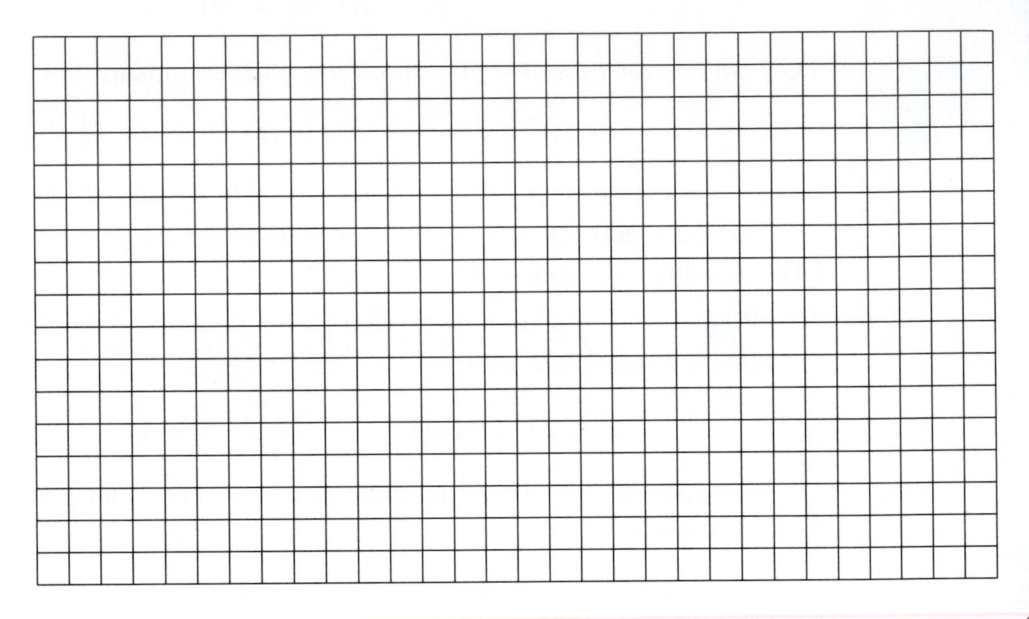

4. Use the distributive property several times to write a sequence of equations which proves that

$$(10 + 3) \cdot (20 + 4) = 10 \cdot 20 \ + \ 10 \cdot 4 \ + \ 3 \cdot 20 \ + \ 3 \cdot 4$$

Your work should be general in the sense that it would remain valid if other numbers were to replace 10, 3, 20, and 4.

5. Relate each step in your sequence of equations in Problem 4 to the subdivided rectangle you created in Problem 3.

6. Find a different expression that is equal to $(A + B) \cdot (C + D + E)$, and explain why the two expressions are equal.

Class Activity 5Q: Using the Distributive Property

1. Use the multiplication facts

$$15 \times 15 = 225$$
$$2 \times 15 = 30$$

to help you mentally calculate

$$17 \times 15$$

Explain how your calculation method is related to the following array, which consists of 17 rows of dots with 15 dots in each row:

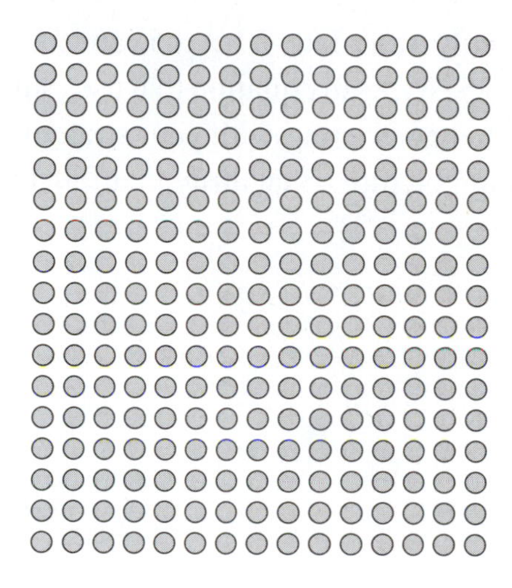

2. Write equations showing how your mental strategy for calculating 17×15 in Problem 3 involves the distributive property.

3. Write an equation that uses subtraction and the distributive property and that goes along with the following array:

4. Mentally calculate 20×15, and use your answer to mentally calculate 19×15. Write an equation that uses subtraction and the distributive property and that goes along with your strategy. Without drawing all the detail, draw a rough picture of an array that illustrates this calculation strategy.

Class Activity 5R: Why Isn't 23×23 Equal to $20 \times 20 + 3 \times 3$?

Kylie has an idea for how to calculate 23×23. She says,

> Twenty times 20 is 400, and 3 times 3 is 9; so 23×23 should be 400 plus 9, which is 409.

Is Kylie's method valid? If not, how could you modify her work to make it correct? Don't just start over in a different way; work with Kylie's idea. Use the large square below, which consists of 23 rows with 23 small squares in each row, to help you explain your answer.

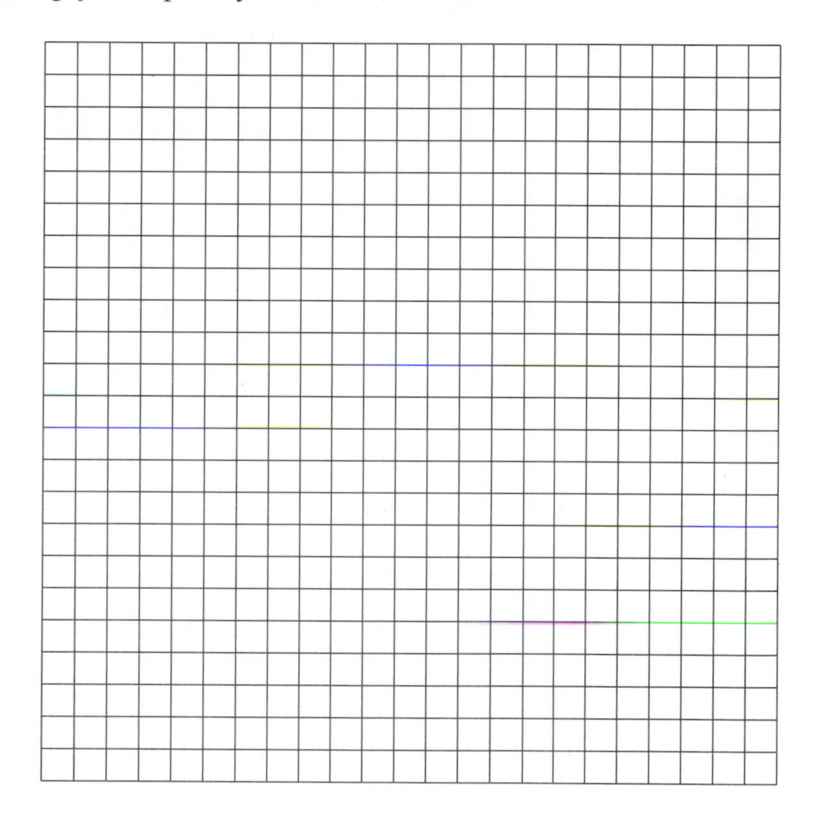

Class Activity 5S: Squares and Products Near Squares

If available, a set of square tiles would be helpful for Problem 2.

The word *square* can refer to a geometric shape or to a related amount of area, such as a "square inch." But there are also square *numbers*. The **square numbers** or **perfect squares** are the counting numbers that can be written as a counting number times itself; for example,

$$1 \times 1, \quad 2 \times 2, \quad 3 \times 3, \quad \ldots, \quad 9 \times 9, \quad \ldots$$

are squares. In the multiplication table, the squares are usually easy to remember; the squares lie on a diagonal, as in the following multiplication table:

×	1	2	3	4	5	6	7	8	9	10
1	1	2	3	4	5	6	7	8	9	10
2	2	4	6	8	10	12	14	16	18	20
3	3	6	9	12	15	18	21	24	27	30
4	4	8	12	16	20	24	28	32	36	40
5	5	10	15	20	25	30	35	40	45	50
6	6	12	18	24	30	36	42	48	54	60
7	7	14	21	28	35	42	49	56	63	70
8	8	16	24	32	40	48	56	64	72	80
9	9	18	27	36	45	54	63	72	81	90
10	10	20	30	40	50	60	70	80	90	100

1. Compare each square in the multiplication table with the entry that is diagonally above and to the right of it (or diagonally below and to the left). For example, compare 49 (which is 7×7) with 48 (which is 6×8). What pattern do you notice?

2. Use arrays to help you explain why the pattern you described in Problem 1 exists. If available, use square tiles to make your arrays; rearrange the tiles in order to relate multiplication facts. If tiles are not available, draw the arrays instead.

3. Now apply the distributive property or FOIL to the expression

$$(A - 1) \cdot (A + 1)$$

to explain why the pattern you discovered in Problem 1 exists.

4. Why are the squares $1 \times 1,\ 2 \times 2,\ 3 \times 3, \ldots$ called squares?

5. Annie is working on the multiplication problem 19×21. Annie says that 19×21 should equal 20×20 because 19 is one less than 20 and 21 is one more than 20.

Annie has a wonderful idea, but is it correct? If not, use the dots shown to help you explain to Annie why not. There are 20 rows of dots with 21 dots in each row.

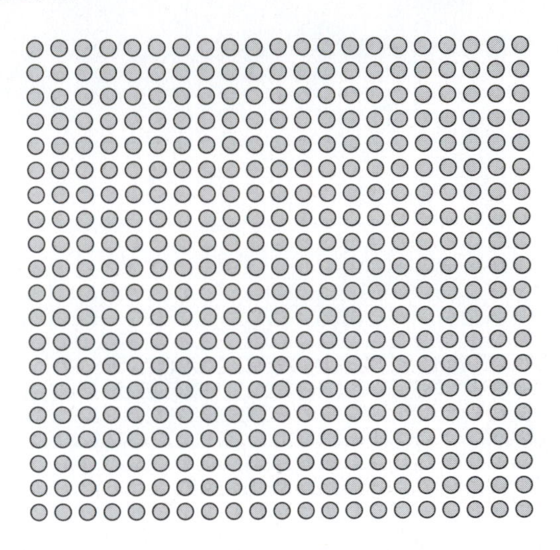

6. Mary is working on the multiplication problem 19×21. Mary says that 19×21 is 21 less than 20×21, and 20×21 is 20 more than 20×20, which she knows is 400. Mary thinks this ought to help her calculate 19×21, but she can't quite figure it out.

Discuss Mary's idea in detail. Can you make her explanation work?

5.6 Mental Math, Properties of Arithmetic, and Algebra

Class Activity 5T: Using Properties of Arithmetic to Aid the Learning of Basic Multiplication Facts

In school, children must learn the basic multiplication facts from $1 \times 1 = 1$ to $10 \times 10 = 100$ (or beyond). These are 100 separate facts, but many of these facts are related by properties of arithmetic. By knowing how some facts are related to other facts, children can structure their understanding of the basic facts in order to learn them better.

$2 \times 2 = 4$	3×2	4×2	5×2	6×2	7×2	8×2	9×2
$2 \times 3 = 6$	$3 \times 3 = 9$	4×3	5×3	6×3	7×3	8×3	9×3
$2 \times 4 = 8$	$3 \times 4 = 12$	$4 \times 4 = 16$	5×4	6×4	7×4	8×4	9×4
$2 \times 5 = 10$	$3 \times 5 = 15$	$4 \times 5 = 20$	$5 \times 5 = 25$	6×5	7×5	8×5	9×5
$2 \times 6 = 12$	$3 \times 6 = 18$	$4 \times 6 = 24$	$5 \times 6 = 30$	$6 \times 6 = 36$	7×6	8×6	9×6
$2 \times 7 = 14$	$3 \times 7 = 21$	$4 \times 7 = 28$	$5 \times 7 = 35$	$6 \times 7 = 42$	$7 \times 7 = 49$	8×7	9×7
$2 \times 8 = 16$	$3 \times 8 = 24$	$4 \times 8 = 32$	$5 \times 8 = 40$	$6 \times 8 = 48$	$7 \times 8 = 56$	$8 \times 8 = 64$	9×8
$2 \times 9 = 18$	$3 \times 9 = 27$	$4 \times 9 = 36$	$5 \times 9 = 45$	$6 \times 9 = 54$	$7 \times 9 = 63$	$8 \times 9 = 72$	$9 \times 9 = 81$

1. Examine the darkly shaded, lightly shaded, and unshaded regions in the preceding multiplication table. Explain how to obtain the unshaded facts quickly and easily from the shaded facts. In doing so, what property of arithmetic do you use?

2. Multiplication facts involving the numbers 6, 7, and 8 are often hard to learn. For each fact in the lightly shaded regions in the table, describe a way to obtain the fact from facts in the darkly shaded region as follows:

a. Draw an array for the fact, and show how to subdivide the array in order to relate the fact to one or more other facts. Try to find several different ways to subdivide the array this way.

b. Write equations that correspond to your subdivided array from Part (a) and that show how the multiplication fact is related to darkly shaded facts. Specify the properties of arithmetic that your equations used.

For example, we could draw either of the two subdivided arrays shown here for the fact 3×7 and write the corresponding equations shown underneath them. Both equations use the distributive property.

3. The *5 × table* is easy to learn because it is "half of the *10 × table*."

$$\begin{array}{ll} 5 \times 1 = 5 & 10 \times 1 = 10 \\ 5 \times 2 = 10 & 10 \times 2 = 20 \\ 5 \times 3 = 15 & 10 \times 3 = 30 \\ 5 \times 4 = 20 & 10 \times 4 = 40 \\ 5 \times 5 = 25 & 10 \times 5 = 50 \\ 5 \times 6 = 30 & 10 \times 6 = 60 \\ 5 \times 7 = 35 & 10 \times 7 = 70 \\ 5 \times 8 = 40 & 10 \times 8 = 80 \\ 5 \times 9 = 45 & 10 \times 9 = 90 \end{array}$$

Write an equation showing the relationship between 10×7 and 5×7 that fits with the statement about the *5 × table* being half of the *10 × table*. Which property of arithmetic do you use?

4. The *9 × table* is easy to learn because you can just subtract a number that is to be multplied by 9 from that number with a 0 placed behind it, as shown next. Explain why this way of multiplying by 9 is valid.

$$9 \times 1 = 10 - 1 = 9$$

$$9 \times 2 = 20 - 2 = 18$$

$$9 \times 3 = 30 - 3 = 27$$

$$9 \times 4 = 40 - 4 = 36$$

$$9 \times 5 = 50 - 5 = 45$$

$$9 \times 6 = 60 - 6 = 54$$

$$9 \times 7 = 70 - 7 = 63$$

$$9 \times 8 = 80 - 8 = 72$$

$$9 \times 9 = 90 - 9 = 81$$

5. What is another pattern (other than the one described in Problem 4) in the *9 × table*?

Class Activity 5U: Solving Arithmetic Problems Mentally

For each of the following arithmetic problems, describe a way to make the problem easy to solve mentally:

1. 4×99

2. 12×125 (Try to find several ways to solve this problem mentally.)

3. $125\% \times 120$

4. $45\% \times 680$

Class Activity 5V: Which Properties of Arithmetic Do These Calculations Use?

The sequences of equations that follow correspond to efficient mental strategies for solving the arithmetic problems of the previous class activity. In each case, describe the strategy in words. That is, describe what a person who is solving the problem mentally in a way that corresponds to the given equations might say to himself. Also, determine which properties of arithmetic were used, and where they were used. Be specific.

1.

$$\begin{aligned} 4 \times 99 &= 4 \times (100 - 1) \\ &= 4 \times 100 - 4 \times 1 \\ &= 400 - 4 \\ &= 396 \end{aligned}$$

2.

$$\begin{aligned} 12 \times 125 &= 10 \times 125 + 2 \times 125 \\ &= 1250 + 2 \times 100 + 2 \times 25 \\ &= 1250 + 200 + 50 \\ &= 1500 \end{aligned}$$

3.

$$\begin{aligned} 12 \times 125 &= (3 \times 4) \times 125 \\ &= 3 \times (4 \times 125) \\ &= 3 \times (4 \times 100 + 4 \times 25) \\ &= 3 \times (400 + 100) \\ &= 3 \times 500 \\ &= 1500 \end{aligned}$$

4.

$$\begin{aligned} 125\% \times 120 &= (100\% + 25\%) \times 120 \\ &= 100\% \times 120 + 25\% \times 120 \\ &= 120 + \frac{1}{4} \times 120 \\ &= 120 + 30 \\ &= 150 \end{aligned}$$

5.

$$\begin{aligned} 45\% \times 680 &= (50\% - 5\%) \times 680 \\ &= 50\% \times 680 - 5\% \times 680 \\ &= \frac{1}{2} \times 680 - \frac{1}{2} \times 10\% \times 680 \\ &= 340 - \frac{1}{2} \times 68 \\ &= 340 - 34 \\ &= 306 \end{aligned}$$

6.

$$\begin{aligned} 16 \times 25 &= (4 \times 4) \times 25 \\ &= 4 \times (4 \times 25) \\ &= 4 \times 100 \\ &= 400 \end{aligned}$$

7.

$$\begin{aligned} 16 \times 25 &= 4 \times 4 \times 5 \times 5 \\ &= 4 \times 5 \times 4 \times 5 \\ &= 20 \times 20 \\ &= 400 \end{aligned}$$

Class Activity 5W: Writing Equations That Correspond to a Method of Calculation

Each arithmetic problem in this activity has a description of the problem solution. In each case, write a sequence of equations that corresponds to the given description. Which properties of arithmetic were used and where? Write your equations in the following form:

$$\begin{aligned} \text{original} &= \text{some expression} \\ &= \vdots \\ &= \text{some expression} \end{aligned}$$

1. $6 \times 800 = ?$

   6 times 8 is 48; then 48 times 100 is 4800.

2. $51 \times 4 = ?$

   50 times 4 is 200, plus another 4 is 204.

3. $35 \times 2 = ?$

   35 twos is two 35s, which is 70.

4. What is 55% of 120?

   Half of 120 is 60. Ten percent of 120 is 12, so 5% of 120 is half of that ten percent, which is 6. So the answer is 60 plus 6, which is 66.

5. What is 35% of 80?

25% is $\frac{1}{4}$, so 25% of 80 is one-fourth of 80, which is 20. Ten percent of 80 is 8. So 35% of 80 is 20 plus 8, which is 28.

6. What is 90% of 350?

10% of 350 is 35. Taking 35 away from 350 leaves 315. So the answer is 315.

7. What is $\frac{7}{8}$ of 2400?

One-eighth of 2400 is 300. Taking 300 away from 2400 leaves 2100. The answer is 2100.

Class Activity 5X: Showing the Algebra in Mental Math

For each arithmetic problem in this activity, find ways to use properties of arithmetic to make the problem easy to do mentally. Describe your method in words, and write equations that correspond to your method. Write your equations in the following form:

$$\begin{aligned} \text{original} &= \text{some expression} \\ &= \vdots \\ &= \text{some expression} \end{aligned}$$

1. 9×99 (Try to find several different ways to solve this problem mentally.)

2. 24×25 (Try to find several different ways to solve this problem mentally.)

3. $\frac{7}{8} \times 128$

4. $26\% \times 840$

5. $5\% \times 48$

6. $15\% \times \$44$

5.7 Why the Procedure for Multiplying Whole Numbers Works

Class Activity 5Y: The Standard Versus the Partial-Products Multiplication Algorithm

1. Use the standard multiplication algorithm and then the partial-products algorithm to solve the following multiplication problems:

$$\begin{array}{r} 495 \\ \times \quad 7 \\ \hline \end{array} \qquad\qquad \begin{array}{r} 495 \\ \times \quad 7 \\ \hline \end{array}$$

$$\begin{array}{r} 234 \\ \times \quad 59 \\ \hline \end{array} \qquad\qquad \begin{array}{r} 234 \\ \times \quad 59 \\ \hline \end{array}$$

2. Compare the two algorithms. How are they the same, and how are they different? Why do both produce the same answer? Compare how "carrying" is handled in the two algorithms. Compare how place value is handled in the two algorithms.

Class Activity 5Z: Why the Multiplication Algorithms Give Correct Answers, Part 1

Graph paper would be helpful for Problem 5 of this activity.

This class activity will help you explain why the partial-products algorithm, and therefore also the standard algorithm, gives correct answers to multiplication problems.

Consider the relatively simple multiplication problem 6×38.

1. Use the partial-products multiplication algorithm to calculate 6×38.

2. The array shown next consists of 6 rows of o's with 38 o's in each row. Subdivide the array into pieces that correspond to the steps in the partial-products algorithm. Use this correspondence to explain why the partial-products algorithm calculates the correct answer to 6×38. Begin your explanation by using the meaning of multiplication to relate the array to the multiplication problem 6×38.

oooooooooo oooooooooo oooooooooo oooooooo

oooooooooo oooooooooo oooooooooo oooooooo

oooooooooo oooooooooo oooooooooo oooooooo

oooooooooo oooooooooo oooooooooo oooooooo

oooooooooo oooooooooo oooooooooo oooooooo

oooooooooo oooooooooo oooooooooo oooooooo

3. Now solve the multiplication problem 6×38 by working with expanded forms and using the distributive property.

$$\begin{aligned} 6 \times 38 &= 6 \times (30 + 8) \\ &= \end{aligned}$$

Relate the equations you just wrote to the steps in the partial-products algorithm. Use this relationship to explain why the partial-products algorithm calculates the correct answer to 6×38.

4. Compare the two explanations you gave in Problems 2 and 3 to show why the partial-products algorithm calculates the correct answer to 6×38.

5. a. Use the partial-products algorithm to calculate 9×26.

b. Draw an array for 9×26 (if it's available, use graph paper; otherwise, indicate the array with a rectangle). Subdivide the array in a natural way so that the parts of the array correspond to the steps in the partial-products algorithm.

c. Solve 9×26 by writing equations that use expanded forms and the distributive property. Relate your equations to the steps in the partial-products algorithm.

Class Activity 5AA: Why the Multiplication Algorithms Give Correct Answers, Part 2

Graph paper would be helpful for Problem 5 of this activity.

This class activity will help you explain why the partial-products multiplication algorithm, and therefore also the standard algorithm, gives correct answers to multiplication problems. It will also help you relate the standard and partial-products algorithms.

Consider the multiplication problem 23×45.

1. Use the partial-products and standard algorithms to calculate

$$\begin{array}{r} 45 \\ \times\ 23 \\ \hline \end{array} \qquad\qquad \begin{array}{r} 45 \\ \times\ 23 \\ \hline \end{array}$$

2. The next array consists of 23 rows of o's with 45 o's in each row. Subdivide the array into pieces that correspond to the steps in the partial-products algorithm. Use this correspondence to explain why the partial-products algorithm calculates the correct answer to 23×45. Begin your explanation by using the meaning of multiplication to relate the array to the multiplication problem 23×45.

oooooooooo oooooooooo oooooooooo oooooooooo ooooo
oooooooooo oooooooooo oooooooooo oooooooooo ooooo
oooooooooo oooooooooo oooooooooo oooooooooo ooooo
oooooooooo oooooooooo oooooooooo oooooooooo ooooo
oooooooooo oooooooooo oooooooooo oooooooooo ooooo
oooooooooo oooooooooo oooooooooo oooooooooo ooooo
oooooooooo oooooooooo oooooooooo oooooooooo ooooo
oooooooooo oooooooooo oooooooooo oooooooooo ooooo
oooooooooo oooooooooo oooooooooo oooooooooo ooooo
oooooooooo oooooooooo oooooooooo oooooooooo ooooo

oooooooooo oooooooooo oooooooooo oooooooooo ooooo
oooooooooo oooooooooo oooooooooo oooooooooo ooooo
oooooooooo oooooooooo oooooooooo oooooooooo ooooo
oooooooooo oooooooooo oooooooooo oooooooooo ooooo
oooooooooo oooooooooo oooooooooo oooooooooo ooooo
oooooooooo oooooooooo oooooooooo oooooooooo ooooo
oooooooooo oooooooooo oooooooooo oooooooooo ooooo
oooooooooo oooooooooo oooooooooo oooooooooo ooooo
oooooooooo oooooooooo oooooooooo oooooooooo ooooo
oooooooooo oooooooooo oooooooooo oooooooooo ooooo

oooooooooo oooooooooo oooooooooo oooooooooo ooooo
oooooooooo oooooooooo oooooooooo oooooooooo ooooo
oooooooooo oooooooooo oooooooooo oooooooooo ooooo

3. Now subdivide the array into pieces that correspond to the steps in the *standard algorithm*. Relate these pieces to the pieces for the partial-products algorithm that you created in Problem 2.

4. Solve the multiplication problem 23×45 by working with expanded forms and using the distributive property several times.

$$\begin{aligned} 23 \times 45 &= (20 + 3) \times (40 + 5) \\ &= \end{aligned}$$

Relate the equations you just wrote to the steps in the partial-products algorithm. Use this relationship to explain why the partial-products algorithm calculates the correct answer to 23×45.

5. Compare the two explanations you gave in Problems 2 and 4 to show why the partial-products algorithm calculates the correct answer to 23×45.

6. Write equations that use the distributive property and that correspond to the steps in the standard algorithm for 23×45. (Put one of the numbers in expanded form.) Relate to Problem 3.

7. **a.** Use the partial-products and standard algorithms to calculate 17×28.

b. Draw an array for 17×28. (Draw your array by drawing a rectangle on graph paper; if you don't have graph paper, draw a rectangle to represent the array rather than drawing 17 rows of 28 items.) Subdivide the array in a natural way so that the parts of the array correspond to the steps in the partial-products algorithm.

c. Now subdivide your array from Part (b) in a natural way so that the parts of the array correspond to the steps in the standard algorithm. Relate these parts to the parts in Part (b).

d. Solve 17×28 by writing equations that use expanded forms and the distributive property. Relate your equations to the steps in the partial-products algorithm.

e. Write equations that use the distributive property and that correspond to the steps in the standard algorithm for 17×28. Relate to Part (c).

Class Activity 5BB: The Standard Multiplication Algorithm Right Side Up and Upside Down

1. Use the standard algorithm to calculate $\begin{array}{r} 347 \\ \times\ 26 \\ \hline \end{array}$

2. Draw a rectangle to represent an array for 26×347. (The rectangle need not be to scale.) Subdivide the rectangle in a natural way so that the parts of the rectangle correspond to the steps of the standard algorithm in Problem 1.

3. Now use the standard algorithm to calculate $\begin{array}{r} 26 \\ \times\ 347 \\ \hline \end{array}$

   Are the steps the same as in Problem 1? Is the answer the same as in Problem 1?

4. Can you show the steps of the standard algorithm in Problem 3 on the rectangle in Problem 2? If so, how?

5. Why must the answers in 1 and 3 be the same even though the steps are different? How is this question related to the rectangle in Problem 2? Which property of arithmetic is relevant here?

6

Multiplication of Fractions, Decimals, and Negative Numbers

6.1 Multiplying Fractions

Class Activity 6A: Writing and Solving Fraction Multiplication Story Problems

Write a story problem for each of the fraction multiplication problems that follow. Then show how to solve your story problem with the aid of pictures. Verify that the solution you obtain from pictures agrees with the solution you obtain by using the fraction multiplication procedure.

1.

$$2 \cdot \frac{1}{3}$$

2.

$$\frac{1}{2} \cdot \frac{1}{3}$$

3.

$$\frac{1}{3} \cdot 2$$

4.

$$3 \cdot \frac{4}{5}$$

Class Activity 6B: Misconceptions with Fraction Multiplication

1. Maisy draws a picture like the one shown to depict $3 \cdot \frac{4}{5}$. Maisy concludes from her picture that

$$3 \cdot \frac{4}{5} = \frac{12}{15}$$

because 12 pieces out of 15 are shaded. Is Maisy right? If not, where is her reasoning flawed?

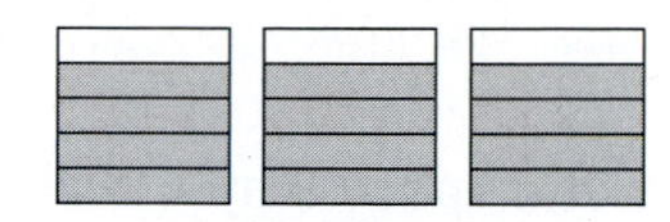

2. Joey, a fourth grader, wanted to figure out how many calories were in a piece of candy. According to the label on the bag, there were 50 calories in 3 pieces. Joey knew he had to find $\frac{1}{3}$ of 50, and he gave the answer $16\frac{2}{6}$. Joey explained his answer this way:

   3 sixteens is 48 and $\frac{1}{3}$ of the other 2 calories is $\frac{2}{6}$.

   Joey drew the following picture to explain the $\frac{2}{6}$. What was Joey confused about?

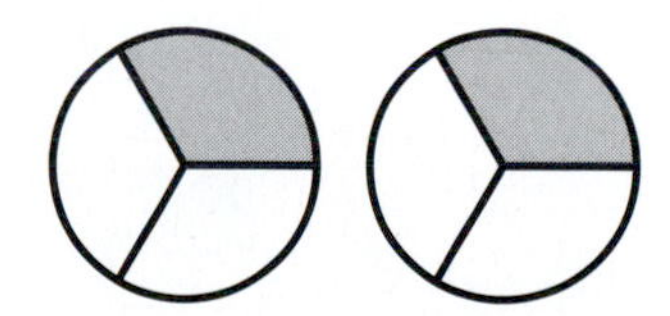

Class Activity 6C: Explaining Why the Procedure for Multiplying Fractions Gives Correct Answers

This class activity will help you explain how the procedure for multiplying fractions comes from the meaning of fractions and the meaning of multiplication.

1. Use the meaning of fractions, the meaning of multiplication, and the next diagram to help you explain why

$$\frac{2}{3} \cdot \frac{5}{8} = \frac{2 \cdot 5}{3 \cdot 8}$$

In particular, explain why it makes sense to multiply the numerators and why it makes sense to multiply the denominators.

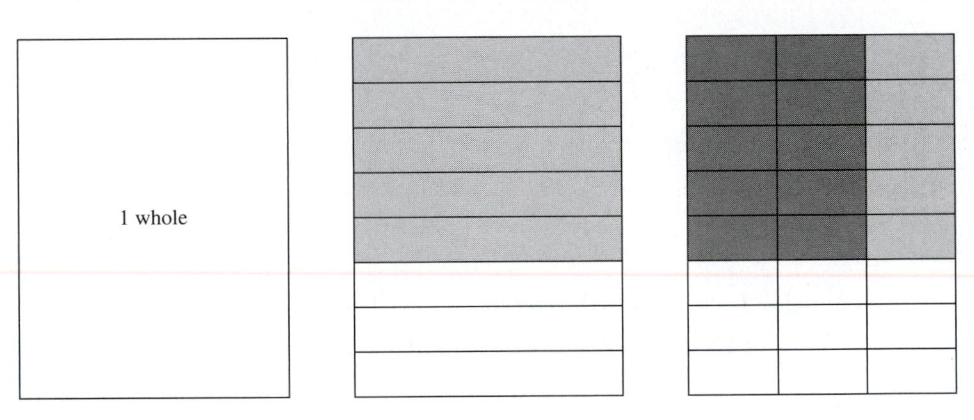

2. Use the meaning of fractions, the meaning of multiplication, and the rectangles in the next figure to help you explain why

$$\frac{2}{5} \cdot \frac{3}{7} = \frac{2 \cdot 3}{5 \cdot 7}$$

In particular, explain why it makes sense to multiply the numerators and why it makes sense to multiply the denominators.

1 whole

Class Activity 6D: When Do We Multiply Fractions?

1. Originally, there were 3 cubic yards of mulch in a mulch pile. Then $\frac{3}{4}$ of the mulch was removed from the pile. What question about the mulch pile can you answer by calculating $\frac{3}{4} \cdot 3$?

2. Shelley reads that one serving of Sweetblaster cereal will provide 80% of her daily value of vitamin C. Shelley is on a diet, so she decides to eat $\frac{2}{3}$ of a serving of Sweetblaster cereal. What question about Shelley's cereal will be answered by calculating $\frac{2}{3} \cdot \frac{80}{100}$?

3. Which of the following problems are story problems for $\frac{2}{3} \cdot \frac{1}{4}$, and which are not? Why?

   a. Joe is making $\frac{2}{3}$ of a recipe. The full recipe calls for $\frac{1}{4}$ cup of water. How much water should Joe use?

   b. $\frac{1}{4}$ of the students in Mrs. Watson's class are doing a dinosaur project. $\frac{2}{3}$ of the children doing the dinosaur project have completed it. How many children have completed a dinosaur project?

   c. $\frac{1}{4}$ of the students in Mrs. Watson's class are doing a dinosaur project. $\frac{2}{3}$ of the children doing the dinosaur project have completed it. What fraction of the students in Mrs. Watson's class have completed a dinosaur project?

   d. There is $\frac{1}{4}$ of a cake left. $\frac{2}{3}$ of Mrs. Watson's class would like to have some cake. What fraction of the cake does each student who wants cake get?

   e. Carla is making snack bags that each contain $\frac{1}{4}$ package of jelly worms. $\frac{2}{3}$ of Carla's grab bags have been bought. What fraction of Carla's jelly worms have been bought?

Class Activity 6E: Multiplying Mixed Numbers

1. Write a story problem for $2\frac{1}{2} \times 1\frac{1}{2}$.

2. Use pictures and the meaning of multiplication to solve your problem from Problem 1.

3. Use the distributive property or FOIL to calculate $2\frac{1}{2} \times 1\frac{1}{2}$ by rewriting this product as $(2 + \frac{1}{2}) \times (1 + \frac{1}{2})$.

4. Identify the four terms produced by the distributive property or FOIL (in Problem 3) in your picture in Problem 2.

5. Now write the mixed numbers $2\frac{1}{2}$ and $1\frac{1}{2}$ as improper fractions, and use the procedure for multiplying fractions to calculate $2\frac{1}{2} \times 1\frac{1}{2}$. How can you see the product of the numerators in your picture in Problem 2? How can you see the product of the denominators in your picture?

Class Activity 6F: What Fraction Is Shaded?

1. For each of the next figures, write an expression that uses both multiplication and addition (or subtraction) to describe the total fraction of the figure that is shaded. (For example, $\frac{5}{7} \cdot \frac{2}{9} + \frac{1}{3}$ is an expression that uses both multiplication and addition). Explain your reasoning. Then determine what fraction of the figure is shaded (in simplest form). In each figure, you may assume that lengths which appear to be equal really are equal.

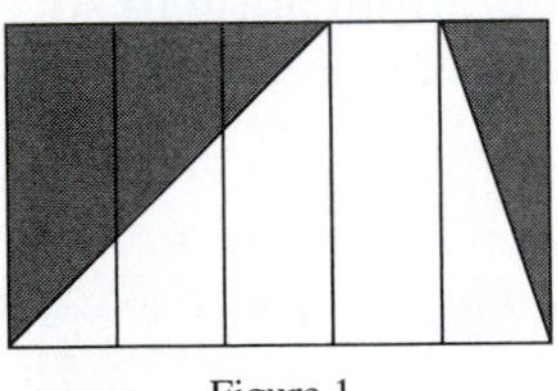

Figure 1

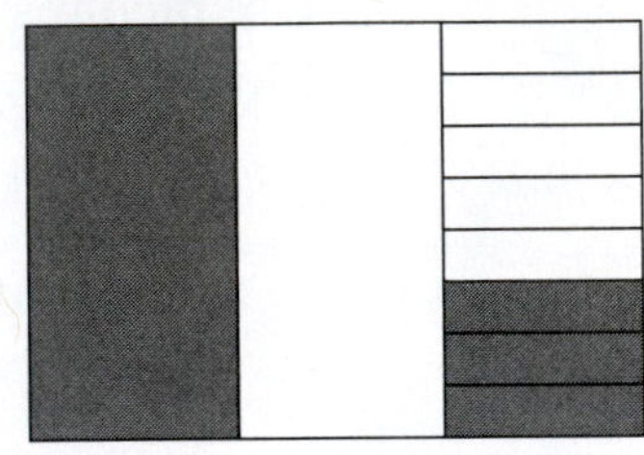

Figure 2

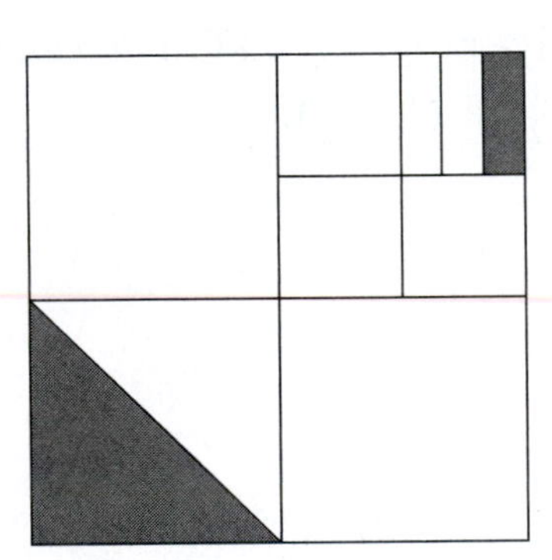

Figure 3

2. Draw a figure in which you shade $\frac{1}{3} \cdot \frac{2}{7} + \frac{1}{2} \cdot \frac{3}{7}$ of the figure.

6.2 Multiplying Decimals

Class Activity 6G: Multiplying Decimals

1. Write a story problem for 2.7×1.35.

2. Ben wants to multiply 3.46×1.8. He first multiplies the numbers by ignoring the decimal points:

$$\begin{array}{r} 3.46 \\ \times \ \ 1.8 \\ \hline 6228 \end{array}$$

Ben knows that he just needs to figure out where to put the decimal point in his answer, but he can't remember the rule about where to put the decimal point. Explain how Ben could reason about the sizes of the numbers in order to determine where to put the decimal point in his answer.

3. Lameisha used a calculator and found that

$$1.5 \times 1.2 = 1.8$$

Lameisha wants to know why the rule about adding the number of places behind the decimal point doesn't work in this case. Why aren't there 2 digits to the right of the decimal point in the answer? Is Lameisha right that the rule about adding the number of places behind the decimal points doesn't work in this case? Explain.

Class Activity 6H: Explaining Why We Place the Decimal Point Where We Do When We Multiply Decimals

1. As indicated in the next diagram, to get from 1.36 to 136, we multiply by 10×10. To get from 2.7 to 27, we multiply by 10. In other words,

$$136 = 10 \times 10 \times 1.36 \quad \text{and} \quad 27 = 10 \times 2.7$$

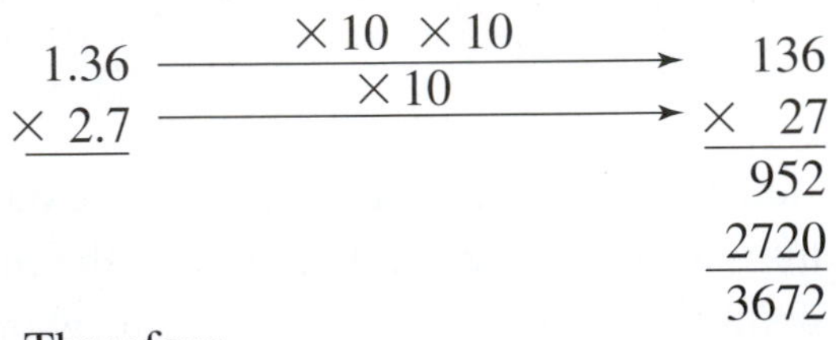

$$\begin{array}{r} 1.36 \\ \times\ 2.7 \\ \hline \end{array} \quad \xrightarrow[\times 10]{\times 10\ \times 10} \quad \begin{array}{r} 136 \\ \times\ 27 \\ \hline 952 \\ 2720 \\ \hline 3672 \end{array}$$

Therefore,

$$\begin{array}{r} 1.36 \\ \times\ 2.7 \\ \hline 952 \\ 2720 \\ \hline \end{array} \quad \begin{array}{c} \xleftarrow{\quad ? \quad} \\ \xleftarrow{\quad ? \quad} \\ \\ \\ \xleftarrow{\quad ? \quad} \end{array} \quad \begin{array}{r} 136 \\ \times\ 27 \\ \hline 952 \\ 2720 \\ \hline 3672 \end{array}$$

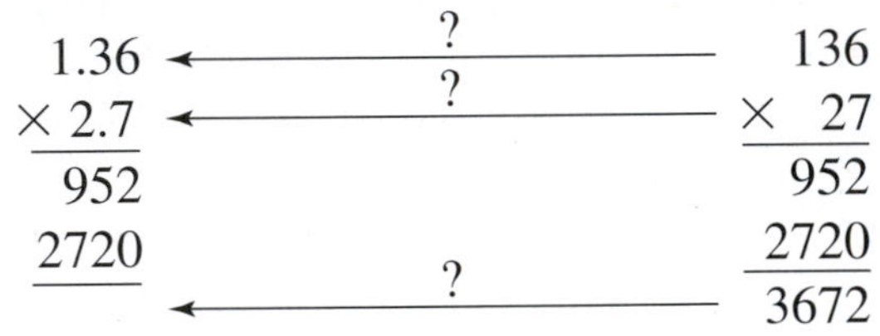

Therefore, what should we do to

$$136 \times 27 = 3672$$

to get back to

$$1.36 \times 2.7?$$

Use your answer to explain the placement of the decimal point in 1.36×2.7.

2. Use the next diagram to help you explain where to put the decimal point in 2.476×1.83.

$$\begin{array}{r} 2.476 \\ \times\ 1.83 \\ \hline \end{array} \longrightarrow \begin{array}{r} 2476 \\ \times\ 183 \\ \hline \end{array}$$

Therefore,

$$\begin{array}{r} 2.476 \\ \times\ 1.83 \\ \hline \end{array} \longleftarrow \begin{array}{r} 2476 \\ \times\ 183 \\ \hline \end{array}$$

3. Working with powers of 10 and using the idea of Problems 1 and 2, explain why the following makes sense: If you multiply a number that has 3 digits to the right of its decimal point by a number that has 4 digits to the right of its decimal point, you should place the decimal point $3 + 4 = 7$ places from the end of the product calculated without the decimal points.

Class Activity 6I: Decimal Multiplication and Areas of Rectangles

In Section 5.3, we decomposed rectangles into groups of squares to explain why we can multiply to find areas of rectangles. Our focus in that section was on rectangles whose side lengths are counting numbers. In this activity, you will explain why we can multiply to find the area of a rectangle whose side lengths are decimals. You will also take apart and recombine a rectangle to determine its area.

Recall that the area of a region, in square units, is the number of 1-unit-by-1-unit squares it takes to cover the region without gaps or overlaps.

1. Explain why we can multiply 2.3×1.8 to find the area of the 2.3-unit-by-1.8-unit rectangle in Figure 6I.1 by using the diagrams in Figure 6I.1 to describe the large rectangle as consisting of 2.3 groups of 1.8 squares.

2. Draw pictures showing how to rearrange portions of the 2.3-unit-by-1.8-unit rectangle in Figure 6I.1 so that you can determine the area of this rectangle. Calculate 2.3×1.8, and verify that it produces the correct area.

3. Discuss the following questions: How is the 2.3-unit-by-1.8-unit rectangle in Figure 6I.1 related to an array for 23×18? How is 2.3×1.8 related to 23×18?

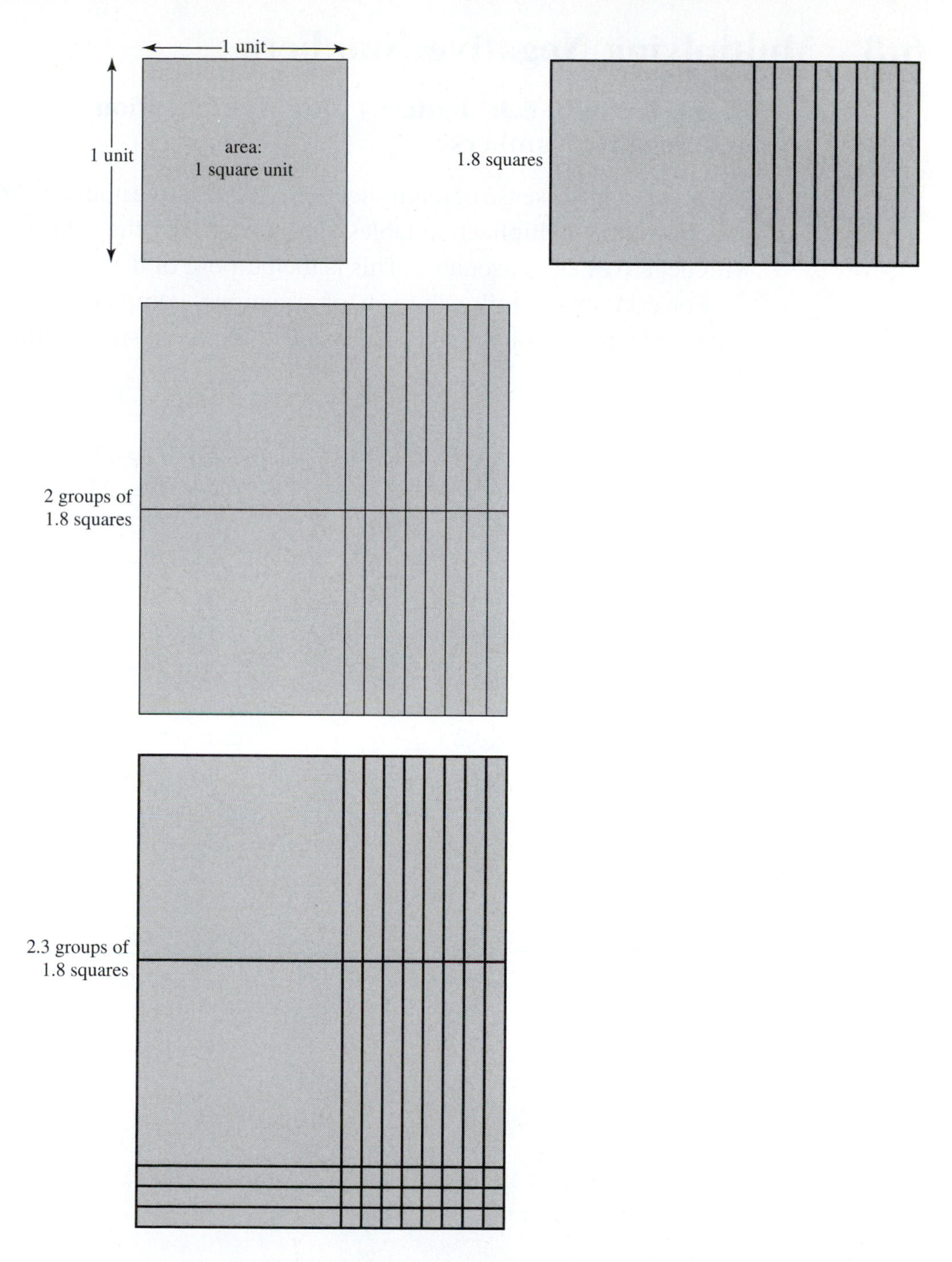

Figure 6I.1
Showing 2.3 Groups with 1.8 Squares in Each Group

6.3 Multiplying Negative Numbers

Class Activity 6J: Patterns with Multiplication and Negative Numbers

How can we make sense of multiplication with negative numbers? By extending patterns in the multiplication tables, we can see why the rules for multiplying with negatives are reasonable. This is the purpose of this class activity.

For each of the following sets of equations, complete the set so that the pattern of numbers on the right-hand side of the equal sign continues. Describe the patterns.

$$6 \times 4 = 24$$
$$6 \times 3 = 18$$
$$6 \times 2 = 12$$
$$6 \times 1 = 6$$
$$6 \times 0 = 0$$
$$6 \times -1 =$$
$$6 \times -2 =$$
$$6 \times -3 =$$
$$6 \times -4 =$$
$$6 \times -5 =$$

$$4 \times 7 = 28$$
$$3 \times 7 = 21$$
$$2 \times 7 = 14$$
$$1 \times 7 = 7$$
$$0 \times 7 = 0$$
$$-1 \times 7 =$$
$$-2 \times 7 =$$
$$-3 \times 7 =$$
$$-4 \times 7 =$$
$$-5 \times 7 =$$

$$-8 \times 4 = -32$$
$$-8 \times 3 = -24$$
$$-8 \times 2 = -16$$
$$-8 \times 1 = -8$$
$$-8 \times 0 = 0$$
$$-8 \times -1 =$$
$$-8 \times -2 =$$
$$-8 \times -3 =$$
$$-8 \times -4 =$$
$$-8 \times -5 =$$

Class Activity 6K: Explaining Multiplication with Negative Numbers (and 0)

1. Write a story problem for 3×-5. Solve the story problem, thereby explaining why 3×-5 is negative. Interpret negative numbers as amounts owed.

2. Explain why the following make sense:

$$\begin{aligned} 0 \times (\text{any number}) &= 0 \\ (\text{any number}) \times 0 &= 0 \end{aligned}$$

3. Assume that you don't yet know what $(-3) \times 5$ is, but you do know that $3 \times 5 = 15$. Use the distributive property to show that the expression

$$(-3) \times 5 + 3 \times 5$$

is equal to 0. Then use that result to determine what $(-3) \times 5$ must be equal to.

4. Assume that you don't yet know what $(-3) \times (-5)$ is, but you do know that $(-3) \times 5 = -15$ from Problem 2, and you know $3 \times 5 = 15$. Use the distributive property to show that the expression

$$(-3) \times (-5) + (-3) \times 5$$

is equal to 0. Then use that result to determine what $(-3) \times (-5)$ must be equal to.

Class Activity 6L: Using Checks and Bills to Interpret Multiplication with Negative Numbers

1. For each of the following transactions, compare the amount of money you will have before and after the transaction takes place:

   a. You get 3 checks, each for $5.

   b. You get 3 bills, each for $5. (Each bill tells you that you owe $5.)

   c. You give away 3 checks, each for $5.

   d. You give away 3 bills, each for $5.

2. Write a consistent set of multiplication problems that describe how the amount of money you have changes due to the transactions in Problem 1. Use negative numbers wherever it makes sense to do so.

Class Activity 6M: Does Multiplication Always Make Larger?

1. In ordinary language, the term "multiply" means "make larger," as in

   Go forth and multiply.

   In mathematics, does multiplying always make a quantity larger? Give some examples to help you explain your answer.

2. For which numbers, N, is $N \times 3$ greater than 3? In other words, for which numbers, N, is $N \times 3 > 3$? Consider all kinds of numbers for N: fractions, decimals, positive, negative. Investigate this question by trying a number of examples and by thinking about the meaning and rules of multiplication.

3. For which numbers, N, is $N \times -3$ greater than -3? In other words, for which numbers, N, is $N \times -3 > -3$? Consider all kinds of numbers for N: fractions, decimals, positive, negative. Investigate this question by trying a number of examples and by thinking about the meaning and rules of multiplication.

6.4 Scientific Notation

Class Activity 6N: Scientific Notation versus Ordinary Decimal Notation

1. A $1.3 trillion tax cut was approved. Write 1.3 trillion in scientific notation and in ordinary decimal notation.

2. The population of the United States is approximately 299 million. Write 299 million in scientific notation and in ordinary decimal notation.

3. Use a calculator to multiply

$$123{,}456{,}789 \times 987{,}654{,}321$$

   a. Does your calculator's display show all the digits in the product $123{,}456{,}789 \times 987{,}654{,}321$? If not, why not? What does your calculator's display mean?

   b. Based on your calculator's display, what can you determine about the ordinary decimal representation of the product $123{,}456{,}789 \times 987{,}654{,}321$?

   c. Can you use the calculator's display to determine how many digits the decimal representation of the product $123{,}456{,}789 \times 987{,}654{,}321$ has? Explain.

   d. Explain how you can "break up" the numbers 123,456,789 and 987,654,321 and use the distributive property or FOIL in conjunction with a calculator in order to determine all the digits in the product $123{,}456{,}789 \times 987{,}654{,}321$. Do not just multiply longhand.

Class Activity 6O: Multiplying Powers of 10

You will need the results of this activity in Class Activity 6Q.

Remember that if A is a counting number, the expression

$$10^A$$

stands for A 10s multiplied together:

$$10^A = \underbrace{10 \times 10 \times \cdots \times 10}_{A \text{ times}}$$

1. Use the meaning of powers of ten to show how to write each of the expressions (a), (b), and (c) as a single power of ten (i.e., in the form 10^A for some exponent A). For example, 10^2 means 10×10, and 10^3 means $10 \times 10 \times 10$; therefore,

$$\begin{aligned} 10^2 \times 10^3 &= (10 \times 10) \times (10 \times 10 \times 10) \\ &= 10^5 \end{aligned}$$

   a. $10^3 \times 10^4$
   b. $10^2 \times 10^5$
   c. $10^3 \times 10^3$

2. In each of (a), (b), and (c) in Problem 1, relate the exponents in the product with the exponent in the answer. (For the example given at the beginning of Problem 1, relate 2 and 3 to 5). In each case, how are the three exponents related?

3. Explain why it is always true that $10^A \times 10^B = 10^{A+B}$ when A and B are counting numbers.

Class Activity 6P: ✻ How Many Digits Are in a Product of Counting Numbers?

When you multiply two counting numbers, how many digits does the product have? This class activity will help you discover the answer to this question.

1. Multiply many 2-digit counting numbers with 3-digit counting numbers. How many digits do the products have?

   Multiply many 3-digit counting numbers with 4-digit counting numbers. How many digits do the products have?

   Multiply many 2-digit counting numbers with 4-digit counting numbers. How many digits do the products have?

   In your examples, how are the number of digits in the two factors related to the number of digits in the product? (Your answer should involve "either . . . or . . . .")

2. Based on your results in Problem 1, predict what will happen in other situations. If you multiply a 9-digit counting number with an 11-digit counting number, how many digits will the product have? In general, if you multiply an n-digit counting number with an m-digit counting number, how many digits will the product have? Give "either . . . or . . ." answers.

Class Activity 6Q: ✻ Explaining the Pattern in the Number of Digits in Products

This class activity continues the previous activity. In this activity, you will explain why the pattern you found in the previous class activity must hold. The key is to relate the number of digits in a counting number to the exponent on the 10 when the number is written in scientific notation.

1. Based on what you discovered in the previous class activity, predict how many digits the product

$$123{,}456{,}789 \times 23{,}456{,}789$$

has.

2. Now use a calculator to compute $123{,}456{,}789 \times 23{,}456{,}789$. Use the calculator's display to determine how many digits the product $123{,}456{,}789 \times 23{,}456{,}789$ has.

3. In general, what is the relationship between the number of digits of a counting number and the exponent on the 10 when that number is written in scientific notation? Explain.

4. Write the numbers 123,456,789 and 23,456,789 and the product 123,456,789 × 23,456,789 in scientific notation. Look carefully at the exponents on the 10s. What relationship do you notice among these three exponents?

5. Take another pair of whole numbers and write them and their product in scientific notation. What is the relationship among the exponents on the 10s? Do this many times, with many pairs of numbers. Do you always get the same relationship as in the last problem, or not?

6. What happens when we multiply numbers that are in scientific notation? Fill in the correct exponents for the 10s on the right in the following equations:

$$(1.2 \times 10^3) \times (3.7 \times 10^5) = (1.2 \times 3.7) \times 10^3 \times 10^5 = (1.2 \times 3.7) \times 10^{—}$$
$$(7.63 \times 10^4) \times (8.14 \times 10^6) = (7.63 \times 8.14) \times 10^4 \times 10^6 = (7.63 \times 8.14) \times 10^{—}$$
$$(4.5 \times 10^7) \times (5.2 \times 10^8) = (4.5 \times 5.2) \times 10^7 \times 10^8 = (4.5 \times 5.2) \times 10^{—}$$

7. Now put the numbers from the previous problem in scientific notation. In some cases, the exponent on the 10 changes. Why?

$$(1.2 \times 3.7) \times 10^{—} =$$
$$(7.63 \times 8.14) \times 10^{—} =$$
$$(4.5 \times 5.2) \times 10^{—} =$$

8. Suppose you have a 7-digit counting number and an 8-digit counting number. What will they look like in scientific notation? Use this to explain why the product of a 7-digit counting number with an 8-digit counting number will have 14 or 15 digits.

7

Division

7.1 The Meaning of Division

Class Activity 7A: The Two Interpretations of Division

1. Use both the "how many in each group?" and the "how many groups?" interpretations of division to explain why $10 \div 2 = 5$. Write a story problem for each case. Draw simple pictures to illustrate.

2. Use both interpretations of division to explain why $14 \div 3 = 4\frac{2}{3}$. Write a story problem for each case. Draw simple pictures to illustrate.

3. Besides $4\frac{2}{3}$, what other answers could you give to the division problem $14 \div 3$? For each answer, write one or two story problems that are best answered that way. In each case, try to write one "how many in each group?" and one "how many groups?" problem.

4. Write a story problem for which you would calculate $14 \div 3$ in order to solve the problem, but which has answer 5.

Class Activity 7B: Why Can't We Divide by Zero?

1. A division problem (exact division, without remainder)

$$A \div B = ?$$

can always be rewritten as a multiplication problem, namely, as

$$? \times B = A$$

or as

$$B \times ? = A$$

Use the fact that every division problem can be rewritten as a multiplication problem to explain why $2 \div 0$ is not defined.

2. Write story problems for the two interpretations of $2 \div 0$. Use your problems to explain why $2 \div 0$ is not defined.

3. Why can't we just say that $2 \div 0$ is 0, remainder 2? If we did, would that fit with the way we defined division with remainder?

4. Write story problems for the two interpretations of $0 \div 2$. Use your problems to explain why $0 \div 2$ *is* defined. Explain the difference between $2 \div 0$ and $0 \div 2$.

5. Explain why $0 \div 0$ is undefined. (We can also say it is "indeterminate.") Can you give the same explanation as for why $2 \div 0$ is not defined? If not, how are the explanations different?

Class Activity 7C: Division Story Problems

1. For each of the following problems, write the corresponding numerical division problem and decide which interpretation of division is involved (the "how many groups?" or the "how many in each group?", exact division or with remainder). Solve the problems.

   a. Gloria has 33 candies to distribute equally among 5 children. How many candies will each child get, and how many candies will be left over?

   b. Bill has a muffin recipe that calls for 2 cups of flour. How many batches of muffins can Bill make if he has 9 cups of flour available? How much flour will be left over? What if Bill is willing to make a fraction of a batch, then how many batches of muffins can Bill make?

   c. If 6 limes cost one dollar, then how much should one lime cost (assuming that all limes are priced equally)?

   d. If 8 cups of ice cream cost $3, then how many cups of ice cream could you buy for $1?

e. One foot is 12 inches. If a piece of rope is 100 inches long, then how long is it in feet?

f. Francine has 20 yards of rope that she wants to cut into 8 equal pieces. How long will each piece be?

g. A gallon of water weighs 8 pounds. How many gallons is 500 pounds of water?

h. If you drive 250 miles at a constant speed and it takes you 4 hours, then how fast did you go?

2. For each of the story problems in Problem 1, label the numerical division problem and its solution with the corresponding units (such as inches, dollars, cups, candies, etc.). For example, for Part (a), write

33	÷	5	=	6	R	3
candies		children		candies		candies

What do the "how many groups?" problems have in common? What do the "how many in each group?" have in common?

Class Activity 7D: Can We Use Properties of Arithmetic to Divide?

In Chapter 5, we often used properties of arithmetic to solve multiplication problems. Think carefully about each of the following:

1. Is division commutative? Explain your answer.

2. Is division associative? Explain your answer.

3. Is the following statement true?

$$200 \div 45 = 200 \div 40 + 200 \div 5$$

Why, or why not? Explain your answer carefully, including diagrams if possible.

4. Is the following statement true?

$$365 \div 7 = 300 \div 7 + 60 \div 7 + 5 \div 7$$

Does it depend on how you express the answer (as a whole number with remainder, a mixed number, or a decimal)?

Class Activity 7E: Reasoning about Division

1. Even though

$$25 \div 12 = 2, \text{ remainder } 1$$

and

$$21 \div 10 = 2, \text{ remainder } 1$$

would it be correct to say that

$$25 \div 12 = 21 \div 10?$$

Explain.

2. **a.** Write and solve a simple story problem for $600 \div 20$.

   b. Use the situation of your story problem in Part (a) to help you solve $600 \div 19$ without a calculator or long division by modifying your solution to $600 \div 20$.

   c. Use the situation of your story problem in Part (a) to help you solve $600 \div 21$ without a calculator or long division by modifying your solution to $600 \div 20$.

Class Activity 7F: Rounding to Estimate Solutions to Division Problems

1. Suppose you want to estimate

$$615 \div 29$$

by rounding 29 up to 30. Both $600 \div 30$ and $630 \div 30$ are easy to calculate mentally. Use reasoning about division to determine which division problem, $600 \div 30$ or $630 \div 30$, should give you a better estimate to $615 \div 29$. Then check your answer by solving the division problems.

2. Suppose you want to estimate

$$615 \div 31$$

by rounding 31 down to 30. Both $600 \div 30$ and $630 \div 30$ are easy to calculate mentally. Use reasoning about division to determine which division problem, $600 \div 30$ or $630 \div 30$, should give you a better estimate to $615 \div 31$. Then check your answer by solving the division problems.

3. Suppose you want to estimate $527 \div 48$. What is a good way to round the numbers 527 and 48 so as to obtain an easy division problem that will give a good estimate to $527 \div 48$? Explain, drawing on what you learned from Problems 1 and 2.

4. Suppose you want to estimate $527 \div 52$. What is a good way to round the numbers 527 and 52 so as to obtain an easy division problem that will give a good estimate to $527 \div 52$? Explain, drawing on what you learned from Problems 1 and 2.

7.2 Understanding Long Division

Class Activity 7G: Dividing without Using a Calculator or Long Division

1. Antrice is working on the following problem: There are 260 pencils to be put in packages of 12. How many packages of pencils can we make, and how many pencils will be left over? Here are Antrice's ideas:

   10 packages will use up 120 pencils. After another 10 packages, 240 pencils will be used up. After 1 more package, 252 pencils are used. Then there are only 8 pencils left, and that's not enough for another package. So the answer is 21 packages of pencils with 8 pencils left over.

   Explain why the equations

$$10 \cdot 12 + 10 \cdot 12 + 1 \cdot 12 + 8 = 260 \tag{7.1}$$

$$(10 + 10 + 1) \cdot 12 + 8 = 260 \tag{7.2}$$

$$21 \cdot 12 + 8 = 260 \tag{7.3}$$

   correspond to Antrice's work, and explain why the last equation shows that $260 \div 12 = 21$, remainder 8.

2. Ashley is working on the division problem $245 \div 15$. She writes

$$15 \times 15 = 225 \qquad \begin{array}{r} 225 \\ +\ \ 15 \\ \hline 240 \\ +\ \ \ 5 \\ \hline 245 \end{array} \qquad 15 + 1 = 16 \text{ remainder } 5$$

   a. Explain why Ashley's strategy makes sense. It may help you to work with a story problem for $245 \div 15$.

   b. Write equations like Equations 7.1, 7.2, and 7.3 that correspond to Ashley's work and which demonstrate that $245 \div 15 = 16$, remainder 5.

3. Maya is working on the division problem $245 \div 15$. She writes the following:

$$\begin{array}{r} 15 \\ \times\ 2 \\ \hline 30 \end{array} \qquad 8 \times 30 = 240 \qquad 8 \times 2 = 16 \qquad \boxed{16 \text{ R } 5}$$

a. Explain why Maya's strategy makes sense. It may help you to work with a story problem for $245 \div 15$.

b. Write equations like Equations 7.1, 7.2, and 7.3 that correspond to Maya's work and which demonstrate that $245 \div 15 = 16$, remainder 5. One side of each equation should be 245.

4. Zane is working on the division problem $245 \div 15$. He writes

$$\begin{array}{r} 15 \\ \times\ \ 2 \\ \hline 30 \\ \times\ \ 2 \\ \hline 60 \\ \times\ \ 4 \\ \hline 240 \\ 5 \text{ left} \end{array} \qquad 2 \times 2 \times 4 = 16 \text{ R } 5$$

a. Explain why Zane's strategy makes sense. It may help you to work with a story problem for $245 \div 15$.

b. Write equations like Equations 7.1, 7.2, and 7.3 that correspond to Zane's work and which demonstrate that $245 \div 15 = 16$, remainder 5.

5. Assume that you don't have a calculator and have forgotten how to do any longhand division method. Explain how you can calculate $2783 \div 125$.

Class Activity 7H: Understanding the Scaffold Method of Long Division

1. Interpret each of the steps in the next scaffold in terms of the following story problem:

   You have 3475 marbles, and you want to put these marbles into bags with 8 marbles in each bag. How many bags of marbles can you make, and how many marbles will be left over?

$$
\begin{array}{r}
4 \\
30 \\
400 \\
8\overline{)3475} \\
-\ 3200 \\
\hline
275 \\
-\ 240 \\
\hline
35 \\
-\ 32 \\
\hline
3
\end{array}
$$

2. Explain how the equations

$$3475 - 400 \times 8 - 30 \times 8 - 4 \times 8 = 3$$
$$3475 - (400 + 30 + 4) \times 8 = 3$$
$$3475 - 434 \times 8 = 3$$

   relate to the scaffold and to the story problem in Problem 1. Then explain why the last equation shows that $3475 \div 8 = 434$, remainder 3.

3. Use the scaffold method to calculate $8321 \div 6$. Interpret each step in your scaffold in terms of the following story problem:

You have 8321 pickles, and you want to put these pickles in packages with 6 pickles in each package. How many packages can you make, and how many pickles will be left over?

4. Write equations like those in Problem 2 for your scaffold in Problem 3. Explain how your equations relate to the scaffold and to the story problem of Problem 3.

Class Activity 7I: Using the Scaffold Method

1. Cassie writes the following scaffold to calculate $7549 \div 12$:

$$
\begin{array}{r}
1 \\
8 \\
20 \\
100 \\
500 \\
12\overline{)7549} \\
\underline{-\ 6000} \\
1549 \\
\underline{-\ 1200} \\
349 \\
\underline{-\ 240} \\
109 \\
\underline{-\ 96} \\
13 \\
\underline{-\ 12} \\
1
\end{array}
$$

Then Cassie writes the following:

$$500 + 100 + 20 + 8 + 1 = 629, \text{ so } 7549 \div 12 = 629, \text{ remainder } 1.$$

a. Show how Cassie could have written a scaffold with fewer steps to solve $7549 \div 12$.

b. Even though Cassie used more steps than she had to, is her method still legitimate? In other words, is her method based on sound reasoning? Explain.

2. Manuel calculates $427 \div 11$, using the following reasoning:

> 10 elevens is 110. After another 10 elevens and another 10 elevens, I'm up to 330. After another 5 elevens, I'm up to 385. One more eleven makes 396. Then one more eleven makes 407. One more eleven makes 418. Then there are 9 left. So the answer is 38, remainder 9.

Write a scaffold for $427 \div 11$ that is based on Manuel's reasoning. The arithmetic in the scaffold will not correspond exactly to Manuel's. Why not?

3. Use the scaffold method and the standard long division method to calculate $3895 \div 14$. Discuss how the two methods are related.

Class Activity 7J: Interpreting Standard Long Division from the "How Many in Each Group?" Viewpoint

1. Use the standard long division algorithm to calculate $1372 \div 3$. Interpret each step in the algorithm in terms of dividing 1372 toothpicks equally among 3 groups, where the toothpicks are arranged in bundles of 1 thousand, 3 hundreds, 7 tens, and 2 individual toothpicks. Show the steps pictorially by drawing how bundles will be unbundled and divided step by step among the 3 groups shown. How is the "bringing down" step in long division related to unbundling toothpicks?

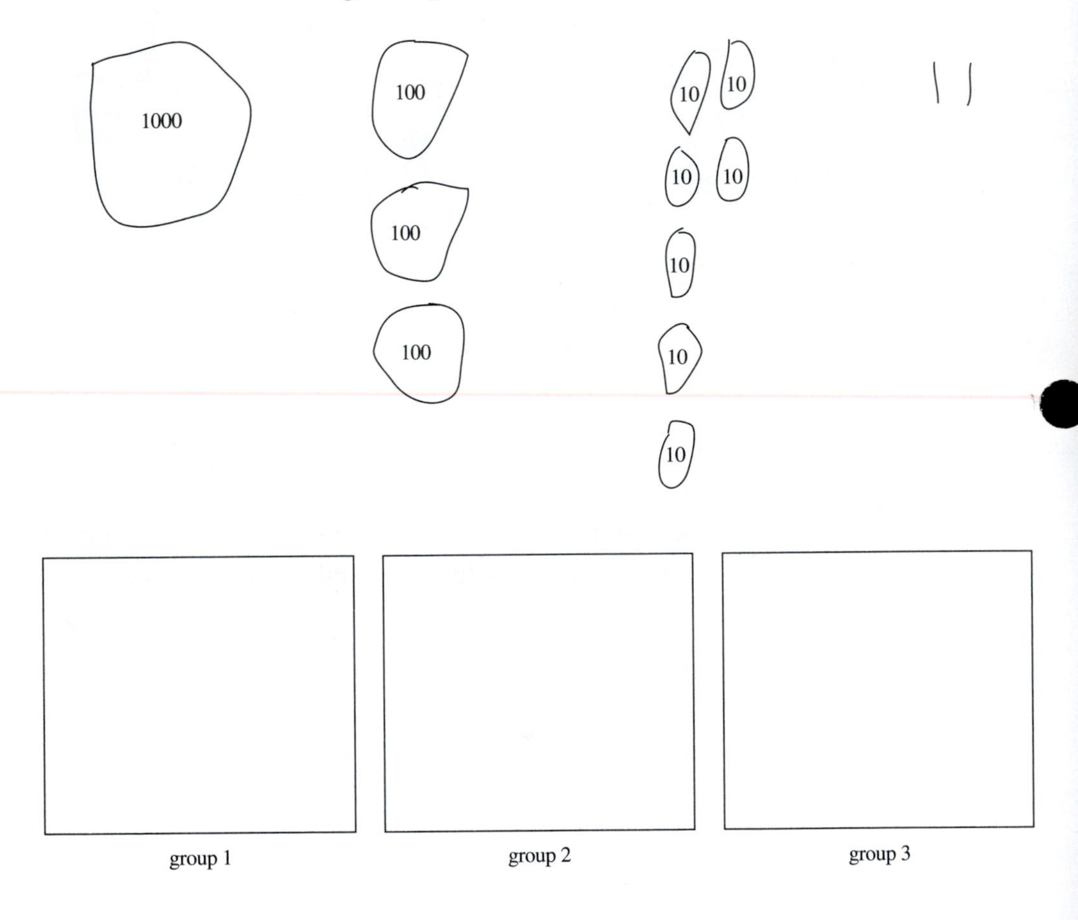

2. Use the standard long division algorithm to calculate $1413 \div 2$. Interpret each step in the algorithm in terms of dividing 1413 toothpicks equally between 2 groups, where the toothpicks are arranged in bundles of 1 thousand, 4 hundreds, 1 ten, and 3 individual toothpicks. Show the steps pictorially by drawing how bundles will be unbundled and divided step by step between the 2 groups shown. How is the "bringing down" step in long division related to unbundling toothpicks?

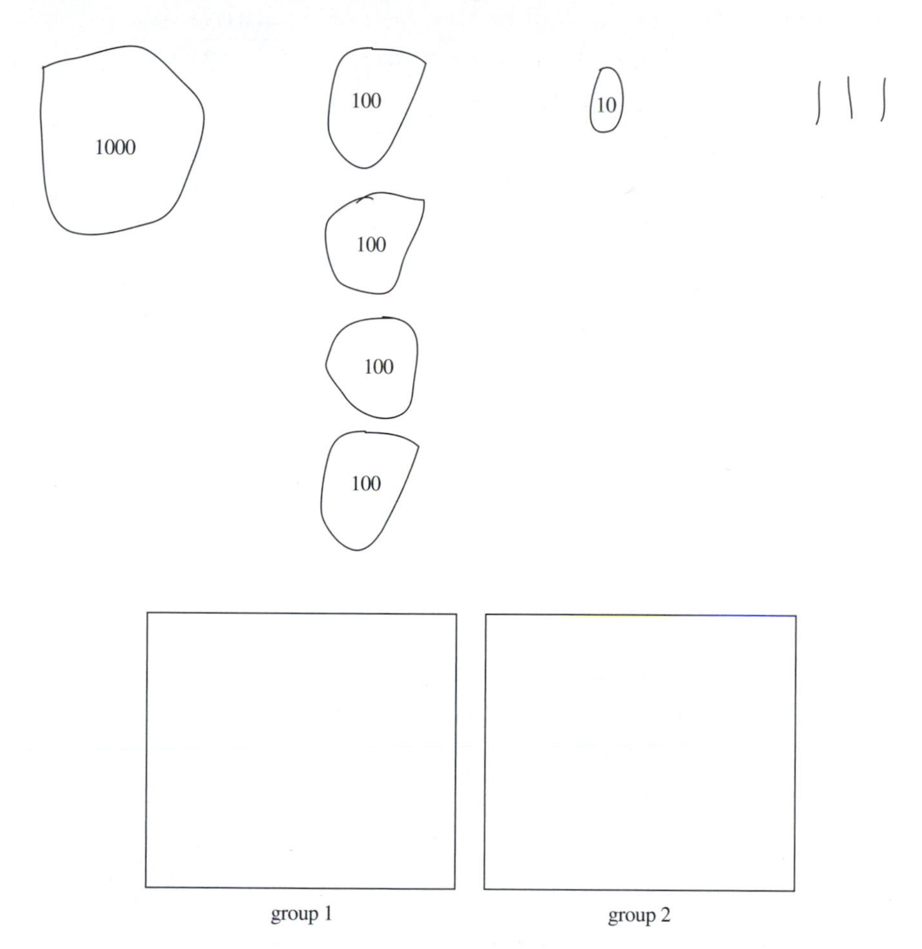

Class Activity 7K: Zeros in Long Division

1. Describe the error in the following division calculation:

```
     39
12)3711
  - 36
    111
  - 108
      3
```

2. Use the scaffold method to calculate 3711 ÷ 12. Discuss why it is easier to avoid the error of Problem 1 when using the scaffold method than when using standard long division.

Class Activity 7L: Using Long Division to Calculate Decimal Number Answers to Whole Number Division Problems

1. Use long division to determine the decimal-number answer to $2674 \div 3$ to the hundredths place.

2. Interpret each step in your long-division calculation for Problem 1 in terms of dividing \$2674 equally among 3 people by imagining that you distribute the money in stages: First distribute hundreds, then tens, then ones, then dimes (tenths), then pennies (hundredths).

Class Activity 7M: Errors in Decimal Answers to Division Problems

1. Given that the answer to a whole number division problem is 4, remainder 1, can you tell what the decimal answer to the division problem is without any additional information? If not, what other information would you need to determine the decimal answer?

2. Ben has been making errors in his division problems. Here are some of Ben's answers:

$$251 \div 6 = 41.5$$
$$269 \div 7 = 38.3$$
$$951 \div 21 = 45.6$$

Use long division to calculate the correct answers to Ben's division problems. Based on your work, determine what Ben is likely to be confused about.

7.3 Fractions and Division

Class Activity 7N: Relating Fractions and Division

1. There are 3 pizzas that will be divided equally among 4 people. How much pizza will each person get? Explain. What does your answer tell you about $3 \div 4$?

2. There are 3 pizzas that will be divided equally among 5 people. How much pizza will each person get? Explain. What does your answer tell you about $3 \div 5$?

3. There are 3 pizzas that will be divided equally among 7 people. How much pizza will each person get? Explain. What does your answer tell you about $3 \div 7$?

4. How are $A \div B$ and $\frac{A}{B}$ related? Explain.

Class Activity 7O: Mixed-Number Answers to Division Problems

1. Describe how to get the mixed-number answer to $8 \div 3$ from the whole-number-with-remainder answer. Explain why your method makes sense by interpreting it in terms of dividing 8 large cookies equally among 3 people.

2. Describe how to use division to write the improper fraction $\frac{23}{4}$ as a mixed number, and explain why the procedure you describe makes sense.

Class Activity 7P: Using Division to Calculate Decimal Representations of Fractions

1. Use long division to show that the decimal representation of $\frac{3}{4}$ is 0.75. Interpret each of your long-division steps in terms of dividing \$3 equally among 4 people, using only dollars, dimes, and pennies.

2. Use long division to determine the decimal representation of $\frac{1}{8}$. Interpret each of your long-division steps in terms of dividing \$1 equally among 8 people, using dollars, dimes, and pennies. (Note that you will not be able to divide evenly without being able to split a penny.)

3. Interpret each of your long-division steps in Problem 4 in terms of dividing the following square into 8 equal parts:

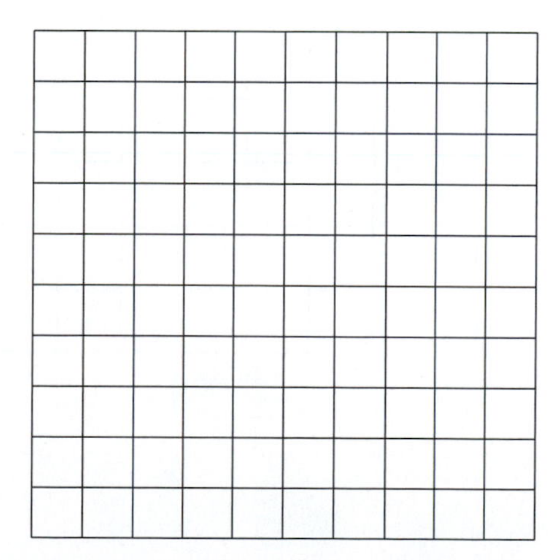

4. Use long division to determine the decimal representation of $\frac{1}{6}$ to the thousandths place. Interpret the steps down to the hundredths place in your long division in terms of dividing $1 equally among 6 people, using dollars, dimes, and pennies.

5. Interpret each of your long-division steps in Problem 4 in terms of dividing the following square into 6 equal parts:

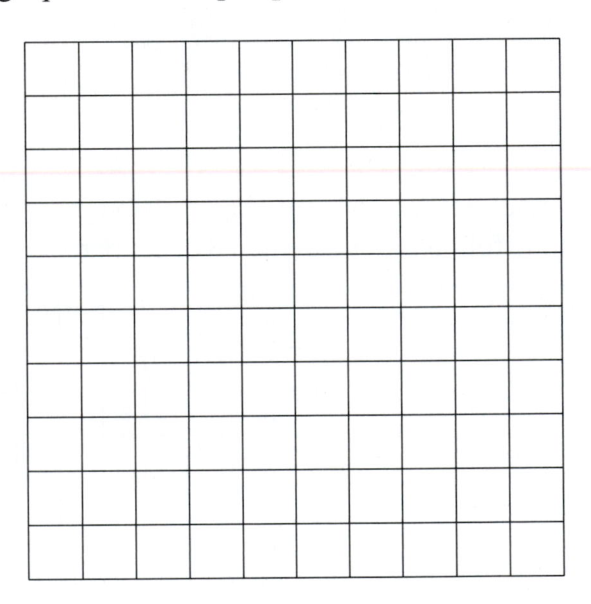

7.4 Dividing Fractions

Class Activity 7Q: "How Many Groups?" Fraction Division Problems

1. a. Write a "how many groups?" story problem for $6 \div 2$.

   b. Write a "how many groups?" story problem for $6 \div \frac{3}{4}$. (If it works, you may modify your story problem from Part (a).)

   c. Assume that you don't know the "invert and multiply" method for dividing fractions. Solve your story problem for $6 \div \frac{3}{4}$ in a simple and concrete way, for example, by drawing pictures. Explain your solution.

2. Tonya and Chrissy are trying to understand $1 \div \frac{2}{3}$ by using the following story problem:

One serving of rice is $\frac{2}{3}$ of a cup. I ate 1 cup of rice. How many servings of rice did I eat?

To solve the problem, Tonya and Chrissy draw a diagram of a large cup divided into three equal pieces, and they shade two of those pieces.

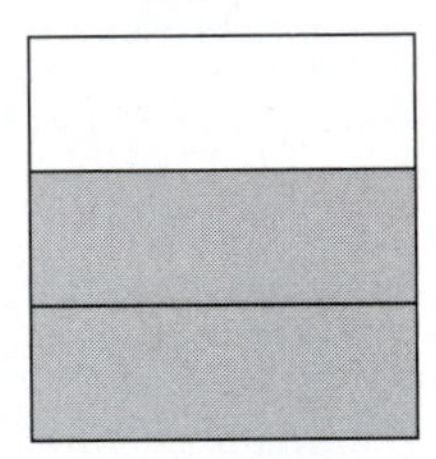

Tonya says, "There is one $\frac{2}{3}$-cup serving of rice in 1 cup, and there is $\frac{1}{3}$ cup of rice left over, so the answer should be $1\frac{1}{3}$."

Chrissy says, "The part left over is $\frac{1}{3}$ cup of rice, but the answer is supposed to be $\frac{3}{2} = 1\frac{1}{2}$. Did we do something wrong?"

Help Tonya and Chrissy.

3. a. Write a “how many groups?” story problem for $8 \div 3$. Choose your story problem so that a mixed-number answer makes sense.

b. Write a “how many groups?” story problem for $1\frac{1}{2} \div \frac{2}{3}$. (If it works, you may modify your story problem from Part (a).)

c. Assume that you don’t know the “invert and multiply” method for dividing fractions. Solve your story problem for $1\frac{1}{2} \div \frac{2}{3}$ in a simple and concrete way, for example, by drawing pictures. Explain your solution.

4. a. Write a "how many groups?" story problem for $\frac{1}{3} \div \frac{3}{4}$.

b. Assume that you don't know the "invert and multiply" method for dividing fractions. Solve your story problem for $\frac{1}{3} \div \frac{3}{4}$ in a simple and concrete way, for example, by drawing pictures. Explain your solution.

Class Activity 7R: "How Many in One Group?" Fraction Division Problems

1. **a.** Verify that the following problems are "how many in one group?" problems for $6 \div 3$:

   i. There are 6 cookies, which will be shared equally among 3 children. How many cookies will each child get?

   ii. 6 cups of flour fill 3 containers completely full (all of which are the same size). How many cups of flour does it take to fill one container?

   iii. Walking at a steady pace, Anna walked 6 miles in 3 hours. How far did Anna walk each hour?

   b. Write a "how many in one group?" story problem for $3 \div \frac{1}{4}$.

   c. Assume that you don't know the "invert and multiply" method for dividing fractions. Solve your story problem for $3 \div \frac{1}{4}$ in a simple and concrete way, for example, by drawing pictures. Explain your solution.

   d. Solve $3 \div \frac{1}{4}$ numerically by "inverting and multiplying"—in other words, by multiplying 3 by $\frac{4}{1}$. Explain how your solution in Part (c) shows this process of multiplying 3 by $\frac{4}{1}$.

2. **a.** Write a "how many in one group?" story problem for $6 \div \frac{3}{4}$.

b. Assume that you don't know the "invert and multiply" method for dividing fractions. Solve your story problem for $6 \div \frac{3}{4}$ in a simple and concrete way, for example, by drawing pictures. Explain your solution.

c. Solve $6 \div \frac{3}{4}$ numerically by "inverting and multiplying"—in other words, by multiplying 6 by $\frac{4}{3}$. Explain how your solution in Part (b) shows this process of multiplying 6 by $\frac{4}{3}$.

3. **a.** Write a “how many in one group?” story problem for $\frac{1}{2} \div \frac{2}{3}$.

b. Assume that you don’t know the “invert and multiply” method for dividing fractions. Solve your story problem for $\frac{1}{2} \div \frac{2}{3}$ in a simple and concrete way, for example, by drawing pictures. Explain your solution.

c. Solve $\frac{1}{2} \div \frac{2}{3}$ numerically by “inverting and multiplying,” in other words, by multiplying $\frac{1}{2}$ by $\frac{3}{2}$. Explain how your solution in Part (c) shows this process of multiplying $\frac{1}{2}$ by $\frac{3}{2}$.

Class Activity 7S: Using "Double Number Lines" to Solve "How Many in One Group?" Division Problems

One helpful aid for solving "how many in one group?" division story problems is a "double number line," such as the one that follows.

1. Will found that $\frac{1}{2}$ pound of nails was $\frac{3}{4}$ of the nails he needed for a project. How many pounds of nails does Will need for the project?

   Use the next double number line to help you solve this problem. Explain your reasoning.

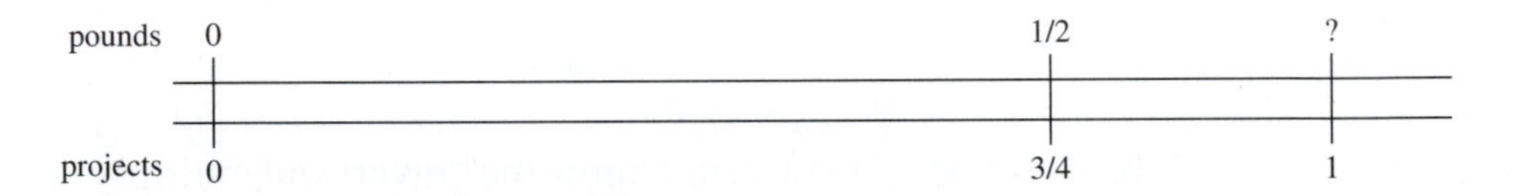

2. It takes Gloria $\frac{1}{3}$ of an hour to walk $\frac{3}{5}$ of a mile. At that rate, how far will Gloria walk in 1 hour? How long does it take for Gloria to walk 1 mile?

   Use double number lines to help you solve these problems. Explain your reasoning.

3. Six liters of juice filled a container $2\frac{1}{2}$ times. How many liters of juice does the container hold?

   Use a double number line to solve this problem. Explain your reasoning.

Class Activity 7T: Explaining "Invert and Multiply" by Relating Division to Multiplication

Every division problem is equivalent to a multiplication problem:

$$A \div B = ? \text{ is equivalent to } ? \times B = A.$$

Assume that you don't know the "invert and multiply" procedure for fraction division, but that you do know how to multiply fractions. Solve each of the following division problems by first rewriting the problem as a multiplication problem:

1. $1 \div \frac{1}{5} = ?$

2. $1 \div 4 = ?$

3. $1 \div \frac{4}{5} = ?$

4. $\frac{A}{B} \div \frac{1}{5} = ?$

5. $\frac{A}{B} \div 4 = ?$

6. $\frac{A}{B} \div \frac{4}{5} = ?$

7. $\frac{A}{B} \div \frac{C}{D} = ?$

Class Activity 7U: Are These Division Problems?

Which of the following are story problems for the division problem $\frac{3}{4} \div \frac{1}{2}$? For those that are, which interpretation of division is used? For those that are not, determine how to solve the problem if it can be solved.

1. Beth poured $\frac{3}{4}$ cup of cereal in a bowl. The cereal box says that one serving is $\frac{1}{2}$ cup. How many servings are in Beth's bowl?

2. Beth poured $\frac{3}{4}$ cup of cereal in a bowl. Then Beth took $\frac{1}{2}$ of that cereal and put it into another bowl. How many cups of cereal are in the second bowl?

3. A crew is building a road. So far, the road is $\frac{3}{4}$ mile long. This is $\frac{1}{2}$ the length that the road will be when it is finished. How many miles long will the finished road be?

4. A crew is building a road. So far, the crew has completed $\frac{3}{4}$ of the road, and this portion is $\frac{1}{2}$ mile long. How long will the finished road be?

5. If $\frac{3}{4}$ cup of flour makes $\frac{1}{2}$ a batch of cookies, then how many cups of flour are required for a full batch of cookies?

6. If $\frac{1}{2}$ cup of flour makes a batch of cookies, then how many batches of cookies can you make with $\frac{3}{4}$ cup of flour?

7. If $\frac{3}{4}$ cup of flour makes a batch of cookies, then how much flour is in $\frac{1}{2}$ of a batch of cookies?

7.5 Dividing Decimals

Class Activity 7V: Quick Tricks for Some Decimal Division Problems

1. Describe a quick way to calculate $32.5 \div 0.5$ mentally. *Hint*: Think in terms of fractions or in terms of money.

2. Describe a quick way to calculate $1.2 \div 0.25$ mentally.

3. Describe a quick way to estimate $7.2 \div 0.333$ mentally.

Class Activity 7W: Decimal Division

1. Write one "how many groups?" story problem and another "how many in one group?" story problem for $23.45 \div 2.7$.

2. The problem

   "How many \$0.25 are in \$12.37?"

   is a story problem for the division problem

   $$12.37 \div 0.25$$

   Modify the story problem so that it retains its same meaning, but becomes a story problem for

   $$1237 \div 25$$

   What can you conclude about how the two division problems

   $$0.25\overline{)12.37} \quad \text{and} \quad 25\overline{)1237}$$

   are related?

3. Explain how the accompanying figure can be interpreted as the following:

$$0.06 \div 0.02 = ?$$

Explain how the same figure can be interpreted as:

$$0.6 \div 0.2 = ? \quad \text{or as} \quad 6 \div 2 = ?$$

What other division problems can the figure illustrate?

How many [figure] are in [figure] ?

4. Fran must calculate $2.45 \div 1.5$ longhand, but she can't remember what to do about decimal points. Instead, Fran solves the division problem $245 \div 15$ longhand and gets the answer 16.33. Fran knows that she must shift the decimal point in 16.33 somehow to get the correct answer to $2.45 \div 1.5$. Explain how Fran could reason about the sizes of the original numbers to determine where to put the decimal point.

7.6 Ratio and Proportion

Class Activity 7X: Comparing Mixtures

1. There are two containers, each containing a mixture of 1 cup red punch and 3 cups lemon-lime soda. The first container is left as it is, but somebody adds 2 cups red punch and 2 cups lemon-lime soda to the second container.

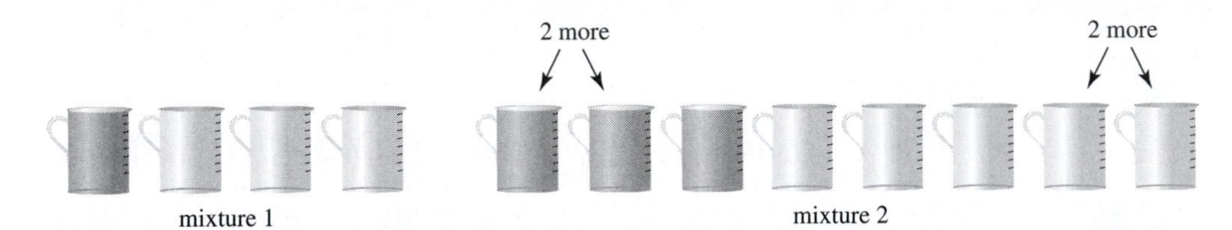

Will the two punch mixtures taste the same? Why or why not? Try to develop several different ways to determine if the two mixtures should taste the same or not.

If the mixtures are available, try them to see if they taste the same or not.

2. There are two containers, each containing a mixture of 2 drops blue paint and 5 drops yellow paint. (All drops are the same size.) The first container is left as it is, but somebody adds 3 drops blue paint and 3 drops yellow paint to the second container.

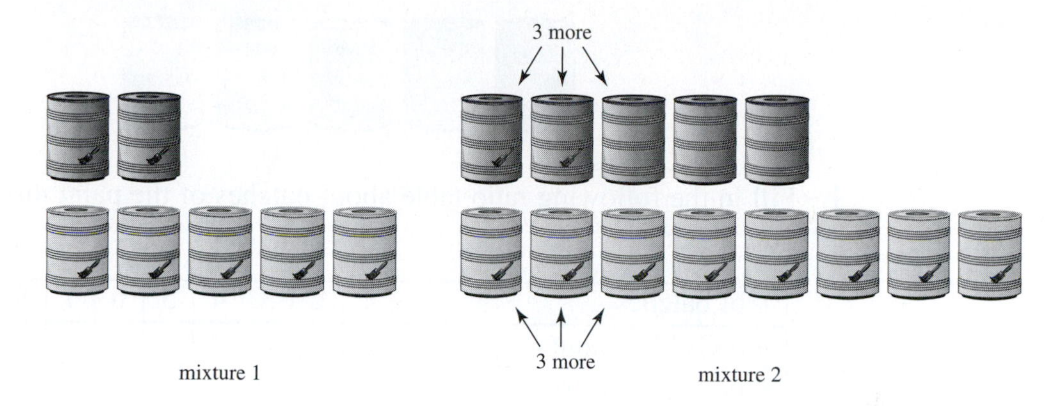

Will the two paint mixtures be the same shade of green? Why or why not? Try to develop several different ways to determine if the two mixtures should look the same or not.

If possible, make the paint mixtures to see if they look the same or not.

Class Activity 7Y: Using Ratio Tables

One batch of a certain shade of green paint is made by mixing 2 pails of blue paint with 3 pails of yellow paint.

1. Fill in the following ratio table about batches of the paint mixture, as just described:

# of batches	1	2	3	4	5	6	7	8	9	10
# pails blue paint	2									
# pails yellow paint	3									
# pails green paint produced	5									

2. Describe the patterns in the 2nd, 3rd, and 4th rows as you go to the right in the ratio table. Explain why those patterns are present.

3. Describe how the entries in the 6th column of the table are related to the entries in the 1st column of the table.

   Describe how entries in the 8th column of the table are related to the entries in the 1st column of the table.

4. If the table were to continue in the same pattern, how would the entries in the 100th column be related to the entries in the 1st column?

How would the entries in the nth column of the table be related to entries in the 1st column of the table? Explain your reasoning.

5. Find relationships among the entries in a column. In particular, describe multiplicative relationships (relationships that use multiplication) among the entries in a column.

6. Use your findings from Problems 4 and 5 to determine in several different ways how to fill in the next ratio table, which is based on the same paint mixture as before.

# of batches	1						
# pails blue paint	2	100			22		
# pails yellow paint	3		45			75	
# pails green paint produced	5			1000			85

7. Make ratio tables for the two punch mixtures of Class Activity 7X, and use your ratio tables to compare the drink mixtures. Will the mixtures taste the same? If not, which mixture will taste more like red punch?

Class Activity 7Z: Using Strip Diagrams to Solve Ratio Problems

Suppose a certain shade of green paint is made by mixing blue paint with yellow paint in a ratio of 2 to 3.

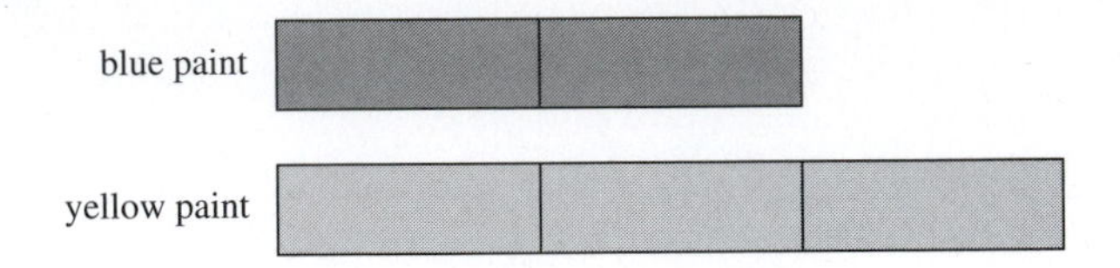

For each of the problems on this page, use the same shade of green paint as above, which is made by mixing blue paint with yellow paint in a ratio of 2 to 3. Explain how to solve the problems by using the strip diagram.

1. If you will use 40 pails of blue paint, how many pails of yellow paint will you need?

2. If you will use 48 pails of yellow paint, how many pails of blue paint will you need?

3. If you want to make 100 pails of green paint, how many pails of blue paint and how many pails of yellow paint will you need?

4. At lunch, there was a choice of pizza or a hot dog. Three times as many students chose pizza as chose hot dogs. All together, 160 students got lunch. How many students got pizza and how many got a hot dog? Draw a strip diagram to help you solve this problem. Explain your reasoning.

5. The ratio of Shauntay's cards to Jessica's cards is 5 to 3. After Shauntay gives Jessica 15 cards, both girls have the same number of cards. How many cards do Shauntay and Jessica each have now? Draw a strip diagram to help you solve this problem. Explain your reasoning.

6. The ratio of Shauntay's cards to Jessica's cards is 5 to 2. After Shauntay gives Jessica 12 cards, both girls have the same number of cards. How many cards do Shauntay and Jessica each have now? Draw a strip diagram to help you solve this problem. Explain your reasoning.

7. Make a new problem for your students by modifying Problem 5 or 6. Change the ratio and change the number of cards that Shauntay gives to Jessica. When you make these changes, which ratios will make the problem easier, and which ratios will make it harder? Once you have chosen a ratio, can the number of cards that Shauntay gives to Jessica be any number, or do you need to take care in choosing this number? Explain.

Class Activity 7AA: Using Simple Reasoning to Find Equivalent Ratios and Rates

1. For a certain shade of pink paint, the ratio of white paint to red paint is 5 to 2. Explain how to use the strip diagram shown next to find at least 6 different quantities of white and red paint that will be in the ratio 5 to 2 (and so will make the same shade of pink). Include at least 3 mixtures where the quantities of paint are not both whole numbers of units. (You may choose any units you like, such as gallons, liters, pails, or something else.)

white paint

red paint

2. Driving at a constant speed, you drove 14 miles in 20 minutes. On the "double number line" that follows, show different distances and times that would give you the same speed. Use simple, logical reasoning to determine these equivalent rates.

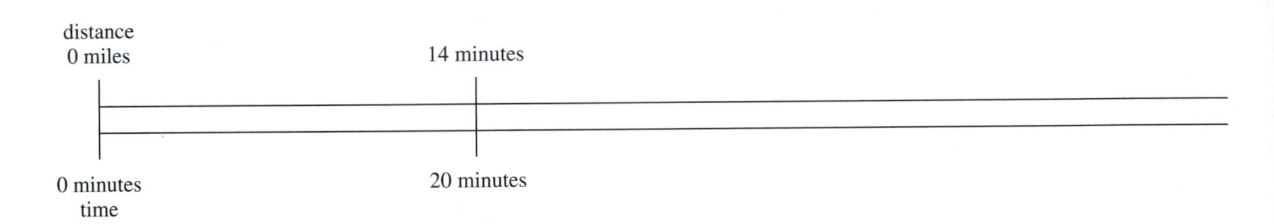

3. Could you use a double number line (as in Problem 2) to show equivalent paint ratios for Problem 1?

   What are some advantages and some limitations of double number lines?

   Would it be a good idea to use a strip diagram (as in Problem 1) to show speeds for Problem 2?

4. A box of Brand A laundry detergent washes 20 loads of laundry and costs $6. A box of Brand B laundry detergent washes 15 loads of laundry and costs $5.

a. In the ratio tables that follow, fill in equivalent rates of loads washed per dollar. Include some examples where the number of loads washed is less than 15 and the cost is less than $5. Explain your reasoning.

Brand A									
loads washed	20								
cost	$6								

Brand B									
loads washed	15								
cost	$5								

If possible, use your tables to make a statement comparing the two brands of laundry detergent.

b. Explain how to use simple logical reasoning to fill in the next tables with equivalent rates. Then use the tables to make statements comparing the two brands of laundry detergent.

Brand A		
loads washed	20	
cost	$6	$1

Brand B		
loads washed	15	
cost	$5	$1

Brand A		
loads washed	20	1
cost	$6	

Brand B		
loads washed	15	1
cost	$5	

c. Robert says that Brand B is less expensive than Brand A because it costs $5 instead of $6. Discuss Robert's reasoning. What would Robert benefit from learning?

5. Traveling at a constant speed, a scooter went $\frac{3}{4}$ of a mile in 4 minutes. Use simple, logical reasoning to help you determine the answers to the next questions.

a. How far did the scooter go in the following amounts of time:
8 minutes? 12 minutes? 10 minutes?
2 minutes? 1 minute?

b. How long did it take the scooter to go 1 mile?

Which aids will be helpful in answering the preceding questions: a strip diagram (as in Problem 1), a double number line (as in Problem 2), or a table (as in Problem 4)?

Class Activity 7BB: Solving Proportions with Multiplication and Division

1. If you mix fruit juice and bubbly water in a ratio of 3 to 5 to make a punch, then how many liters of fruit juice and how many liters of bubbly water will you need to make 24 liters of punch?

   a. Explain how to use a strip diagram to solve this problem.

   b. Suppose you will give this paint problem to students, but you decide to change 24 liters to 10 liters. Will the problem be just as easy to solve or not? How can you change the 24 liters so that the problem is still easy to solve with a strip diagram? How can you change the 24 liters so that the problem becomes harder to solve with a strip diagram?

2. If you mix fruit juice and bubbly water in a ratio of 3 to 5 to make a punch, then how many liters of fruit juice and how many liters of bubbly water will you need to make 10 liters of punch?

   a. Use multiplication, division, and logical reasoning to explain how to fill in the blanks in the following table of equivalent ratios, thereby solving this punch problem:

# liters juice	3		
# liters bubbly water	5		
# liters punch	8	1	10

   b. Describe the strategy for solving the punch problem that the table in Part (a) helps you use. What is the idea behind the way the table was created?

3. If you mix $\frac{3}{4}$ cup of red paint with $\frac{2}{3}$ cup of yellow paint to make an orange paint, then how many cups of red paint and how many cups of yellow paint will you need if you want to make 15 cups of the same shade of orange paint?

   a. Use multiplication, division, and logical reasoning to explain how to fill in the blanks in the following table with equivalent ratios, thereby solving this paint problem:

# cups red	$\frac{3}{4}$			
# cups yellow	$\frac{2}{3}$			
# cups orange		17	1	15

   b. Describe the strategy for solving the paint problem that the table in Part (a) helps you use. What is the idea behind the way the table was created?

4. Chandra made a milkshake by mixing $\frac{1}{2}$ cup of ice cream with $\frac{3}{4}$ cup of milk. Use the most elementary reasoning you can to determine how many cups of ice cream and milk Chandra should use if she wants to make the same milkshake (i.e., using the same ratios) in the following amounts:

   a. using 3 cups of ice cream

   b. to make 3 cups of milkshake

5. Russell was supposed to mix 3 tablespoons of weed killer concentrate with $1\frac{3}{4}$ cups of water to make a weed killer. By accident, Russell put in an extra tablespoon of weed killer concentrate, mixing 4 tablespoons of weed killer concentrate with $1\frac{3}{4}$ cups of water. How much water should Russell add to his mixture so that the ratio of weed killer concentrate to water will be the same as in the correct mixture? Use the most elementary reasoning you can to solve this problem.

Class Activity 7CC: Ratios, Fractions, and Division

For a certain shade of orange paint, the ratio of red to yellow is 3 to 5. For each of the following fractions and division problems, write a question about the orange paint that will be answered by the given fraction (viewed as parts of a whole) or by solving the associated division problem:

1. $\frac{3}{8}$ or $3 \div 8$

2. $\frac{5}{8}$ or $5 \div 8$

3. $\frac{3}{5}$ or $3 \div 5$

4. $\frac{5}{3}$ or $5 \div 3$

5. $\frac{8}{3}$ or $8 \div 3$

6. $\frac{8}{5}$ or $8 \div 5$

Class Activity 7DD: Solving Proportions by Cross-Multiplying Fractions

Read the following recipe problem:

> A recipe that serves 6 people calls for $2\frac{1}{2}$ cups of flour. How much flour will you need to serve 10 people, assuming that the ratio of people to cups of flour remains the same?

One familiar way to solve this problem is by setting up and solving a proportion, as follows: First, we let x be the amount of flour we need to serve 10 people. Then we set two fractions equal to each other:

$$\frac{x}{10} = \frac{2\frac{1}{2}}{6}$$

In setting these fractions equal to each other, we may say "x is to 10 as $2\frac{1}{2}$ is to 6." Next, we "cross-multiply" to obtain the equation

$$6 \cdot x = 10 \cdot 2\frac{1}{2}$$

Finally, we solve for x by dividing both sides of the equation by 6. Therefore,

$$x = \frac{10 \cdot 2\frac{1}{2}}{6} = \frac{10 \cdot \frac{5}{2}}{6} = \frac{25}{6} = 4\frac{1}{6}$$

and we see that $4\frac{1}{6}$ cups of flour are needed to serve 10 people.

This class activity will help you understand the rationale for this method of solving proportions.

1. In the solution we just found, we worked with two fractions:

$$\frac{x}{10} \quad \text{and} \quad \frac{2\frac{1}{2}}{6}$$

Interpret the meaning of these fractions in terms of the recipe problem at the beginning of this activity. Explain why these two fractions should be equal.

2. After setting two fractions equal to each other, the next step in solving the proportion was to "cross-multiply." Why does it make sense to cross-multiply? What is the rationale behind the procedure of cross-multiplying?

3. In the preceding solution, we set up the proportion

$$\frac{x}{10} = \frac{2\frac{1}{2}}{6}$$

What is another way to set up a proportion so that the unknown amount of flour, x, is in the numerator of one of the fractions? Interpret the two fractions in your new proportion in terms of the recipe problem. Use your interpretations to explain why the two fractions should be equal.

4. Now solve the recipe problem in a different way by using logical thinking and by using the most elementary reasoning you can. Explain your reasoning clearly.

Class Activity 7EE: Can You Always Use a Proportion?

Sometimes a problem that looks as if it could be solved by setting up a proportion actually can't be solved that way. Before you set up a proportion to solve a problem, ask the following question about quantities in the problem: If I double one of the quantities, should the other quantity also double? If the answer is "no," then you cannot solve the problem by setting up a proportion.

1. Ken used 3 loads of stone pavers to make a circular (i.e., circle-shaped) patio with a radius of 10 feet. Ken wants to make another circular patio with a radius of 15 feet, so he sets up the proportion

$$\frac{3 \text{ loads}}{10 \text{ feet}} = \frac{x \text{ loads}}{15 \text{ feet}}$$

Is this correct? If not, why not? Is there another proportion that Ken could set up to solve the problem? (The area of a circle that has radius r units is πr^2 square units.)

2. In a cookie factory, 4 assembly lines make enough boxes of cookies to fill a truck in 10 hours. How long will it take to fill the truck if 8 assembly lines are used? Is the proportion

$$\frac{10 \text{ hours}}{4 \text{ lines}} = \frac{x \text{ hours}}{8 \text{ lines}}$$

appropriate for this situation? Why or why not? If not, can you solve the problem another way? (Assume that all assembly lines work at the same steady rate.)

3. In the cookie factory of Problem 2, how long will it take to fill a truck if 6 assembly lines are used? (If you get stuck here, move on to the next problem and come back.)

4. Robyn used the following reasoning to solve the previous problem:

> Since 4 assembly lines fill a truck in 10 hours, 8 assembly lines should fill a truck in half that time, namely, in 5 hours. Since 6 assembly lines is halfway between 4 and 8, it ought to take halfway between 10 hours and 5 hours, or $7\frac{1}{2}$ hours, to fill a truck.

Robyn's reasoning seems quite reasonable, but is it really correct? Let's look carefully.

Fill in the following table by using logical thinking about the assembly lines:

# of assembly lines	**# of hours to fill a truck**
1	
2	
4	10 hours
8	
16	
32	

Now apply Robyn's reasoning again, but this time to 1 assembly line versus 32. Sixteen assembly lines is approximately halfway between 1 and 32. But is the number of hours it takes to fill a truck by 16 assembly lines approximately halfway between the number of hours it takes to fill a truck by 1 assembly line versus by 32 assembly lines?

What can you conclude about Robyn's reasoning?

Class Activity 7FF: * The Consumer Price Index

Sometimes news reports compare how much was spent on items in the past with how much is spent on the same items now. For example, a report in *Newsweek* magazine on January 24, 2005, published a table like the following one, listing costs of recent presidential inaugurals:

Inaugural Price Tags (Amount spent, adjusted to 2004 $)		
Nixon	1973	$17 million
Carter	1977	11
Reagan	1981	34
Reagan	1985	35
Bush	1989	46
Clinton	1993	30
Clinton	1997	35
Bush	2001	43
Bush	2005	40 (goal)

1. Why do you think the table shows the amounts spent on inaugurals in 2004 dollars instead of the actual amounts spent?

   How do you think the table would be different if the actual dollar amounts spent on inaugurals were listed instead of 2004 dollars?

The following table shows the value of the CPI for some selected years:

Consumer Price Index (selected years)			
year CPI	**year CPI**	**year CPI**	**year CPI**
1940: 14.0	1960: 29.6	1980: 82.4	2000: 172.2
1950: 24.1	1970: 38.8	1990: 130.7	2004: 188.9

2. What salary in the year 2000 would have had the same buying power as $20,000 did in 1950?

3. If a gallon of milk cost $3 in 2000, what should it have cost in 1940, if you use the CPI to adjust for inflation?

   Does this necessarily give you the actual cost of a gallon of milk in 1940? Why not?

4. If a certain item cost $2 in 1980 and $3.50 in 2000, would it be reasonable to argue that its price had gone down from 1980 to 2000? How so?

5. In both 1980 and 2000, gasoline cost about $1.20 per gallon. Put the price of gasoline in both years into 2000 dollars, and compute the percent by which inflation-adjusted gasoline prices fell from 1980 to 2000.

8

Geometry

8.1 Visualization

Class Activity 8A: What Shapes Do These Patterns Make?

You will need scissors and tape for this activity.

The next picture shows small versions of patterns for shapes. Figure A.2 on page 597 has large versions of these patterns that can be cut out, folded, and taped to make closed, solid shapes.

Before you cut, fold, and tape the large patterns on page 597, *visualize* the folding process in order to help you visualize the final shapes. Be patient and make several attempts. Remember that the goal of this section is to improve your visualization skills. Discuss and debate your ideas with some of your classmates. Then cut out the large patterns on page 597 along the solid lines, fold down along the dotted lines (or fold up—but be consistent), and tape sides with matching labels together.

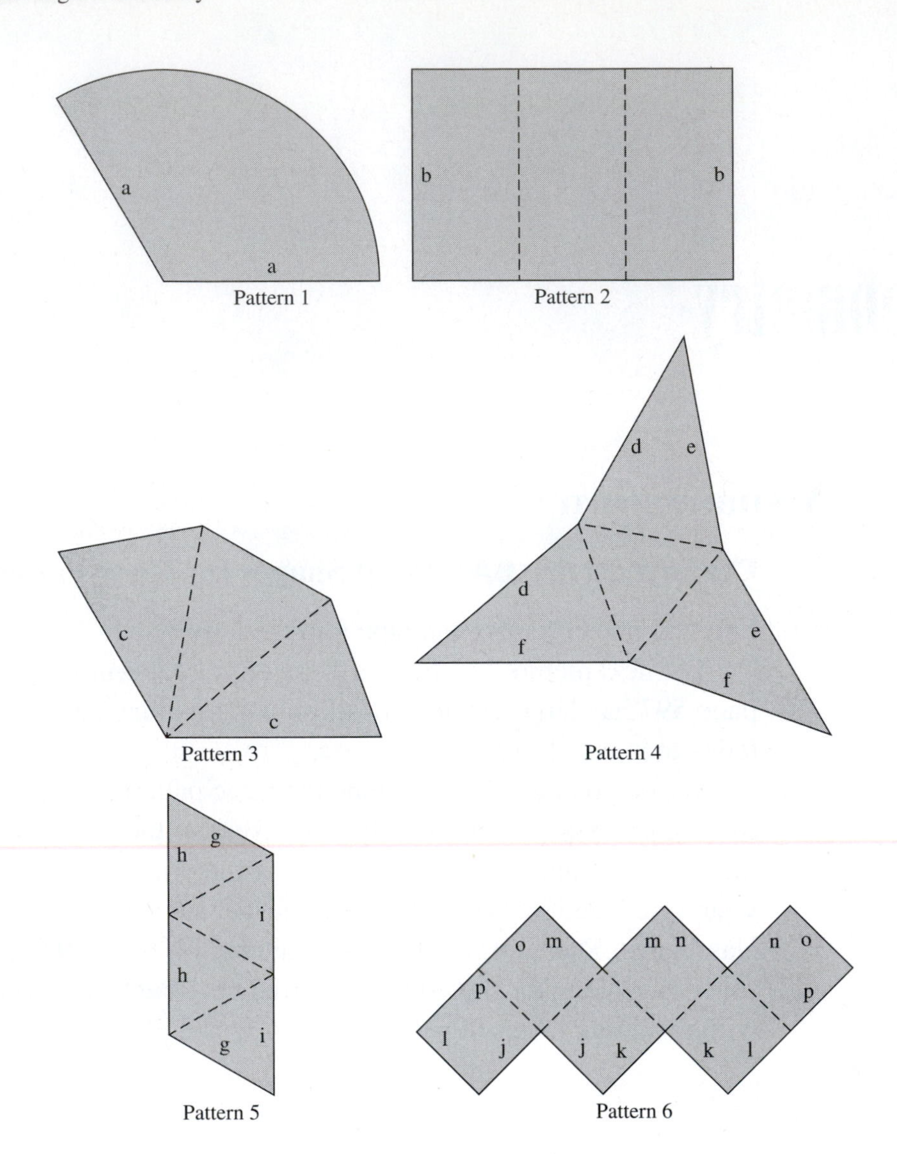

Pattern 1

Pattern 2

Pattern 3

Pattern 4

Pattern 5

Pattern 6

Class Activity 8B: Parts of a Pyramid

You will need paper, scissors, and tape for Problem 1 of this activity. You will need modeling dough or modeling clay and dental floss for Problem 3.

The following picture shows a pyramid (like an Egyptian pyramid):

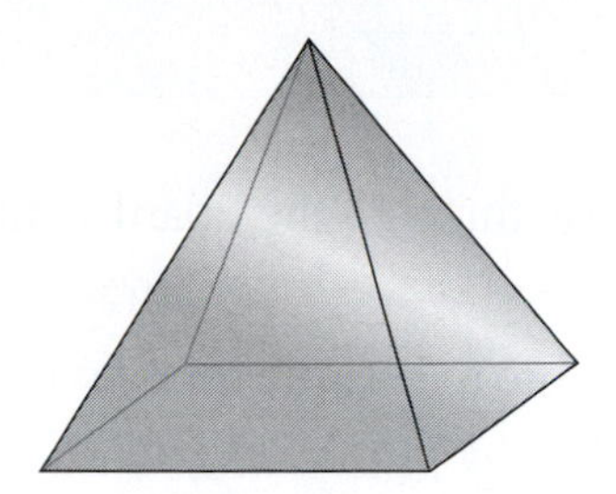

1. What shapes do you need to make the outer surface of the pyramid out of paper?

   Make a paper model of the pyramid, including the bottom. Be willing to "get messy," make mistakes, and modify your initial ideas until you find shapes that you can cut out and tape together to make a pyramid.

2. Now visualize a plane slicing through the pyramid. The places where the plane passes through the pyramid form a shape in the plane. Which shapes in the plane can be made this way, by slicing the pyramid? Use your model to help you, but also visualize each case *without* using your model.

3. Use modeling dough or modeling clay to make a pyramid. Slice straight through your pyramid with dental floss, as if you were slicing the pyramid with a plane. The place where the dental floss cuts through the pyramid should make a plane shape. What plane shape did you get this way? Put the pyramid back together, and try slicing it in a different way. Now what shape do you get? Find as many different plane shapes as you can.

4. Are there any plane shapes that can't arise from slicing the pyramid? Give some examples. How do you know they can't arise?

Class Activity 8C: Slicing through a Board

You will need modeling clay or modeling dough and dental floss for this activity.

Suppose you have an ordinary 2″-by-4″ board, as pictured here:

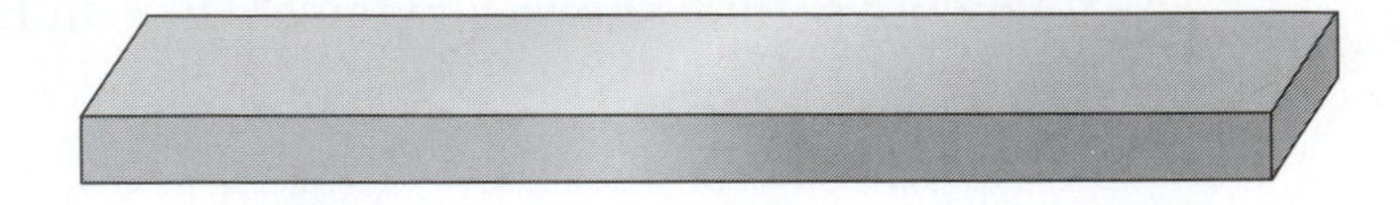

If you saw through the board with a straight cut, the place where the board was cut makes a shape in a plane.

1. Is it possible to saw the board with a straight cut in such a way that the shape formed by the cut is a square? Try to visualize if this is or isn't possible.

2. Is it possible to saw the board with a straight cut in such a way that the shape formed by the cut is *not* a rectangle? (A square *is* a kind of rectangle.) Try to visualize if this is or isn't possible. If the answer is yes, what kind of shape other than a rectangle can you get?

3. Use modeling dough to make a model of a 2″-by-4″ board. Make a straight slice through your model with dental floss. Describe the plane shape that the cut made. Now restore your model of the board to its original shape, and make a different slice through your model. Keep trying different ways to slice your model. Describe all the different shapes you get where the model was cut.

4. Are there any plane shapes that can't arise from slicing the board? Give some examples. How do you know they can't arise?

Class Activity 8D: Visualizing Lines and Planes

1. Visualize a line in a plane. The line divides the plane into two disjoint pieces—visualize these two pieces as well. Visualize a line in space. Does a line in space divide space into disjoint pieces? The word **disjoint** means *distinct and not meeting*.

2. Visualize *two* lines in a plane. How many disjoint pieces do the two lines divide the plane into? The answer depends on how the lines are positioned relative to each other—explain. Draw pictures to aid your explanation, but be sure to also "see" this in your "mind's eye."

3. Visualize *three* lines in a plane. How many disjoint pieces do the three lines divide the plane into? Again, the answer depends on how the lines are positioned relative to each other. Try to *visualize* the distinct configurations. Then draw pictures to help you show all the different types of configurations and to see how many disjoint pieces there are in each case.

4. Visualize a plane in space. A single plane in (3-dimensional) space divides space into two disjoint regions. If the plane you are thinking of is horizontal, then these two regions are the region above the plane and the region below the plane. Visualize a nonhorizontal plane and the two distinct regions into which it divides space.

5. Visualize *two* planes in space. Into how many disjoint regions do two planes divide space? The answer depends on how the planes are arranged relative to each other—explain. Use two pieces of paper to help you, but be sure to try to see the planes in your mind's eye.

6. Visualize *three* planes in space. Into how many disjoint regions do three planes divide space? As always, the answer depends on how the planes are arranged in space. Describe all the different types of configurations, and make paper models to help you. (You will probably want to cut the paper and use tape for this.) Be sure to also *visualize* all these different types of configurations. How can you tell whether you've found all the different possibilities for disjoint regions there can be?

Class Activity 8E: The Rotation of the Earth and Time Zones

A table-tennis ball and a flashlight would be useful for Problem 2. A larger ball such as a basketball, or better yet, a globe, can be used to demonstrate to the whole class.

The surface of the earth is divided into different time zones. For example, the continental United States has four time zones: Eastern, Central, Mountain, and Pacific. This activity will help you explain why it makes sense to have different time zones and how times vary as you move about the world.

1. Visualize the earth in space, rotating about the axis through the North and South Poles. The earth completes a full rotation about its axis every 24 hours (approximately). Visualize the sun, far away, sending light rays to illuminate the portion of the earth that is facing the sun. Figures 8E.1 and 8E.2 may help.

2. If available, a flashlight and a table-tennis ball can be used to simulate the sun and the earth. Hold the table-tennis ball between two fingers, more or less vertically, at points representing the North and South Poles. Shine the flashlight on the side of the ball. If you have a globe, picture the sun rays coming from one side of the room, illuminating one side of the globe. On your table-tennis ball, or globe, where is it about noon? Where is it about midnight?

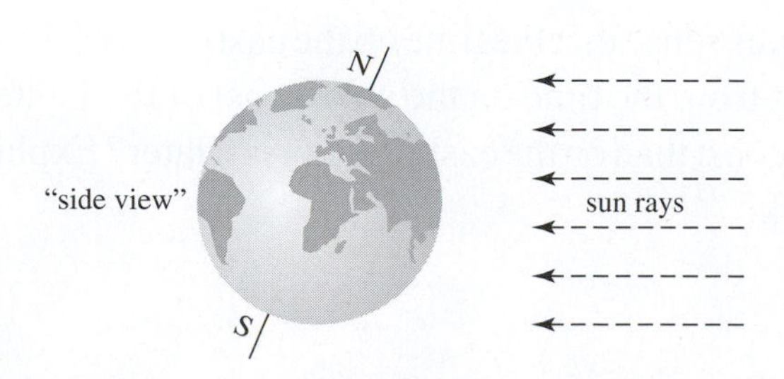

Figure 8E.1
The Earth

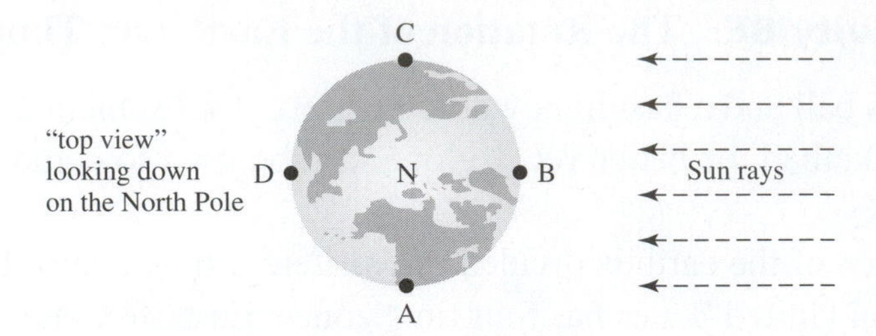

Figure 8E.2
The Earth, Seen from above the North Pole

3. Given that the sun rises in the east, approximately what time of day is it at locations A, B, C, and D in Figure 8E.2, in which the earth is pictured as seen from above the North Pole? Explain.

4. Now go back to your globe or table-tennis ball as in Problem 2. Where is it sunrise, and where is it sunset? Explain.

5. Viewed from outer space, looking down on the North Pole, which way does the earth rotate, clockwise or counter-clockwise? Use Problems 3 and 4 to explain your answer.

6. Explain why it makes sense that the time on the east coast of the United States should be different from the time on the west coast of the United States. Is it earlier on the west coast than on the east coast, or is it later? Explain why, using your previous work.

Class Activity 8F: Explaining the Phases of the Moon

Every month the moon goes through phases, waxing from a new moon to a full moon and then waning from a full moon back to a new moon, as shown in the following diagram:

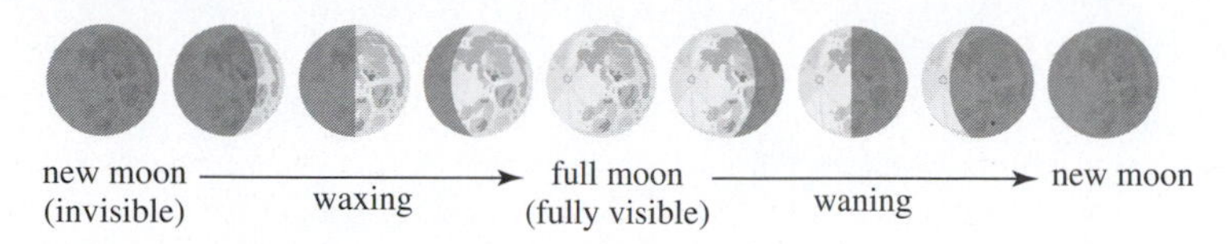

The phases of the moon are caused by its rotation about the earth every month. The moon gives off no light of its own, so we can only see it because light from the sun reflects off the moon. Therefore, *we only see that portion of the moon which faces both toward the sun and toward us on the earth.* As the moon rotates about the earth, different portions of the moon become visible to people on the earth, depending on the positions of the earth, moon, and sun.

1. Imagine floating far above the earth and moon in outer space, looking straight down at the plane of the path of the moon around the earth, as in Figure 8F.1, Parts (A) and (B). Explain why the shaded portion of the moon shown in Part (B) represents the part of the moon that is visible from the earth. Use the dashed lines to help you.

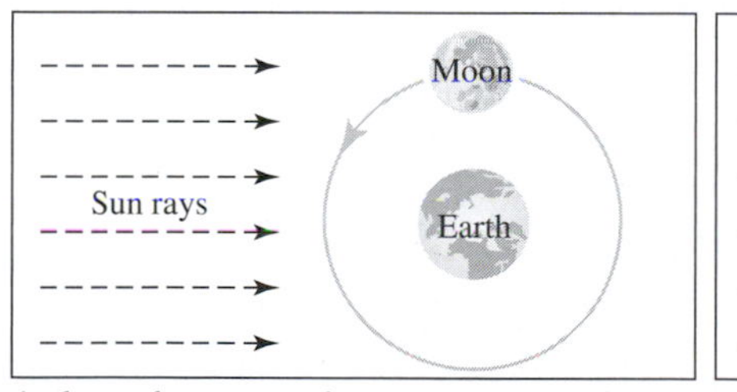

A: the earth, moon, and sun rays, not to scale

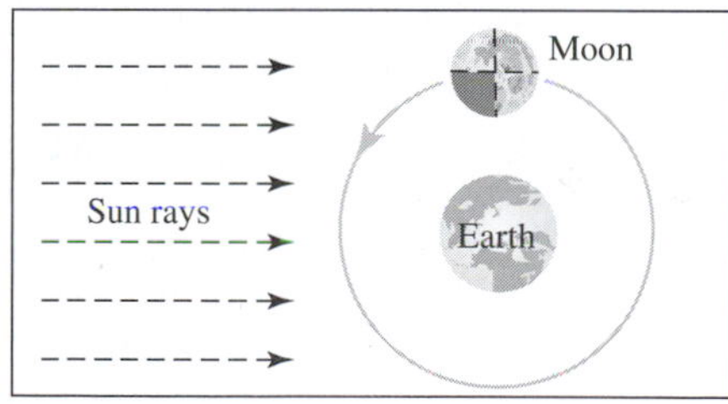

B: the portion of the moon visible from earth

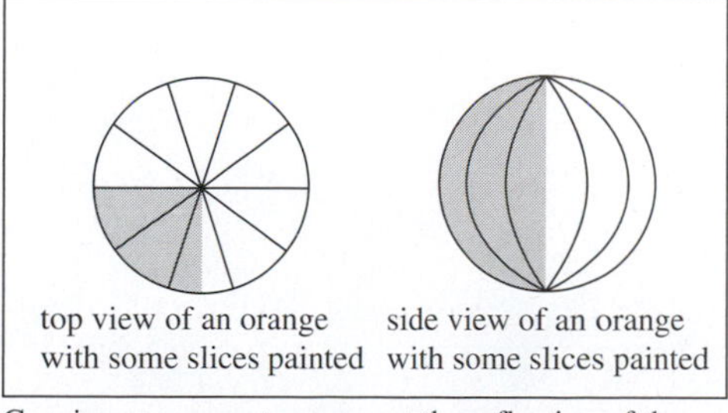

top view of an orange
with some slices painted
side view of an orange
with some slices painted

C: using an orange to represent the reflection of the sun's rays from the moon

Figure 8F.1

In What Phase Is the Moon?

2. Describe how the moon appears to a person on the earth when the earth, moon, and sun's rays are positioned as shown in Figure 8F.1, Parts (A) and (B). To help you, use the idea of looking down on the top of an orange and at the side of an orange, as indicated in Part (C).

3. Describe how the moon appears to a person on the earth in each of the diagrams in Figure 8F.2. In each case, explain why the moon appears that way. Parts (A)–(D) are not to scale, but Part (E) shows the earth and moon, and the distance between them, to scale.

4. Determine the phase of the moon in Figure 8F.1 (A) on page 253, and in Figure 8F.2 (A–D). In which diagram is the moon new? In which is it full? In which is it waxing? In which is it waning? Explain your answers.

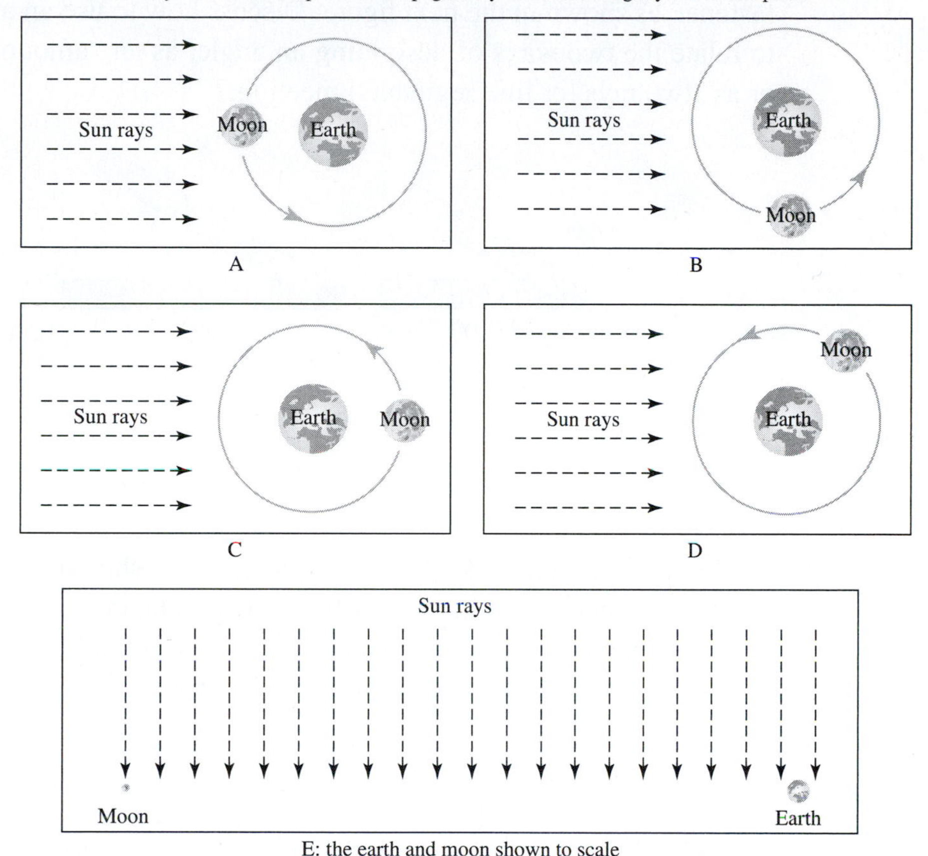

Figure 8F.2
In What Phase Is the Moon?

You might be wondering why there aren't eclipses at every new moon and full moon. A solar eclipse occurs when the moon passes right in front of the sun, as observed from the earth, temporarily obscuring our view of the sun. A lunar eclipse occurs when the earth is directly between the sun and moon and casts a shadow on the moon. At a new moon, the moon is between the sun and the earth. At a full moon, the earth is between the sun and the moon. So why isn't there a solar eclipse at every new moon and lunar eclipse at every full moon? The reason is that the sun is not in the plane in which the moon revolves around the earth, so that the earth, moon, and sun rarely lie in a straight line. You should think of the sun rays shown in Figure 8F.2 as coming from slightly *above* (or below) the plane formed by the page.

8.2 Angles

Class Activity 8G: Angle Explorers

1. An "angle explorer" is made by fastening two cardboard strips with a brass fastener, as shown in the next figure. Discuss how to use an angle explorer to relate the two ways of describing an angle: as an "amount of turning," or as "two rays (or line segments) meeting."

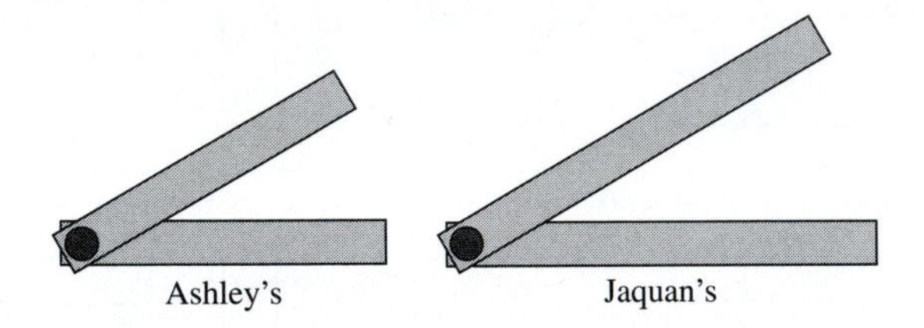

2. Ashley and Jaquan make "angle explorers," as shown in the preceding figure, but Ashley uses shorter cardboard strips than Jaquan. Ashley says that her angle explorer shows a smaller angle than Jaquan's. What is Ashley's misconception, and how could you help her correct it?

3. When we show an angle, do we have to make one of the rays or line segments horizontal, or is it okay to show an angle as on the angle explorer here?

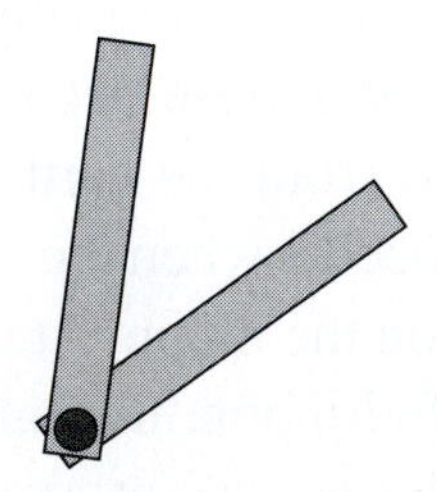

Class Activity 8H: Angles Formed by Two Lines

1. The next figure shows three pairs of lines meeting (or you may wish to think of this as showing one pair of lines in three situations, when the lines are moved to different positions). In each case, how do angles a and c appear to be related, and how do angles b and d appear to be related?

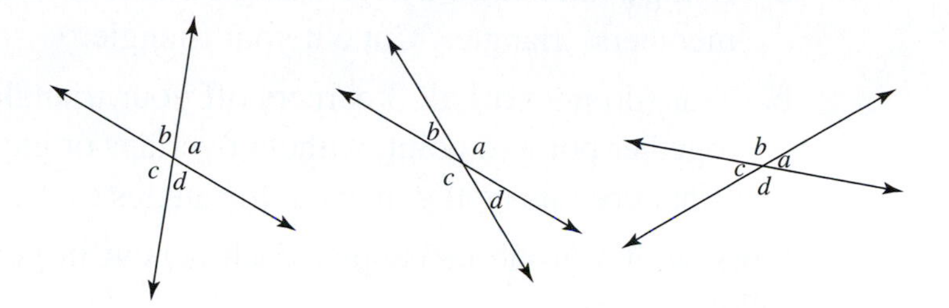

2. Do you think the same phenomenon you observed in Problem 1 will hold for *any* pair of lines meeting at a point? Do you have a convincing reason why or why not?

3. Explain *why* what you observed in Problem 1 must always be true by using the fact that an angle formed by a straight line is 180°. Use this fact to tell you something about several pairs of angles.

Class Activity 8I: Seeing that the Angles in a Triangle Add to 180°

You will need a ruler and scissors for this activity.

1. Work with a group of people. Each person in your group should do the following:
   a. Using a ruler, draw a large triangle that looks different from other group members' triangles. Cut out your triangle.
   b. Tear (do not cut) all 3 corners off your triangle. Then put the corners together point to point, without overlaps or gaps. What do you notice? What does this tell you about the angles of the triangle?
2. Work with a group of people. Each person in your group should do the following:
   a. Using a ruler, draw a large triangle that looks different from the other group members' triangles. Cut out your triangle.
   b. As indicated next, fold the corner that is opposite the longest side (or a longest side) of your triangle down to meet the longest side. Do this in such a way that the fold line is parallel to the longest side of the triangle.

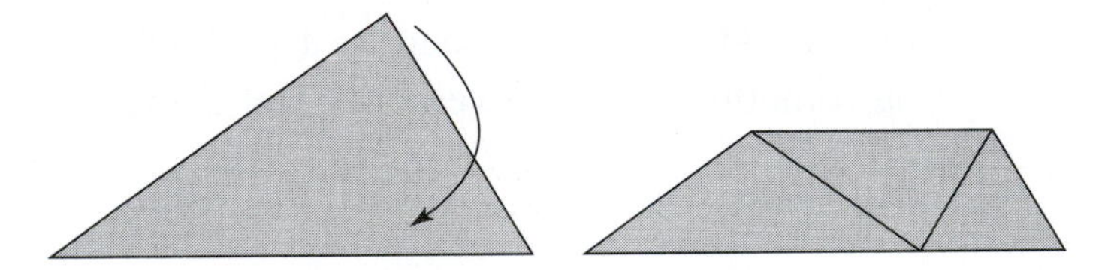

   c. Fold the other two corners of the triangle in to meet the corner that is now along the longest side of the triangle.

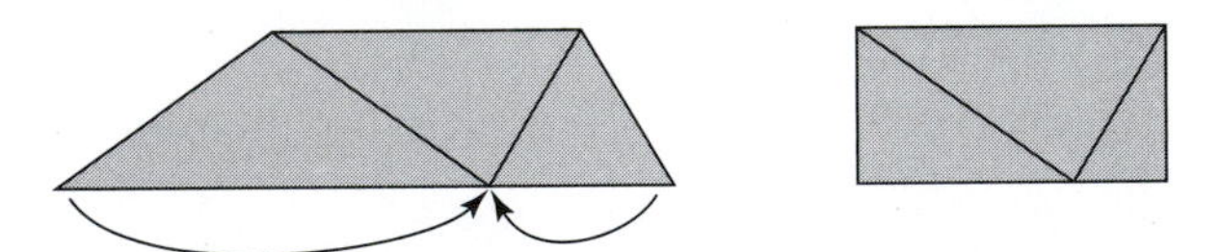

   d. What does this way of folding the triangle show you about the angles of the triangle? Do the other people in your group reach the same conclusion, even though they started with different triangles?

These are neat ways to see that the angles in a triangle add to 180°, but it is not so easy to explain *why* they must always work.

Class Activity 8J: Using the Parallel Postulate to Prove that the Angles in a Triangle Add to 180°

This activity will show you a way to prove that the angles in a triangle add to 180°.

1. Each of the pictures in the diagram that follows shows a pair of parallel lines and third line that crosses the parallel lines. In each case, how do the angles labeled a and b appear to be related? Use the Parallel Postulate to prove your answer.

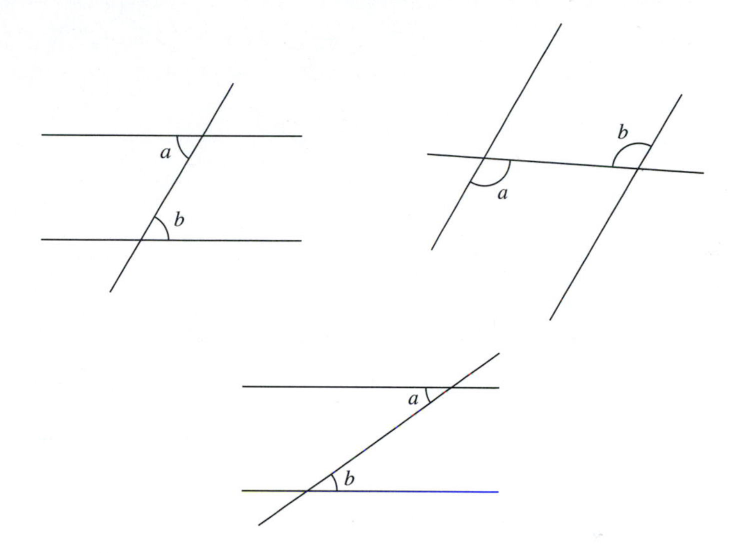

2. Given any triangle with corner points A, B, and C, let a, b, and c be the angles of the triangle at A, B, and C, respectively. Consider the line through A parallel to the side BC that is opposite A.

Based on what you discovered in Problem 1, what can you say about the three adjacent angles at A that are formed by the triangle and the line through A?

What can you conclude about the sum of the angles in the triangle?

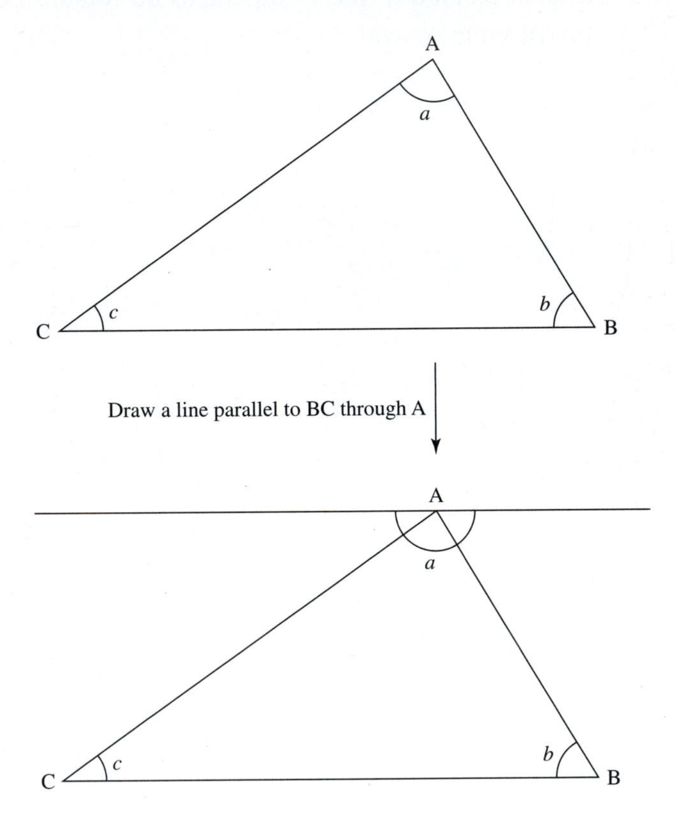

3. What if you used a different triangle in Problem 2? Would you still reach the same conclusion?

Class Activity 8K: Describing Routes, Using Distances and Angles

You will need a protractor and a ruler for this activity.

1. The map in the next figure shows a route along which Dave, who is blindfolded, is to walk. On the map, 1 inch represents 5 of Dave's paces. Dave starts at point A, facing the route, so that he is ready to start walking along it. Use a protractor to help you tell Dave how to get to point H. For example, here's what Dave should do to get from point A to point C:

   Starting at A, go 5 paces to B. At B, turn clockwise (to your right) 60°. Go another 10 paces to C.

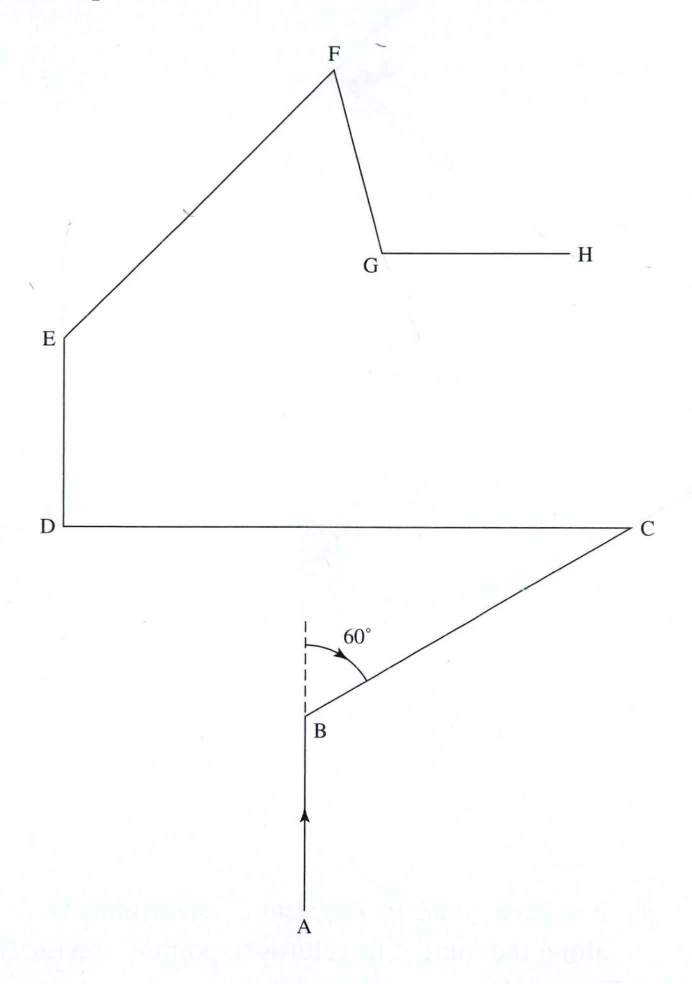

2. For each of the routes in the next figure, tell Dave, who is blindfolded, how to walk around the route. One inch on the map represents 5 of Dave's paces. In each case, Dave will start at the point labeled A, facing the route, so that he is ready to start walking along it.

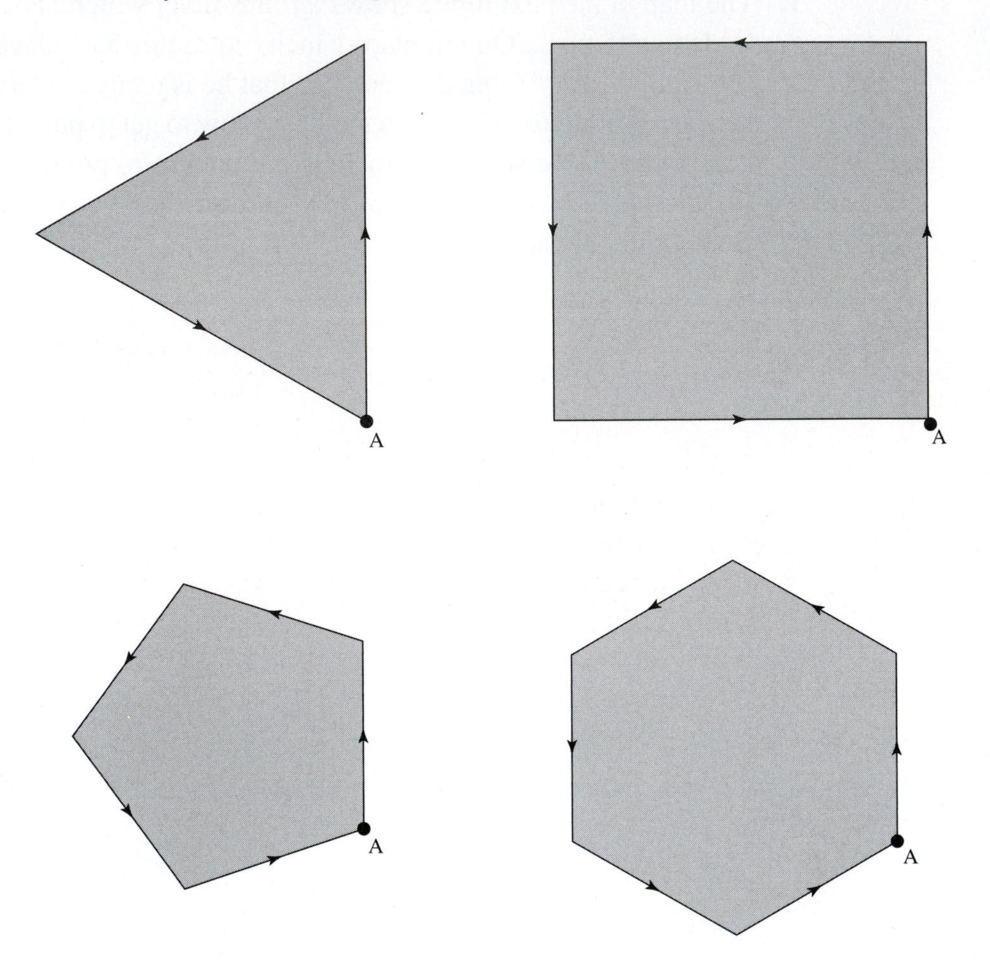

3. For each route in Problem 2, determine Dave's total amount of turning along the route if he returns to point A and *turns to face the same direction in which he started.*

Class Activity 8L: Explaining Why the Angles in a Triangle Add to 180° by Walking and Turning

You will need sticky notes and, if available, masking tape.

This activity will show you a way to understand why the angles in a triangle add to 180°. It is best done as a demonstration for the whole class.

1. Put three "dots" (sticky notes) labeled A, B, and C on the floor to create the corners of a triangle. If possible, connect them with masking tape to make a triangle you can see. Label a point P on the line segment between A and B.

2. Pick two people: one to be a *walker* and one to be a *turner*. The rest are *observers*.

   The walker's job: Stand at point P, facing point B. Walk all the way around the triangle, returning to point P.

   The turner's job: Stand at one fixed spot, and face the same direction that the walker faces at all times. This means that when the walker turns at a corner, you should turn in the same way.

   The observers' job: Observe the walker and the turner, and make sure that they really are facing the same direction at all times.

   Repeat the walking and turning described above until everyone can confidently answer the following questions:

   a. Let's say that the walker and turner were facing north when the walker began walking around the triangle. Which directions did the turner face during the experiment? Were any directions left out? Were any directions repeated?

   b. What was the full angle of rotation of the turner when the walker walked once all the way around the triangle, returning to point P?

3. On the next diagram, show which angles the walker turned through at the corners of the triangle. Label these angles d, e, and f.

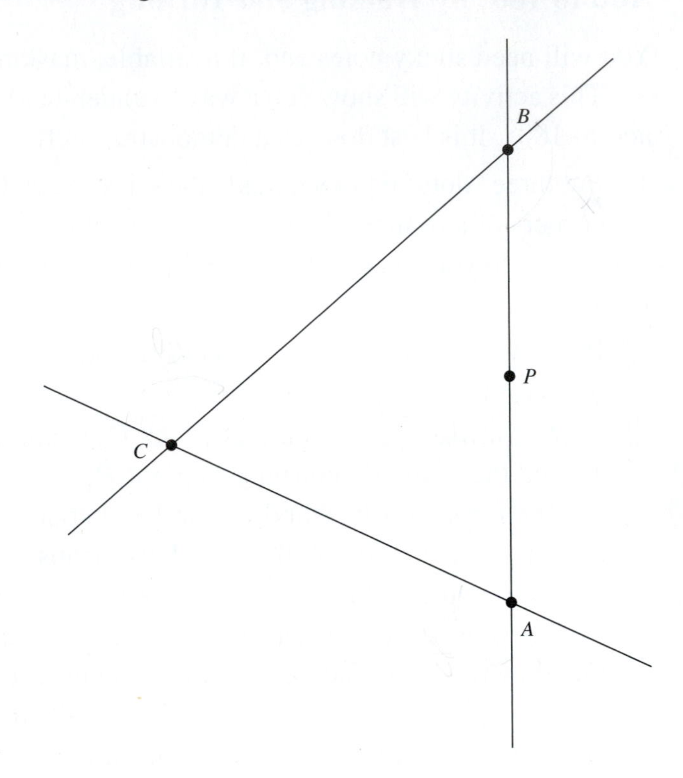

4. Based on your answer to Problem 2(b), what can you say about the value of $d + e + f$?

5. Check your answer to the previous problem by using a protractor to find the angles d, e, f and then adding them.

What if you used a different triangle? The values of d, e, and f might be different, but what about $d + e + f$?

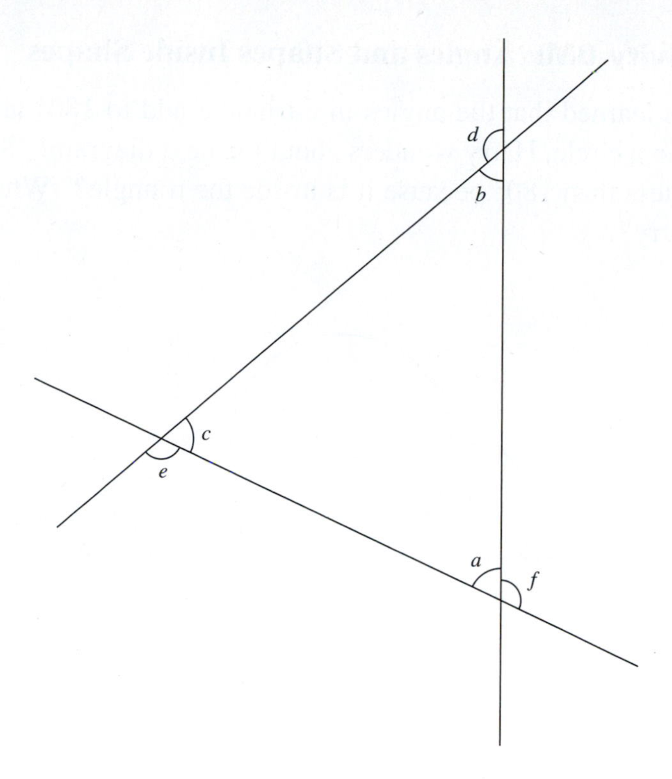

Figure 8L.1
The Exterior Angles of a Triangle Add to 360°

6. You should have just found that

   the sum of the **exterior angles** of a triangle is 360°

   In other words, $d+e+f = 360°$, where d, e, and f are the exterior angles, as shown in Figure 8L.1.

   Use the formula $d + e + f = 360°$ to explain why the sum of the **interior angles** in a triangle is equal to 180°. In other words, show that $a+b+c = 180°$, where a, b, and c are the interior angles of a triangle, as shown in Figure 8L.1.

   Hint: What do you notice about $a + f$, $b + d$, and $c + e$? Can you use this somehow?

7. What if you used a different triangle in this activity? Would you still reach the same conclusion?

Class Activity 8M: Angles and Shapes Inside Shapes

1. Harry has learned that the angles in a triangle add to 180° and that there are 360° in a circle. Harry wonders about the next diagram: "Shouldn't the circle be less than 180° because it is inside the triangle?" What could you say to Harry?

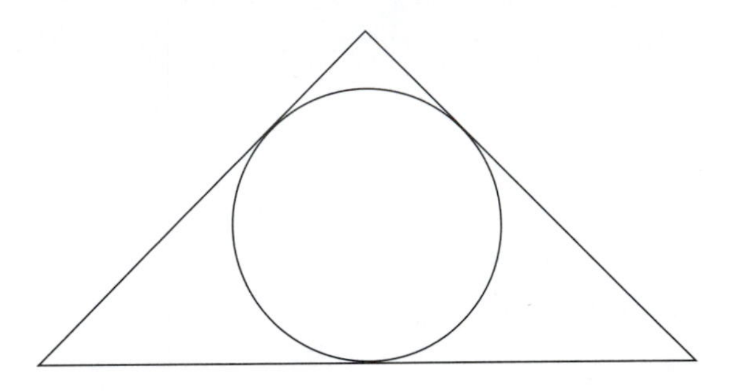

2. The next picture shows a large triangle subdivided into three smaller triangles. If the angles in each of the three smaller triangles add to 180°, then why don't the angles in the larger triangle add to

$$180° + 180° + 180° = 540°$$

instead of 180°? Relate the 540° to the sum of the angles in the large triangle.

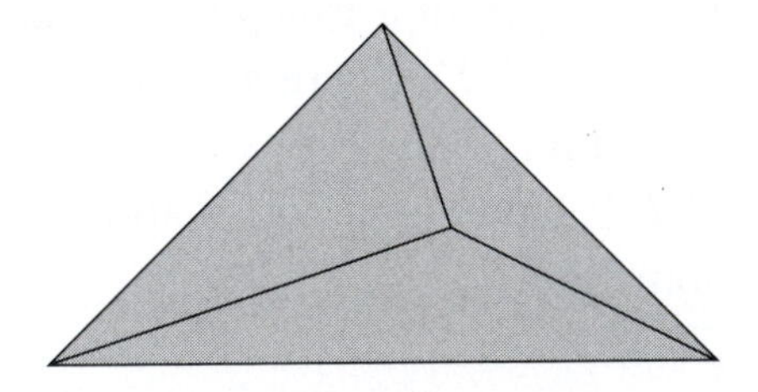

Class Activity 8N: Angles of Sun Rays

1. How are the following related?

   a. The height of the sun in the sky

   b. The length of the shadows of telephone poles (or other objects)

   c. The angle that the sun's rays make with horizontal ground

   Draw diagrams to help you describe and explain the relationships.

2. Because the sun is so far away, light rays coming from the sun form virtually parallel straight lines near the earth. When light rays from the sun strike a location on the earth, they form an angle with the horizontal ground. Figure 8N.1 shows a diagram of a portion of the earth seen from above the North Pole, marked N, and sun rays traveling toward the earth. Assume that these sun rays are parallel to the page.

   a. At each location A, B, C, and D in Figure 8N.1, show the angle that a sun ray makes with the horizontal ground there.

   b. At each location A, B, C, and D in Figure 8N.1, describe how the sun would appear to a person standing there. Is the sun high in the sky or low in the sky? Is the sun in the east, in the west, or directly overhead?

   c. At each location A, B, C, and D in Figure 8N.1, describe how long the shadow of a telephone pole would be. Is the shadow of a telephone pole long or short compared to the other locations? Show the shadows of the telephone poles on Figure 8N.2.

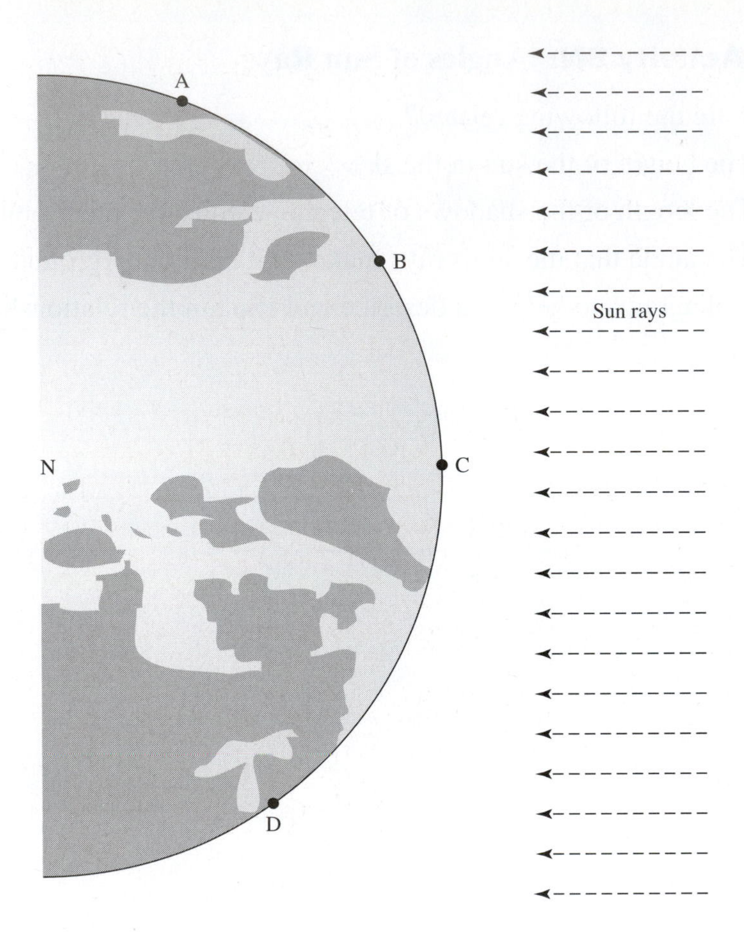

Figure 8N.1

Parallel Sun Rays Traveling toward the Earth

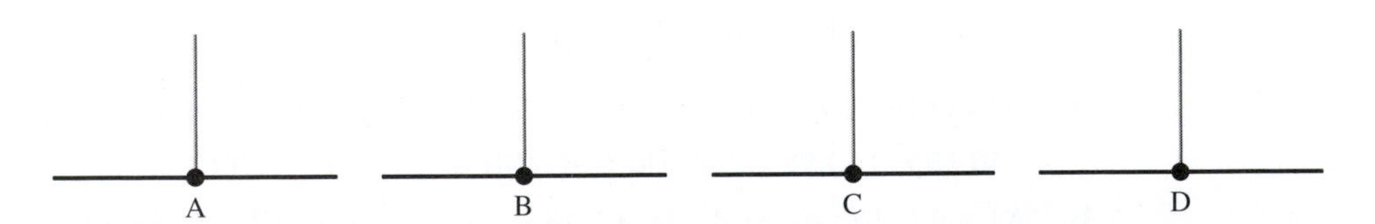

Figure 8N.2

Show the Shadows Formed by the Telephone Poles at Points A, B, C, D in Figure 8N.1

Class Activity 8O: How the Tilt of the Earth Causes Seasons

If available, a globe or a ball with labels representing the North and South Poles would be useful for this activity.

What causes the seasons? Some people mistakenly believe that the seasons are caused by the earth's varying distance from the sun. In fact, the distance from the earth to the sun varies only a little during the year. The seasons are caused by the tilt of the earth's axis.

1. Visualize the earth rotating around the sun over the course of a year. As the earth travels around the sun, the axis between the North and South Poles of the earth remains parallel to itself in space; in other words, the tilt of the earth's axis does not change. The earth's axis is tilted 23.5° from the perpendicular to the plane in which the earth rotates about the sun, as shown in the following diagram:

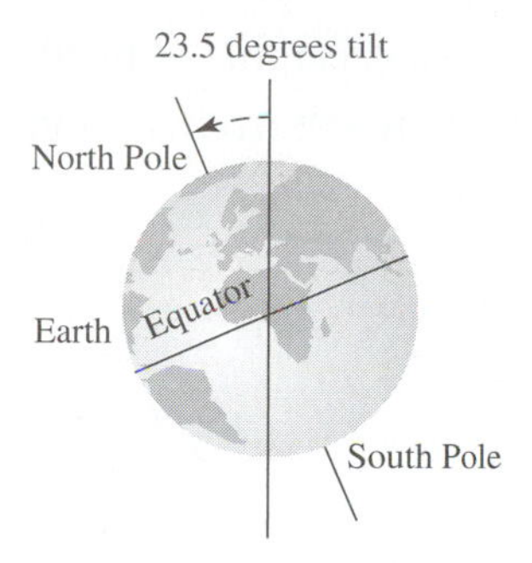

Simulate the earth's yearly journey around the sun either with a globe or as follows: Make a fist around a pencil. The pencil represents the earth's axis (through the North and South Poles), and your fist represents the earth. Imagine the sun in front of you. Move your fist around your imagined sun, keeping your pencil tilted to represent the tilt of the earth. *Keep the pencil parallel to its original position* as you move your "earth" around the "sun." Notice that as the earth travels around the sun, the North Pole is sometimes tilted toward the sun and sometimes tilted away from the sun.

2. The following diagram shows the earth at two different times of year. In diagram 1, the North Pole is tilted directly toward the sun. In diagram 2, the South Pole is tilted directly toward the sun, and the North Pole is tilted directly away from the sun. In both diagrams, assume that the sun rays are parallel to the page. To picture other times of year, visualize the sun's rays coming from outside of the plane of the page.

   a. Points A and B in diagrams 1 and 2 are both shown at around noon. In diagram 1, is the sun higher in the sky at point A or point B? In diagram 2, is the sun higher in the sky at point A or point B?

   b. During the day, the earth rotates on its axis. In both diagrams 1 and 2, visualize how the locations of points A and B will change throughout the day. Which of the points, A or B, will receive more sunlight over the course of a day in diagram 1? Which of the points, A or B, will receive more sunlight over the course of a day in diagram 2?

   c. What season is it in the northern hemisphere in diagram 1? What season is it in the northern hemisphere in diagram 2? Why?

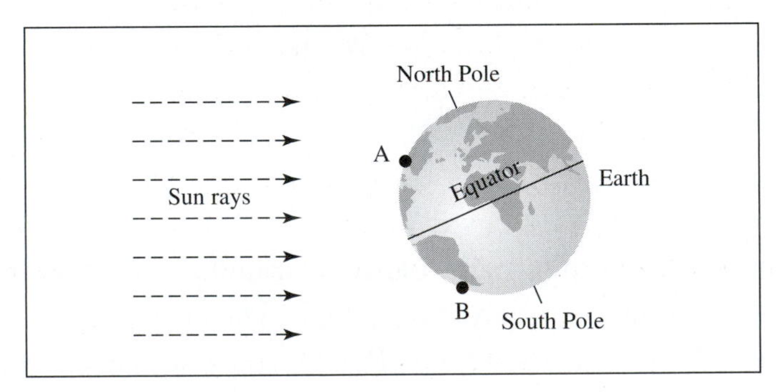

Diagram 1

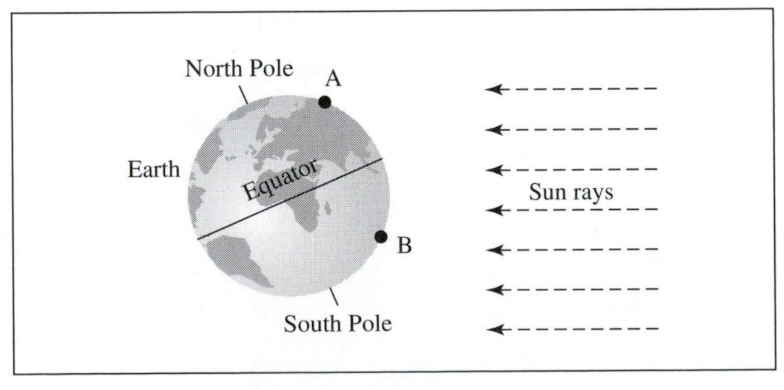

Diagram 2

Figure 8O.1

The Earth's Position around the Sun at Different Times of Year

Class Activity 8P: How Big Is the Reflection of Your Face in a Mirror?

You will need a mirror and a ruler for the first part of this activity.

How big is the reflection of your face in a mirror? Is it the same size as your face, or is it larger or smaller, and if so, how much? The answers to these questions may surprise you.

1. Use a ruler to measure the length of your face from the top of your forehead to your chin. Now hold a mirror *parallel* to your face and measure the length of your face's reflection in the mirror, from the top of the forehead to the chin. Measure carefully!

   Hold the mirror closer to your face or farther away (but always *parallel* to your face), and repeat the measuring processes just described. The *position* of your reflection will probably change, but does the *size* of your reflection change or not? The answer may surprise you.

   Compare the length of your face and the length of your reflected face in the mirror. How do these lengths appear to be related?

2. Figure 8P.1 shows a side view of a person looking into a mirror. Using the laws of reflection, determine what the person will see at each of the points A, B, and C. We see objects by seeing the light that travels from the object to our eyes. So a person looking at a particular point sees the light that travels in a straight line from that point to the person's eye. To determine what a person sees at a point in a mirror, you must determine where the light at that point came from. For this you will need the laws of reflection.

3. Use the reflection laws to show where the person looking into the mirror in Figure 8P.2 will see the top of her forehead and where she will see the bottom of her chin. Measure the length of the person's face and the length of her reflected face in the mirror. How do these lengths appear to be related? Your result should fit with what you discovered in Problem 1.

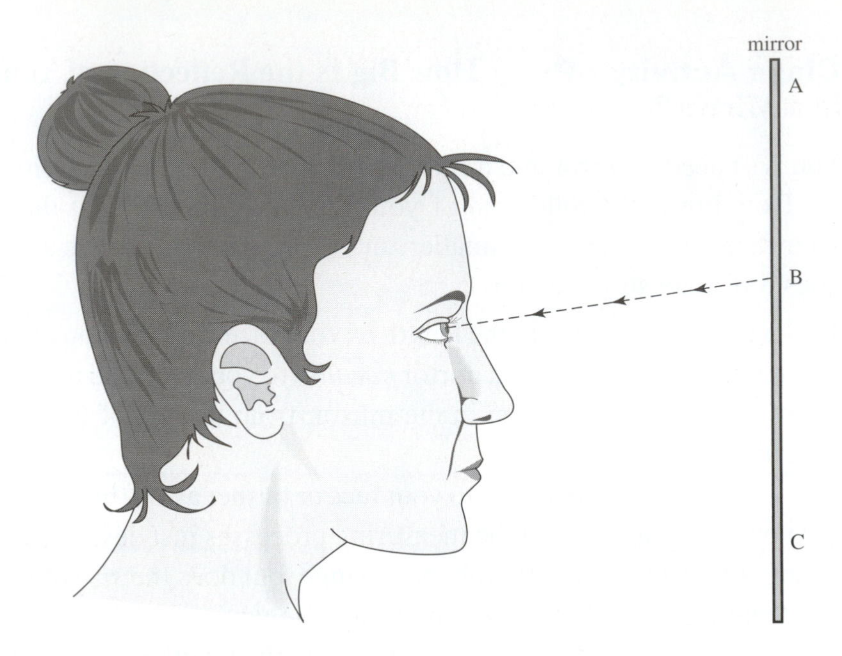

Figure 8P.1
Looking into a Mirror

Figure 8P.2
Where Is the Person's Reflection in the Mirror?

Class Activity 8Q: Why Do Spoons Reflect Upside Down?

A large, reflective spoon would be helpful for this activity.

When you look at your reflection in the bowl of a spoon, you will notice that (in addition to looking quite distorted) your image will appear upside down. Have you ever wondered why? We will explore this now.

Figure 8Q.1 is a diagram of a person looking into the bowl of a (very large) spoon. Only a cross section of the spoon is shown. The lines shown are the *normal lines* to the spoon at the points A, B, and C. Use these normal lines and the laws of reflection to determine what the person sees when she looks at points A, B, and C. (Assume that the person sees light that enters the center of her eye.)

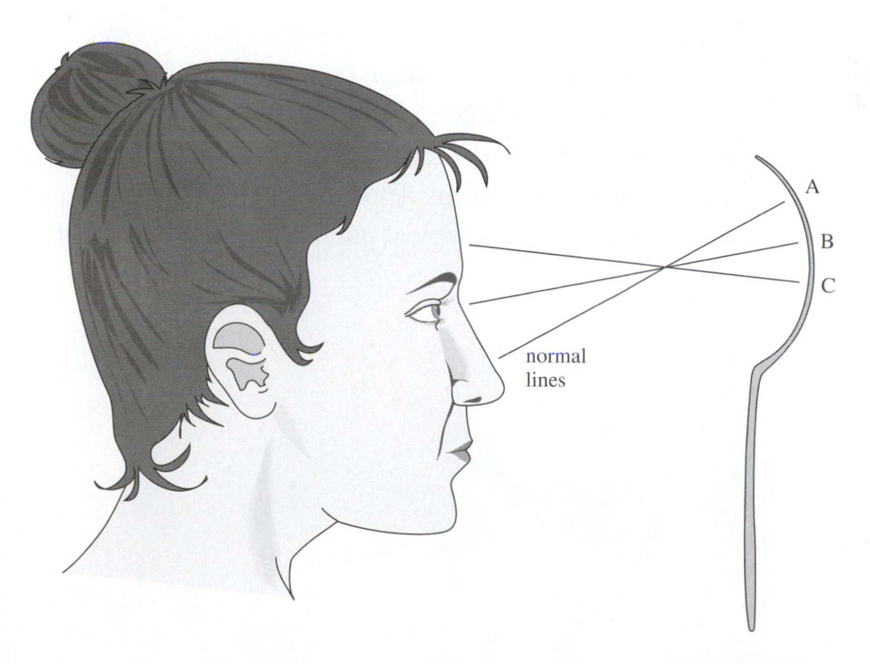

Figure 8Q.1
Looking into a Spoon

Class Activity 8R: The Special Shape of Satellite Dishes

Satellite dishes are shaped in a special way so as to efficiently capture incoming radio waves broadcast from a satellite. The top diagram of Figure 8R.1 shows the cross section of a satellite dish. The curve formed by this cross section is called a **parabola**.

1. Verify that all incoming radio waves hitting a parabolic satellite dish (top diagram of Figure 8R.1) reflect onto the receiver. The receiver is shown as a small circle inside and above the center of the satellite dish. (Actual satellite dishes usually do not extend above their receiver, as this one does, but this exaggerated diagram will be easier to work with.)

2. The cross section of a cone-shaped satellite dish forms a v-shaped curve, as shown in the bottom diagram of Figure 8R.1. What kind of receiver would a cone-shaped satellite dish need in order to capture all the incoming radio waves reflected from the inside of the satellite dish? Could the receiver be as small as the one for a typical parabolic satellite dish? Explain.

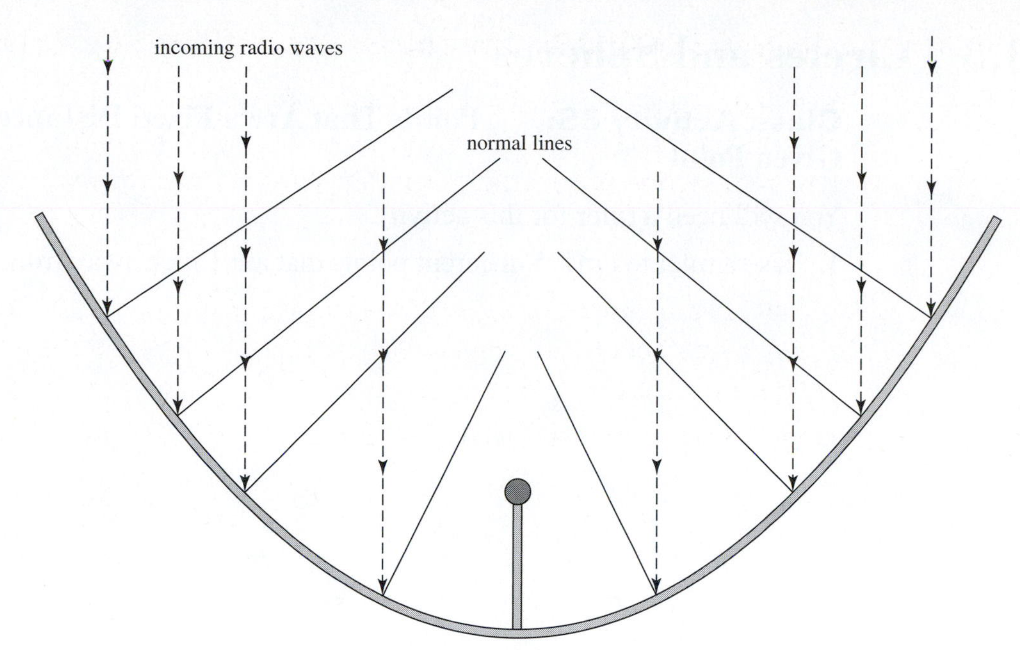

cross section of a parabolic satellite dish

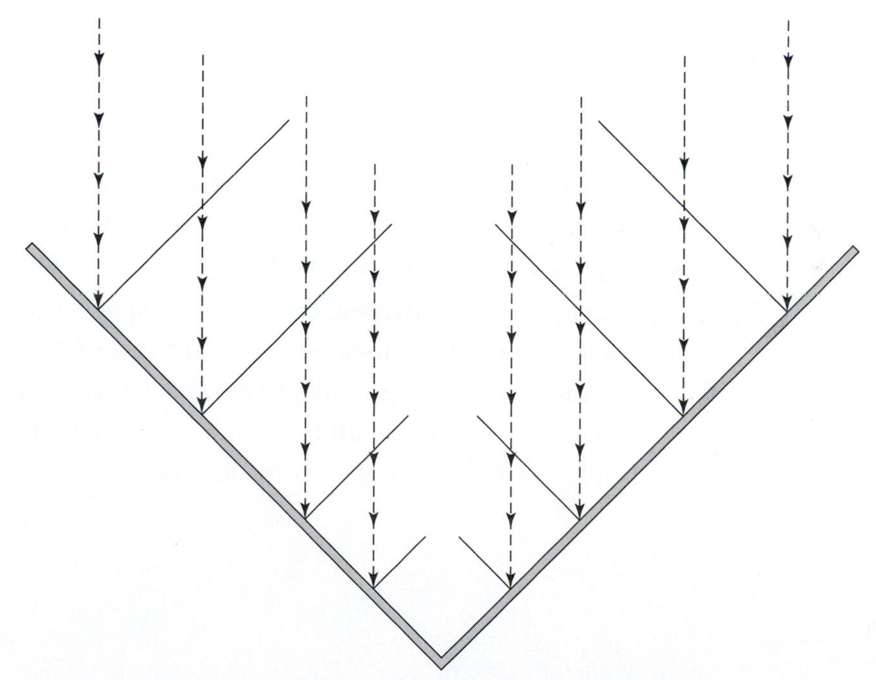
cross section of a hypothetical cone-shaped satellite dish

Figure 8R.1
Cross Sections of Satellite Dishes

8.3 Circles and Spheres

Class Activity 8S: Points That Are a Fixed Distance from a Given Point

You will need a ruler for this activity.

1. Use a ruler to draw 5 different points that are 1 inch away from the point P:

P

   Now draw 5 more points that are 1 inch away from the point P. If you could keep drawing more and more points that are 1 inch away from point P, what shape would this collection of points begin to look like?

2. Have one person point to a particular point in space. Call that point P. Using a ruler, find several other locations in space that are one ruler-length away from point P. (A ruler-length might be 12 inches or 6 inches, depending on your ruler.) Try to visualize all the points in space that are one ruler-length away from your point P. What shape do you see?

Class Activity 8T: Using Circles

You will need a compass for this activity. Some string would also be useful.

1. Use the definition of a circle to explain why a compass draws a circle. How is the radius of the circle related to the compass?

2. Explain how to use a piece of string and a pencil to draw a circle. Use the definition of a circle to explain why your method will draw a circle.

3. Suppose that, 1 hour and 45 minutes ago, a prisoner escaped from a prison located at point P, shown on the map that follows. Due to the terrain and the fact that no vehicles have left the area, police estimate that the prisoner can go no faster than 4 miles per hour. Show all places on the map where the prisoner might be at this moment. Explain.

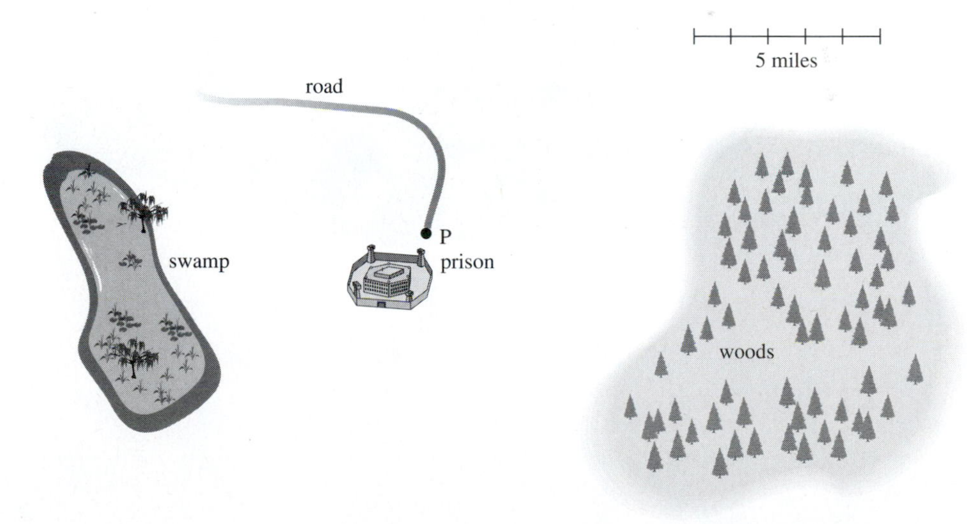

4. The treasure map for this exercise shows an X and an O marking two spots. The treasure is described as buried under a spot that is 30 feet from X and 50 feet from O. Use this information to help you show where the treasure might be buried. Is the information enough to tell you *exactly* where the treasure is buried? Explain.

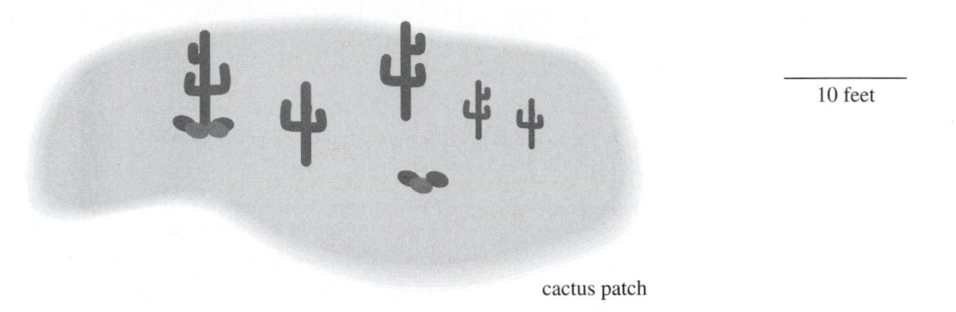

X O

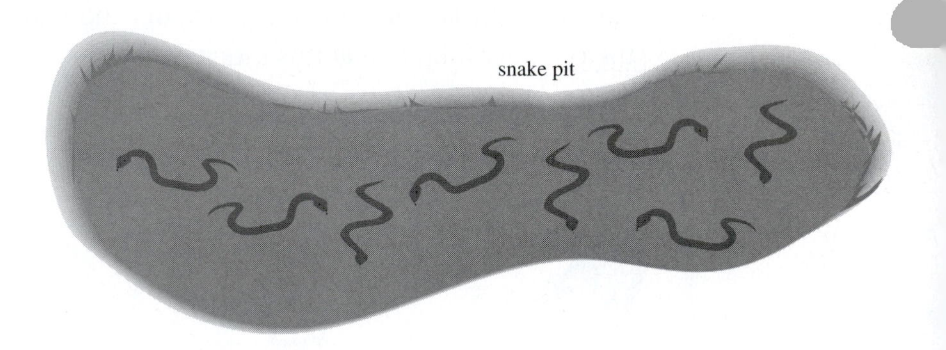

Class Activity 8U: The Global Positioning System (GPS)

You will need string for this activity.

The Global Positioning System (GPS) is a navigation system designed and maintained by the U.S. Department of Defense for military use, but is also available for civilian use. With a hand-held GPS unit, it is possible to determine your location on the earth, accurate to within about 10 meters. The GPS system uses a network of 24 satellites that orbit the earth. Figure 8U.1 shows a GPS satellite. A GPS unit receives information on its distance from each satellite within its range. The GPS unit can then figure its location by using those distances, based on the geometry of spheres.

Suppose a GPS unit learns that it is a certain distance from satellite 1 and another certain distance from satellite 2. The following activity will help you describe all possible locations for the GPS unit by simulating the situation with people and string:

1. Pick two people to represent satellites 1 and 2, and pick a third person who will show all possible locations of the GPS unit. The GPS person should stand between satellites 1 and 2.
2. Cut two pieces of string, representing the distances from satellites 1 and 2 to the GPS unit.

Figure 8U.1
A GPS Satellite

3. Satellites 1 and 2 should each hold one end of their piece of string, and the GPS person should hold the other ends of the two pieces of string in one hand, pulling the strings tight. (Everyone may have to adjust positions so that it is possible to do this. Once suitable positions are found, everyone should stay fixed in his or her position.)
4. The designated GPS person will now be able to show all possible locations of the GPS unit by moving the strings, while keeping them pulled tight and held in one hand (and while satellites 1 and 2 remain fixed in their positions). See Figures 8U.2 and 8U.3.
5. Describe the shape of all possible locations of the GPS unit. How is this related to the intersection of two spheres?

Figure 8U.2
Showing a Position of a GPS Unit

Figure 8U.3
Showing Another Position of a GPS Unit

6. Now suppose that there is also a third satellite beaming information to the GPS unit. Pick a person to represent sallelite 3, and cut a piece of string to represent the distance of satellite 3 to the GPS unit.
7. Satellite 3 should hold one end of her or his string while the GPS person holds the other end in the same hand with the strings from satellites 1 and 2. By pulling all three strings tight, the GPS person can show all possible locations of the GPS unit. In general, there will be two such locations. See Figures 8U.4 and 8U.5.

Figure 8U.4
One Position of a GPS Unit, Given Information from 3 Satellites

Figure 8U.5
Another Position of a GPS Unit, Given Information from 3 Satellites

How is this exercise related to the intersection of three spheres?

As you've seen, with information from three satellites, a GPS unit can narrow its location to one of two points. If one of those two locations can be recognized as being in outer space, and not on the surface of the earth, then the GPS unit can report its location on the earth. This is the idea behind the GPS system.

Class Activity 8V: Circle Curiosities

You will need various coins for Problem 1 and a compass for Problem 2.

1. How many pennies can you place around a single penny, each touching the inner penny? Try it and see. Do the pennies fit snugly around the inner penny or not?

   Now try placing as many pennies as you can around a single nickel so that each penny touches the inner nickel. Do the pennies fit snugly around the inner nickel or not?

   How many nickels can you place around a single nickel, each touching the inner nickel? Do the nickels fit snugly around the inner nickel or not?

2. Although the accompanying design looks complex, it is surprisingly easy to make. Use a compass to draw a design like the one below. Explain how to draw the design. If you look closely, you will also see the phenomenon you discovered about arranging pennies in Problem 1.

8.4 Triangles, Quadrilaterals, and Other Polygons

Class Activity 8W: Using a Compass to Draw Triangles and Quadrilaterals

You will need a compass and ruler for this activity. Focus on the definition of a circle throughout the activity.

1. Use a compass to help you draw an isosceles triangle. Without measuring any side lengths, explain why your triangle must be isosceles.

2. Try to draw an equilateral triangle by using only a ruler and pencil, no compass. Why does this not work so well?

   Now use a compass to help you draw an equilateral triangle. Without measuring any side lengths, explain why your triangle must be equilateral.

3. To make an equilateral triangle, follow the steps outlined in the next figure. Notice that you really need to draw only the top portions of the circles. Use this method to make examples of several different equilateral triangles.

Explain *why* this method must always produce an equilateral triangle.

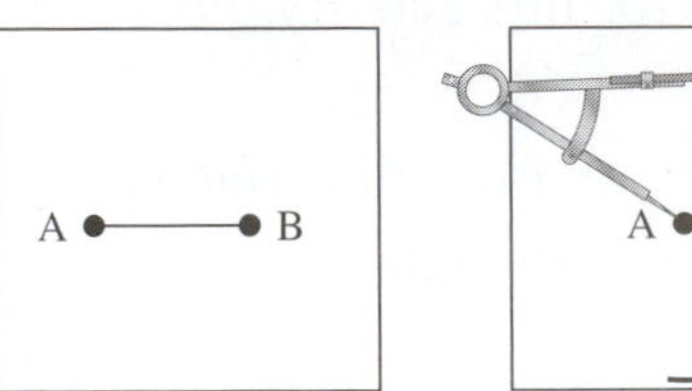

Step 1: Start with any line segment AB.

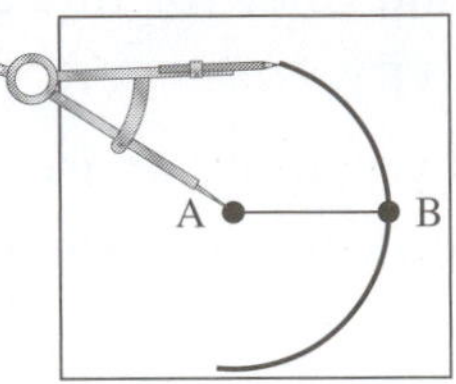

Step 2: Draw a circle centered at A, passing through B.

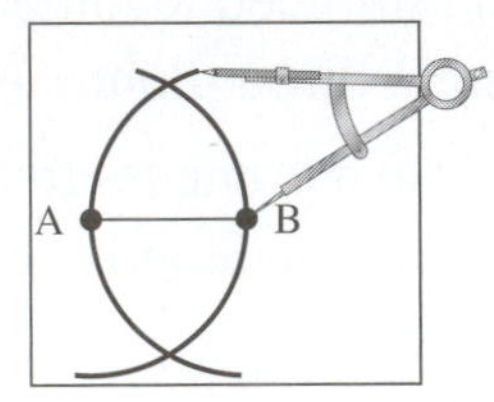

Step 3: Draw a circle centered at B, passing through A.

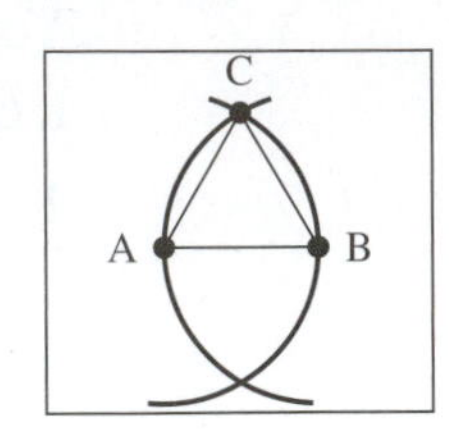

Step 4: Label one of the two points where the circles meet C. Connect A, B, and C with line segments.

4. In the next picture, the line segments AB and AC have the same length. Use a ruler and compass to draw a rhombus that has AB and AC as two of its sides. (You may wish to try drawing the rhombus without a compass first. Notice that it is difficult and clumsy to do so.)

Hint: To create the rhombus, you will need to construct a fourth point, D, such that the distance from D to B is equal to the distance from D to C, and such that these two distances are also equal to the common distance from A to B and A to C. Think about the *definition of circles* in order to help you figure out how to construct the point D.

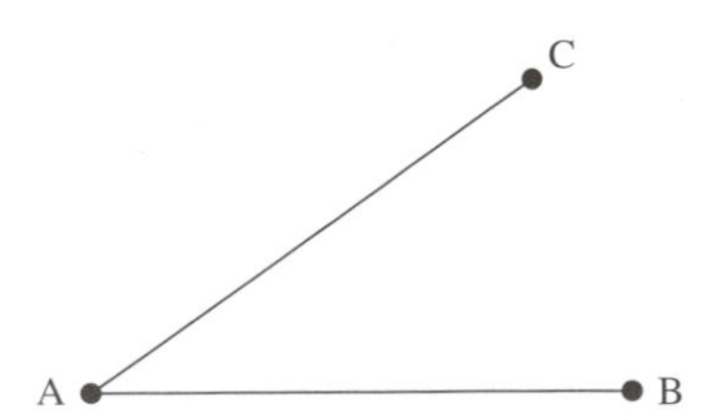

5. The line segment AB shown in the next figure is 4 inches long. Use a ruler and compass to construct a triangle that has AB as one of its sides, has a side that is 3 inches long, and has another side that is 2 inches long. Describe how you constructed your triangle, and explain why your construction must produce the desired triangle. *Hint*: Modify the construction of an equilateral triangle shown in Problem 4 of Activity 8W by drawing circles of different radii.

A ——————————————— B

6. Take a blank piece of paper. Use a ruler and compass to construct a triangle that has one side of length 6 inches, one side of length 5 inches, and one side of length 3 inches.

7. Is it possible to make a triangle that has one side of length 6 inches, one side of length 3 inches, and one side of length 2 inches? Explain.

Class Activity 8X: Making Shapes by Folding Paper

You will need paper, scissors, and a ruler for this activity.

1. To create an isosceles triangle, follow the next set of instructions.

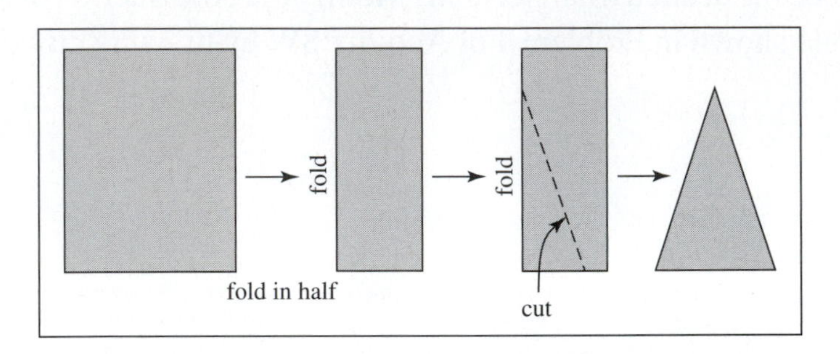

Figure 8X.1

A Method for Making Isosceles Triangles

a. By referring to the definition of isosceles triangle, explain *why* the method described must always create an isosceles triangle.

b. What properties does your isosceles triangle have? Find as many as you can. Explain if you can.

2. To create a rhombus, follow the next set of instructions.

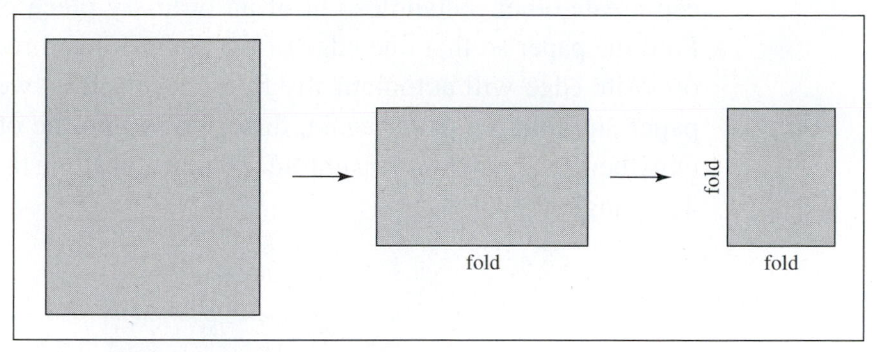

Step 1: Fold a rectangular piece of paper in half and then in half again, creating perpendicular fold lines.

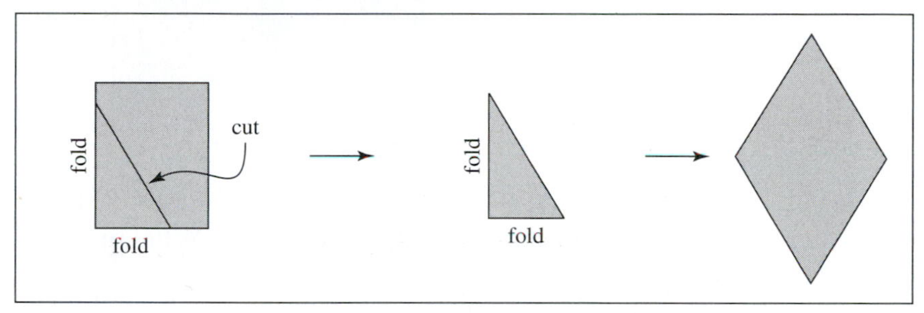

Step 2: Draw a line from anywhere on one fold to anywhere on the perpendicular fold. Cut along the line you drew. When you unfold, you will have a rhombus.

a. By referring to the definition of rhombus, explain *why* the method described must always create a rhombus.

b. What properties does your rhombus have? Find as many as you can. Explain if you can.

3. An ordinary rectangular piece of paper is one example of a rectangle. You can create other rectangles out of an ordinary piece of paper as follows: Fold the paper so that one edge of the paper folds directly onto itself. The opposite edge will automatically fold onto itself as well. Now unfold the paper and fold the paper again, this time so that the other two edges fold onto themselves. When you unfold, you can cut along the fold lines to create 4 rectangles.

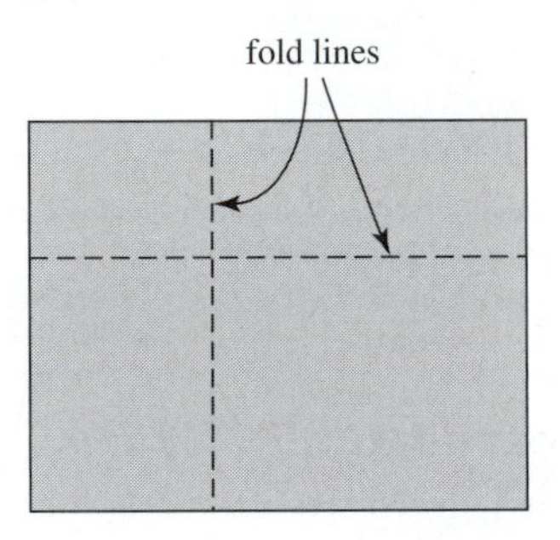

By referring to the definition of a rectangle, explain why the method just described must always create rectangles. What properties do your rectangles have?

4. To create a parallelogram, start with a rectangular piece of paper of any size. Follow the next set of steps.

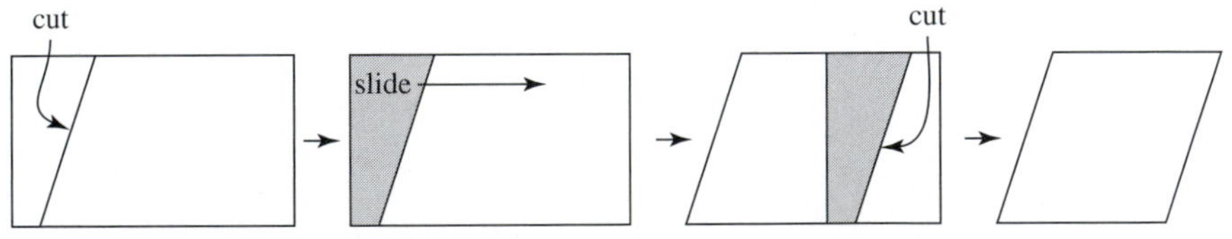

Draw a line segment connecting two opposite sides of a rectangle. Cut along the line segment. Put the piece back (shown shaded) and slide it over as shown. Mark and cut at the leading edge of the slid-over piece.

What properties does your parallelogram have?

Class Activity 8Y: Constructing Quadrilaterals with Geometer's Sketchpad

In this activity, you will construct various kinds of quadrilaterals. Save your sketches for use in a subsequent activity.

To do this activity, you must be able to do the following in Geometer's Sketchpad:

- Construct points and line segments with the construction tools.
- Use the *Construct* menu to construct a line segment between two given points.
- Use the *Construct* menu to construct a line that is parallel to another line and goes through a given point.
- Use the *Construct* menu to construct a line that is perpendicular to another line and goes through a given point.
- Use the *Construct* menu to construct a circle that goes through a given point and has a radius of the same length as a given line segment.
- Hide objects by using the *Display* menu.

1. Construct a "general" quadrilateral. You should be able to move the vertices so as to vary the quadrilateral and show (theoretically) all possible quadrilaterals (if your computer screen were infinitely large).
2. Construct a "general" trapezoid. Use the definition of trapezoid so that your construction is guaranteed to be a trapezoid, even when you move the vertices. This means that you will need to use the *Construct* menu to construct a parallel line to a given line segment. By moving the vertices of your trapezoid, you should be able to show (theoretically) all possible trapezoids (if your computer screen were infinitely large).
3. Construct a "general" parallelogram. Use the definition of parallelogram so that your construction is guaranteed to be a parallelogram, even when you move the vertices. By moving the vertices of your parallelogram, you should be able to show (theoretically) all possible parallelograms (if your computer screen were infinitely large).

4. Construct a "general" rectangle. Use the definition of rectangle so that your construction is guaranteed to be a rectangle, even when you move the vertices. By moving the vertices of your rectangle, you should be able to show (theoretically) all possible rectangles (if your computer screen were infinitely large).
5. Construct a "general" square. Use the definition of square so that your construction is guaranteed to be a square, even when you move the vertices. To do this construction, it will help you to remember the definition of a circle. By moving the vertices of your square, you should be able to show (theoretically) all possible squares (if your computer screen were infinitely large).
6. Construct a "general" rhombus. Use the definition of rhombus so that your construction is guaranteed to be a rhombus, even when you move the vertices. To do this construction, it will help you to remember the definition of a circle. By moving the vertices of your rhombus, you should be able to show (theoretically) all possible rhombuses (if your computer screen were infinitely large).

Class Activity 8Z: Relating the Kinds of Quadrilaterals

1. In each of Parts (a)–(d), describe the relationships among the given kinds of shapes. In each case, write as many sentences as you can of the following forms:

   Every _______ is a _______.
   There are _______ that are not _______.

   a. Describe the relationship between squares and rectangles.

   b. Describe the relationship between squares and rhombuses.

   c. Describe the relationship between rectangles and trapezoids.

   d. Describe the relationship between rhombuses and parallelograms.

2. **a.** Explain why squares and rectangles are related the way they are.

   b. Explain why squares and rhombuses are related the way they are.

   c. When you relate rectangles and trapezoids, are you relying *only* on information that is stated directly in the definitions of these shapes, or are you relying on information that is not stated directly in the definitions, such as visual information?

   d. When you relate rhombuses and parallelograms, are you relying *only* on information that is stated directly in the definitions of these shapes, or are you relying on information that is not stated directly in the definitions, such as visual information?

Class Activity 8AA: Venn Diagrams Relating Quadrilaterals

1. Draw a Venn diagram relating the set of squares and the set of rectangles.

2. Draw a Venn diagram relating the set of rhombuses, the set of squares, and the set of rectangles.

3. Draw a Venn diagram relating the set of parallelograms and the set of trapezoids.

4. Our definition of a trapezoid is a quadrilateral with *at least one* pair of parallel sides. Some books define "trapezoid" as a quadrilateral with *exactly one* pair of parallel sides. How would the Venn diagram relating parallegrams and trapezoids be different if we used this other definition of "trapezoid?"

Class Activity 8BB: Investigating Diagonals of Quadrilaterals with Geometer's Sketchpad

You will need sketches for a general quadrilateral, rectangle, and rhombus from Class Activity 8Y for this activity.

To do this activity, you will need to be able to measure angles and lengths of line segments. Both of these can be done with the *Measure* menu. To measure an angle, you will need to select three points *in the following order*: a point on one line segment, the point where the line segments meet, and a point on the other line segment.

1. Choose a vertex of your general quadrilateral. Construct a diagonal of the quadrilateral that has your chosen point on it. The diagonal splits the angle at your chosen point into two angles. Measure these two angles, and compare them. Observe how these two angles change as you move various vertices of the quadrilateral around to make different quadrilaterals.

Now choose a vertex of your general rhombus. Construct a diagonal of the rhombus that has your chosen point on it. As before, the diagonal splits the angle at your chosen point into two angles. Measure these two angles, and compare them. Observe how these two angles change as you move various vertices of the rhombus around to make different rhombuses.

Compare what you observed for the quadrilateral with what you observed for the rhombus.

2. Construct the two diagonals in each of your sketches of a general quadrilateral, rectangle, and rhombus. Now move the vertices in your sketches around in different ways so as to see many different quadrilaterals, rectangles, and rhombuses. As you move the vertices, observe the diagonals, and do the following:

- Measure the angles that the diagonals make with each other.
- Find where the diagonals meet—where is this point located on the diagonals?
- Compare the lengths of the two diagonals.

What do the diagonals of rectangles all have in common? What do the diagonals of rhombuses all have in common? What about diagonals of general quadrilaterals?

Class Activity 8CC: Investigating Diagonals of Quadrilaterals (Alternate)

1. Figure 8CC.1 shows examples of different kinds of quadrilaterals and their diagonals. Look carefully at the diagonals and do the following:
   - Observe and (if possible) measure the angles that the diagonals make with each other.
   - Observe where the diagonals meet—where is this point located on the diagonals?
   - Compare the lengths of the two diagonals.

   What do the diagonals of rectangles all have in common? What do the diagonals of rhombuses all have in common? What about diagonals of general quadrilaterals?

2. Shown in Figure 8CC.2 are some rhombuses and other quadrilaterals with one of the two diagonals drawn in. The two angles created by the diagonal are labeled a and b. In each case, compare angle a with angle b. Compare what you observed for quadrilaterals with what you observed for rhombuses.

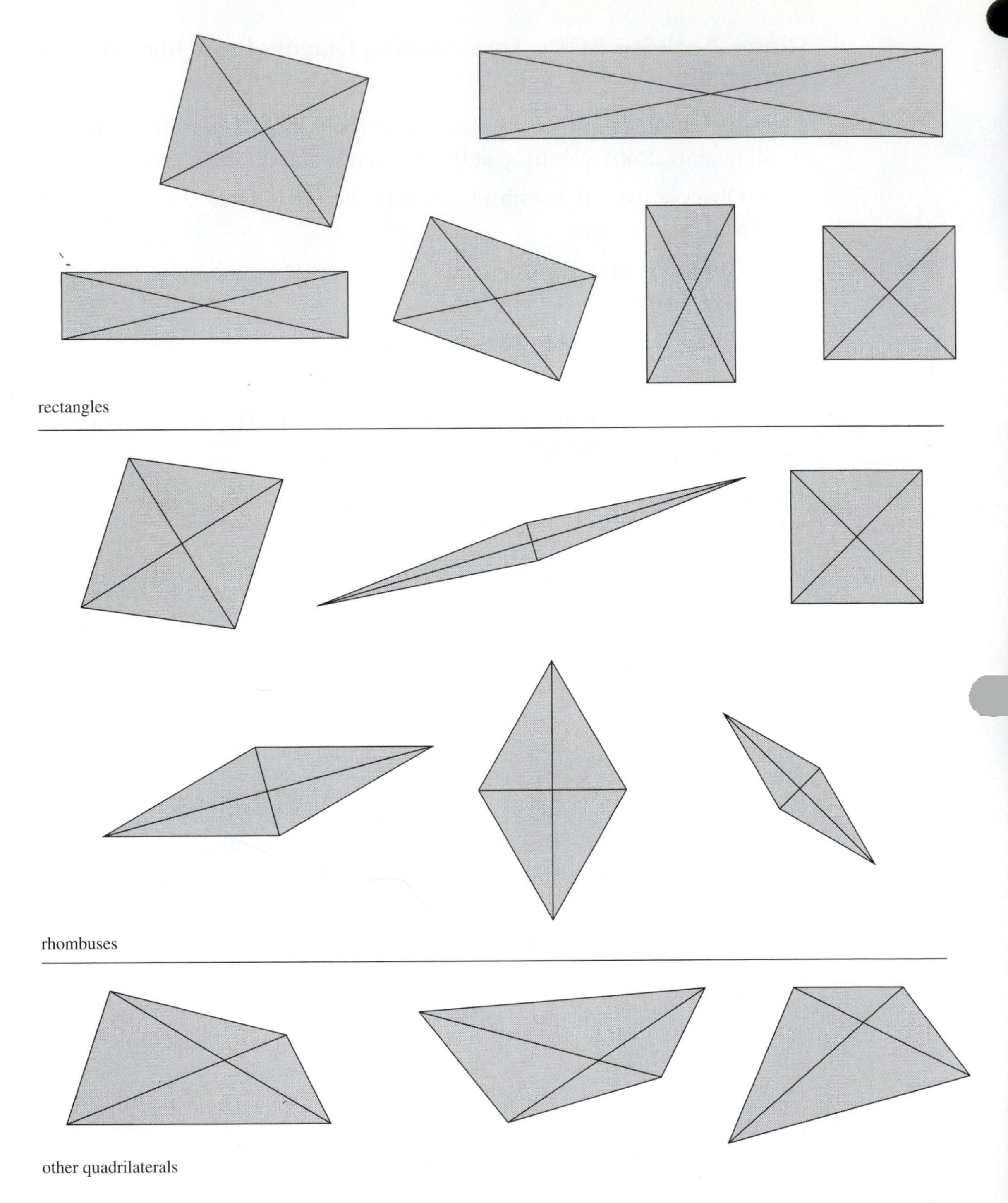

Figure 8CC.1
Some Quadrilaterials and Their Diagonals

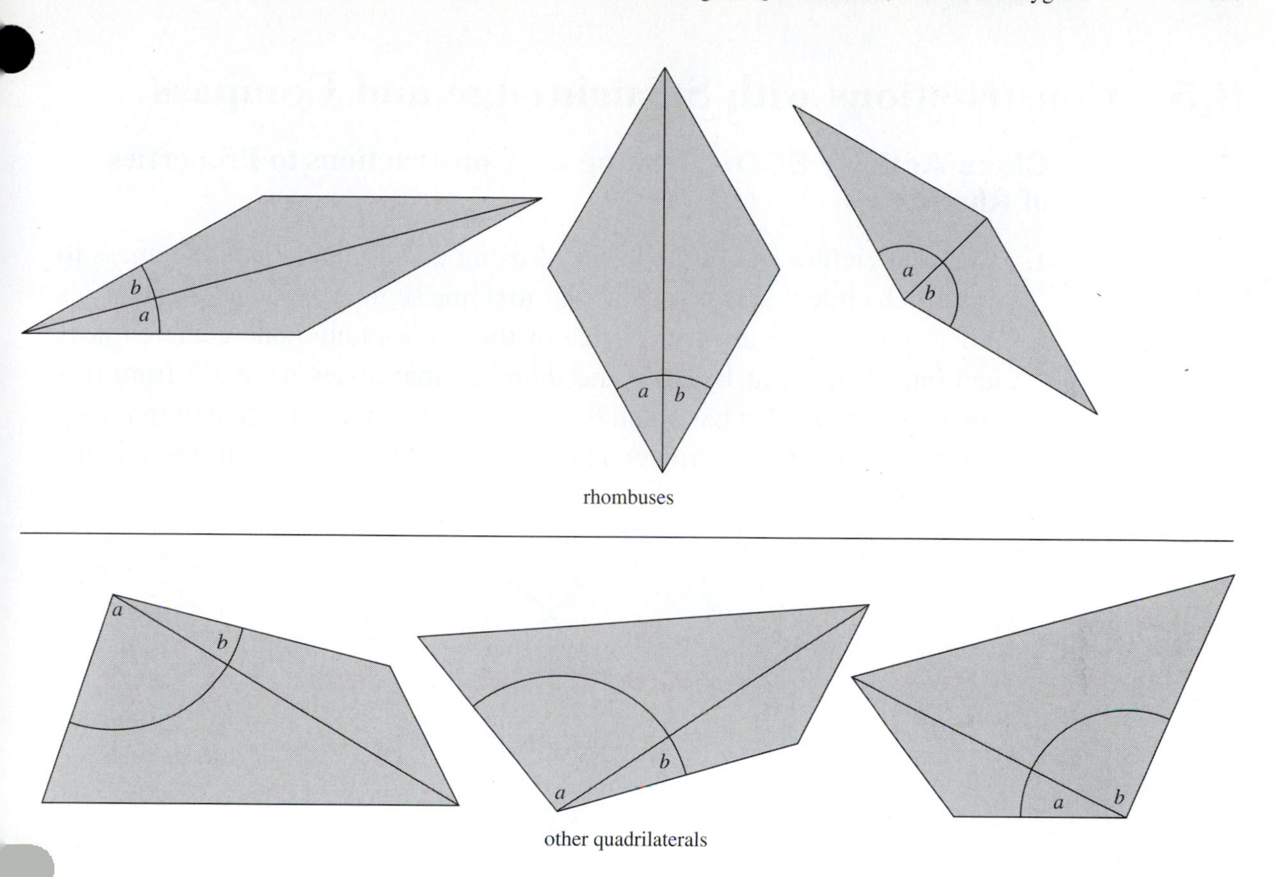

Figure 8CC.2

Some Quadrilaterals and the Angles Made by Their Diagonals

8.5 Constructions with Straightedge and Compass

Class Activity 8DD: Relating the Constructions to Properties of Rhombuses

1. The next picture shows the result of using a straightedge and compass to construct a line that is perpendicular to a line segment AB and that divides AB in half. There are two circles of the same radius, one centered at A and one centered at B. Draw the rhombus that arises naturally from this construction and that has A and B as vertices. Use the definition of rhombus to explain *why* the quadrilateral that you identify as a rhombus really must be a rhombus.

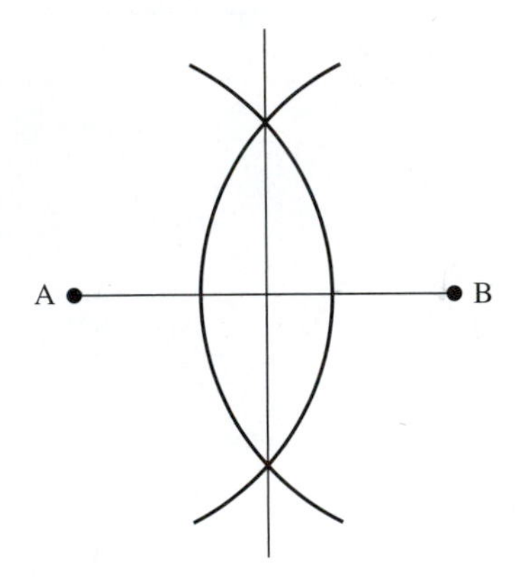

2. Which special properties of rhombuses are related to the construction of a line that is perpendicular to a line segment and divides the line segment in half? Explain.

3. The next picture shows the result of using a straightedge and compass to construct a ray that divides an angle in half. There are three circles of the same radius, one centered at P, one centered at Q, and one centered at R. Draw the rhombus that arises naturally from this construction. Use the definition of rhombus to explain *why* the quadrilateral that you identify as a rhombus really must be a rhombus.

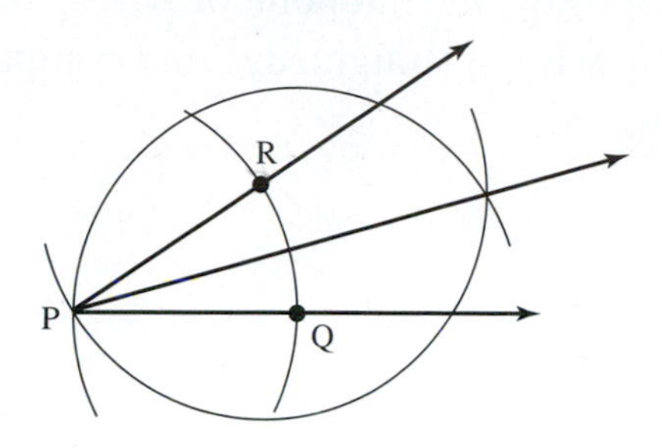

4. Which special properties of rhombuses are related to the construction of a ray that divides an angle in half? Explain.

Class Activity 8EE: Constructing a Square and an Octagon with Straightedge and Compass

Use a straightedge and compass to do the following constructions:

1. Using a straightedge and compass, construct a line that is perpendicular to the line segment AB shown in the next figure and that passes through point A (*not* through the midpoint of AB!). *Hint*: First extend the line segment AB.

Now use a straightedge and compass to construct a square that has AB as one side.

2. Using a straightedge and compass, construct an octagon whose vertices all lie on the circle in the next figure and whose sides all have the same length.

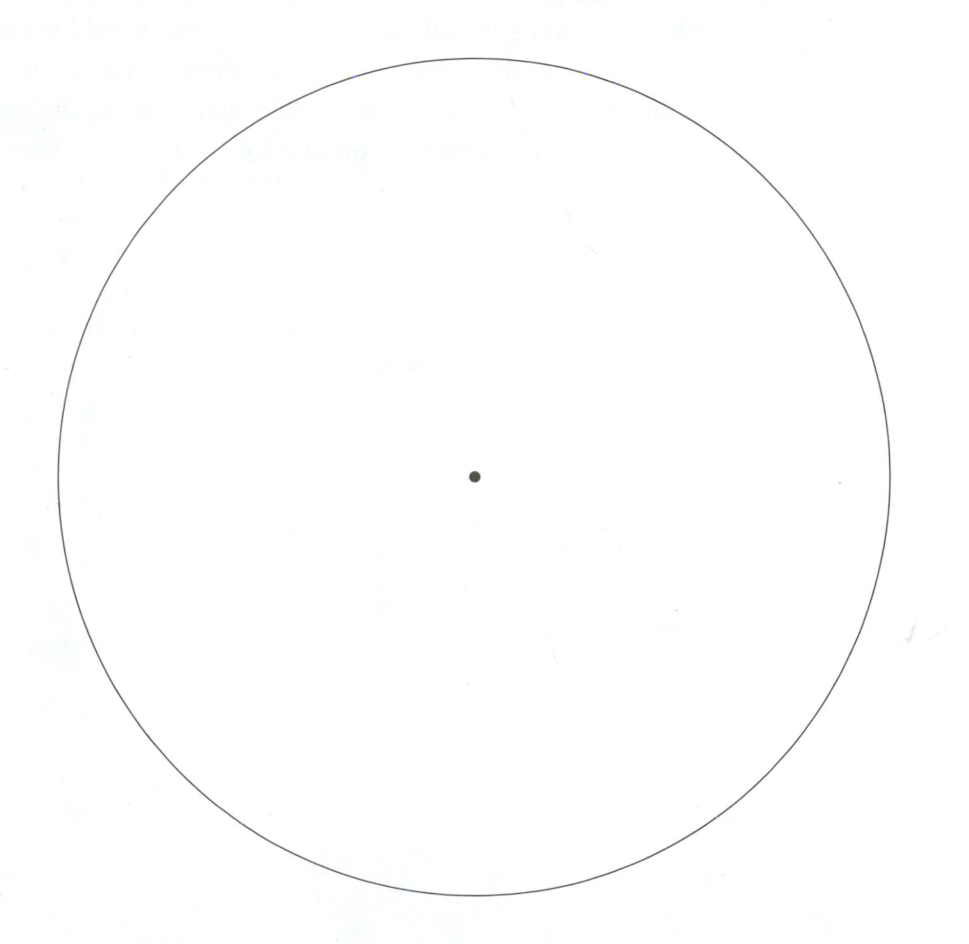

8.6 Polyhedra and Other Solid Shapes

Class Activity 8FF: Patterns for Prisms, Cylinders, Pyramids, and Cones

You will need scissors, tape, a compass, a ruler, graph paper, and blank paper for Problems 2–5 of this activity.

1. Examine the patterns in the next set of figures. Try to visualize what would happen if you were to cut these patterns out on the heavy lines, fold them on the dotted lines, and tape various sides together. What kinds of polyhedra would they make? Which parts of the patterns would make the bases?

   Now cut out the large versions of these patterns on pages 599, 601, 603, and 605 on the heavy lines, fold them on the dotted lines, and tape (or hold) the sides together to form various polyhedra. Were your predictions correct?

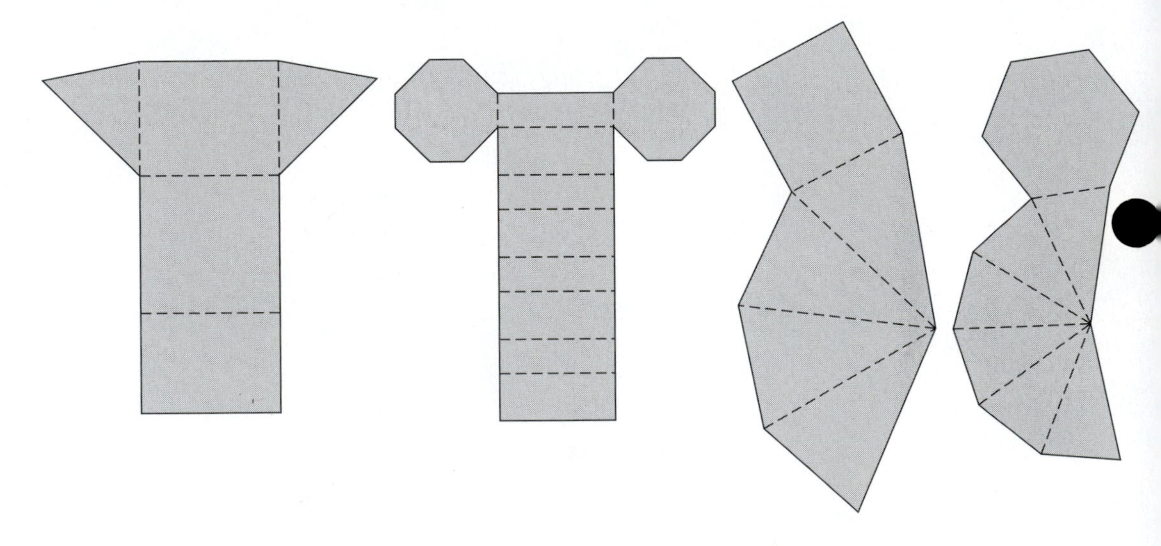

2. Make a pattern for a cylinder with a circular base. Make your pattern so that when you cut it out you will not need to overlap any paper to make your cylinder. You may leave the bases off your pattern. Cut out your pattern and see if it works.

3. Use graph paper to make a pattern for a prism whose base is a rectangle that is not a square. Include the bases in your pattern. Cut out your pattern and see if it works.

4. Make a pattern for a cone with a circular base. Construct your pattern so that when you cut it out you will not need to overlap any paper to make your cone. You may leave the base off your pattern. Cut out your pattern and see if it works.

5. Use a straightedge and compass to make a pattern for a pentagon-based pyramid. You may leave the pentagon base off your pattern. Cut out your pattern and see if it works.

Class Activity 8GG: Making Prisms and Pyramids

You will need either toothpicks and miniature marshmallows or modeling clay, or straws and regular-size marshmallows or modeling clay for this activity. If it is available, you could also use soap bubble liquid.

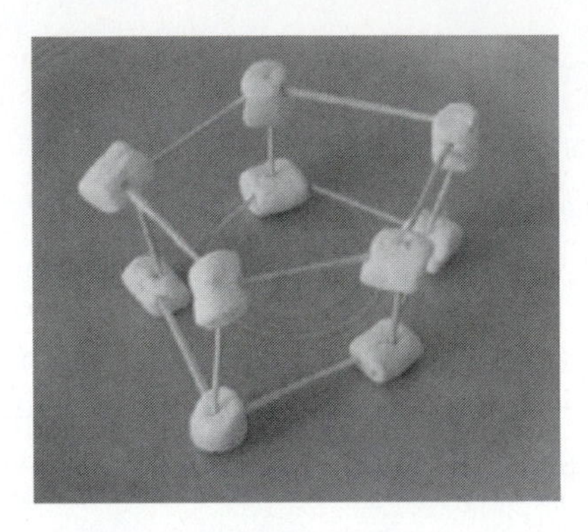

As in the photograph, stick toothpicks into miniature marshmallows or into small balls of modeling clay to make the shapes listed. To make larger shapes, stick straws into balls of clay or into regular-size marshmallows. In each case, visualize the shape first and predict how many toothpicks and marshmallows you will need for the following:

1. a rectangular prism

2. a triangular prism

3. a pyramid with a triangle base

4. a pyramid with a square base

Some teachers like to dip these shapes into a soap bubble solution to show the faces of the shapes.

Class Activity 8HH: Analyzing Prisms and Pyramids

1. Answer the next group of questions *without* using a model. Use your visualization skills and look back at other models. (When you are done, verify with a model if one is available.)

   How many faces does a pentagonal prism have? Why? Describe the faces of a pentagonal prism. What kinds of shapes are they? How many of each kind of shape are there?

   How many edges and how many vertices does a pentagonal prism have? Explain.

2. Answer the next group of questions *without* using a model. Use your visualization skills and look back at other models. (When you are done, verify with a model if one is available.)

   How many faces does a pyramid with a hexagonal base have, assuming that you count the base? Why? Describe the faces of a pyramid with a hexagonal base. What kinds of shapes are they? How many of each kind of shape are there?

   How many edges and how many vertices does a pyramid with a hexagonal base have? Explain.

Class Activity 8II: What's Inside the Magic 8 Ball?

You will need a Magic 8 Ball for this activity. (One or more can be shared by a class.) Magic 8 Balls are typically available where popular children's toys are sold. You will also need scissors and tape or snap-together plastic triangles.

1. Play with a Magic 8 Ball for a little while. Generally speaking, how does the Magic 8 Ball work? What sorts of things are inside it? Why do we see different "answers" at different times?

2. Other than breaking the Magic 8 Ball, how might you determine the shape that is inside the Magic 8 Ball?

3. Describe some properties that you think the shape inside the Magic 8 Ball should have.

4. Use snap-together plastic triangles, or use the paper triangles on page 607 to create various closed 3-dimensional shapes. Which of these shapes could possibly be the shape inside the Magic 8 Ball?

Class Activity 8JJ: Making Platonic Solids with Toothpicks and Marshmallows

You will need toothpicks and either miniature marshmallows or modeling clay for this activity.

Make all five Platonic solids by sticking toothpicks into miniature marshmallows or small balls of clay. (Your dodecahedron may sag a little, especially if you use marshmallows.) Refer to the following descriptions of the Platonic solids:

Tetrahedron is made of 4 equilateral triangles, with 3 triangles coming together at each vertex.

Cube is made of 6 squares, with 3 squares coming together at each vertex.

Octahedron is made of 8 equilateral triangles, with 4 triangles coming together at each vertex.

Dodecahedron is made of 12 regular pentagons, with 3 pentagons coming together at each vertex.

Icosahedron is made of 20 equilateral triangles, with 5 triangles coming together at each vertex.

Use your models to fill in the following table:

SHAPE	# AND TYPE OF FACES	# OF EDGES	# OF VERTICES
Tetrahedron	4 equilateral triangles		
Cube	6 squares		
Octahedron	8 equilateral triangles		
Dodecahedron	12 regular pentagons		
Icosahedron	20 equilateral triangles		

Class Activity 8KK: Why Are There No Other Platonic Solids?

You will need scissors and tape for this activity. Snap-together plastic triangles, squares, and pentagons could also be helpful.

1. Each vertex of a tetrahedron has 3 equilateral triangles coming together. Each vertex of an octahedron has 4 equilateral triangles coming together. Each vertex of an icosahedron has 5 equilateral triangles coming together. Can there be a convex polyhedron whose faces are equilateral triangles and for which 6 equilateral triangles come together at each vertex? Investigate this by working with paper or plastic equilateral triangles. (Page 607 has triangles you can cut out.) Explain your conclusion.

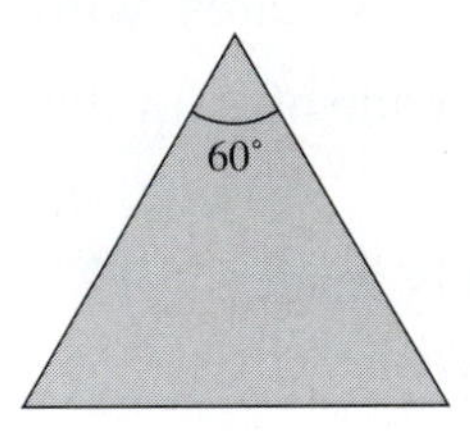

2. What if there were a polyhedron for which 7 or more equilateral triangles came together at each vertex? Could such a polyhedron be convex? Investigate this by working with paper or plastic equilateral triangles. Explain your conclusion.

3. At each vertex of a cube, there are 3 squares coming together. Can there be a convex polyhedron whose faces are squares and for which 4 squares come together at each vertex? Is it possible to make a convex polyhedron in such a way that 5 or more squares come together at each vertex? Investigate these questions by working with paper or plastic squares. (Page 609 has squares you can cut out.) Explain your conclusions.

4. At each vertex of a dodecahedron, there are 3 regular pentagons coming together. Is it possible to make a convex polyhedron in such a way that 4 or more regular pentagons come together at each vertex? Investigate this question by working with paper or plastic regular pentagons. (Page 609 has pentagons you can cut out.) Explain your conclusions.

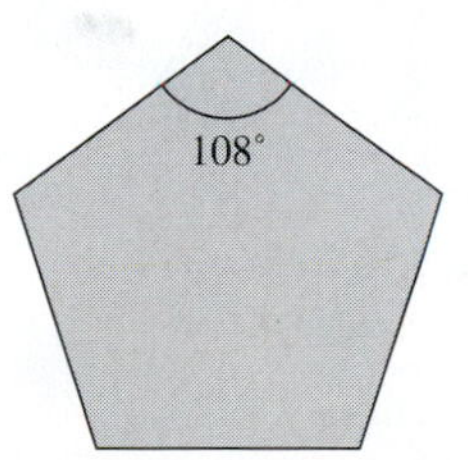

5. Can there be a convex polyhedron whose faces are all regular hexagons and for which 3 hexagons come together at each vertex? Investigate this question by working with paper or plastic regular hexagons. (Page 611 has hexagons you can cut out.) Explain your conclusion.

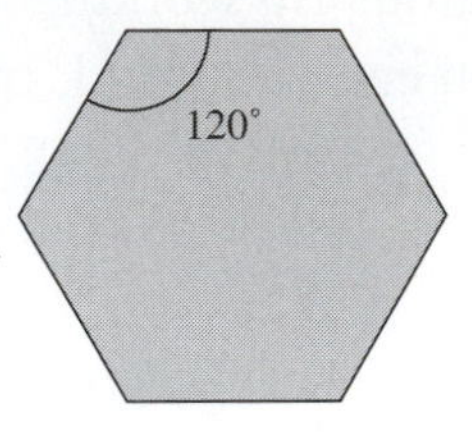

6. Is it possible to make a convex polyhedron whose faces are all regular hexagons? Consider the same question for regular 7-gons, 8-gons, etc. (Page 611 has hexagons, 7-gons, and octagons that you can cut out.)

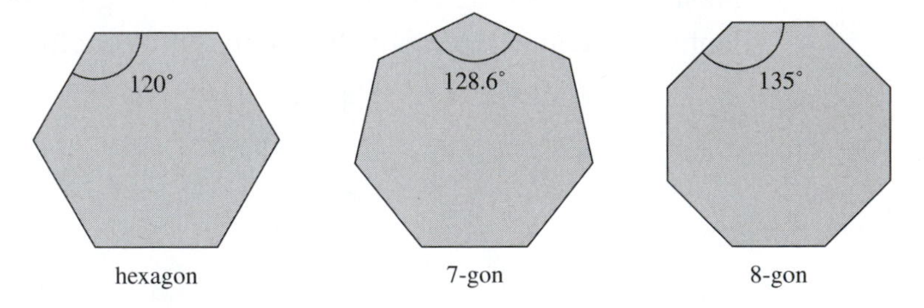

Class Activity 8LL: * Relating the Numbers of Faces, Edges, and Vertices of Polyhedra

To do this activity, you will need to have made a polyhedron, either out of paper; out of plastic, snap-together triangles, squares, or pentagons; or out of a combination of these polygons.

1. Count the number of faces, edges, and vertices of your polyhedron. Following is a table of numbers of faces, edges, and vertices of shapes that other students have made in the past, where you can record your data and the data of some of your classmates in the blank spaces:

	# FACES	# EDGES	# VERTICES
yours			
partner			
partner			
partner			
partner			
partner			
shape 1	14	21	9
shape 2	10	15	7
shape 3	24	36	14
shape 4	4	6	4
shape 5	8	12	6
shape 6	20	30	12

2. There is an interesting relationship between the numbers of faces, edges, and vertices. Can you find it?

   The previously recorded data (for shapes 1–6) are from polyhedra made only out of triangles. For polyhedra made only out of triangles, there is a special relationship between the number of faces and the number of edges that other polyhedra do not share. Can you find it? Challenge: Explain why this relationship must hold.

9

Geometry of Motion and Change

9.1 Reflections, Translations, and Rotations

Class Activity 9A: Exploring Rotations

You will need scissors for this activity.

1. Cut out the large circle and the two copies of a shape on page 613.
2. Place one copy of the shape on one side of your circle. You will be rotating your circle by 180° (a half turn) about its center. But before you rotate your circle, try to visualize the final position and orientation of your shape. Place the second copy of the shape on the circle to show how you think the first shape will be oriented after you rotate the circle 180°.

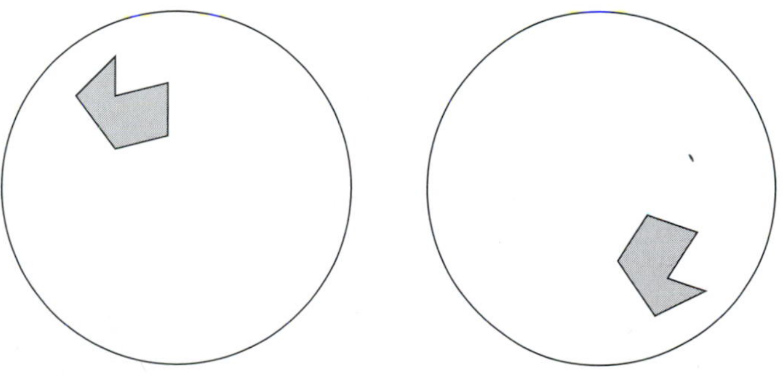

Will a half turn produce this?

3. Leave the first shape on the circle. Without changing the orientation of the second shape, move it to the side of the circle. Hold the center of the circle down with the tip of a pencil and rotate the circle 180° about its center. Did you correctly predict the final orientation of your shape? If so, the first and second shapes will be oriented in the same way.

4. Repeat the previous steps, but this time with your shape in a different orientation or in a different location on the circle. Keep repeating until you can accurately predict the final orientation of the shape after the circle is rotated.
5. Repeat the previous steps, but this time with a 90° rotation (a quarter turn), either clockwise or counterclockwise, instead of a 180° rotation.

Class Activity 9B: Exploring Reflections

You will need scissors, and reusable adhesive clay or paper clips, for this activity.

1. Cut out the large circle and the two copies of a shape on page 613. You may want to shade the backs of the shapes so that they look the same (more or less) on the front and back. Fold your circle in half to form a crease line that goes through the center of the circle.
2. Place one copy of the shape somewhere on the circle. You will be reflecting the shape across the crease line in the circle. But before you reflect the shape, try to visualize its final position and orientation. Place the second copy of the shape on the circle to show where you think the first shape will be located and how you think the first shape will be oriented after you reflect it across the line. Attach the second shape to the circle with adhesive clay.

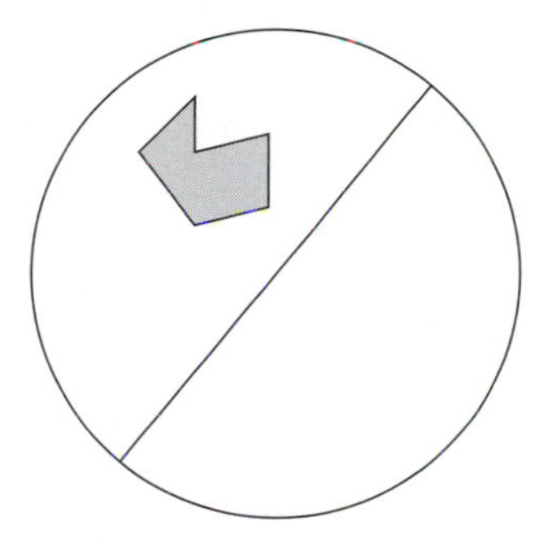

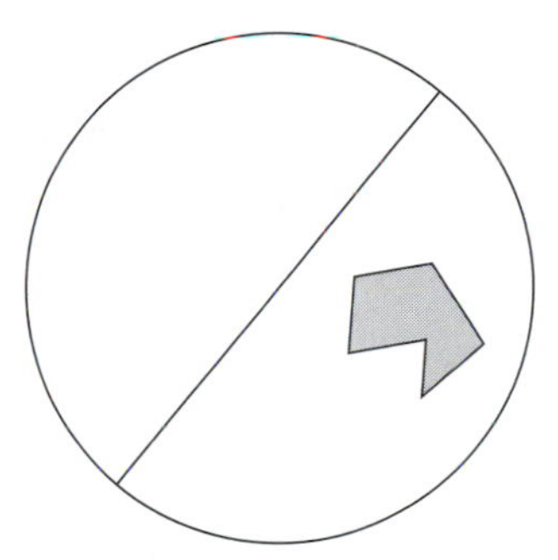

Will reflecting across the line produce this?

3. Now reflect your first shape across the line by holding the shape in place, folding the circle along the line (with the shapes on the inside), and transferring the shape to the other side of the circle. Did you correctly predict the final position and orientation of your shape? If so, the two shapes will match one another exactly.
4. Repeat the previous steps, but this time with your shape in a different orientation or in a different location. Also, change the orientation of the crease line: Make it horizontal, vertical, and diagonal.

Class Activity 9C: Exploring Reflections with Geometer's Sketchpad

For this activity, you will need to use the drawing tools for drawing circles, line segments, lines, rays, and points in Geometer's Sketchpad. You will also need to select objects with the selection tool.

1. Draw a shape or design that is not symmetrical, such as the one shown in the next figure. Then, separately, draw a line, a line segment, or a ray. Select the line, line segment, or ray, and then choose *Mark Mirror* from the *Transform* menu.

2. Your next step will be to reflect your shape across your line. Before you continue, try to visualize the location and orientation of the reflected shape.
3. Select your shape, and then choose *Reflect* from the *Transform* menu. Geometer's Sketchpad shows you both the initial location of your shape and its final location after reflection across your marked line.
4. Move your original shape and your marked line around on the screen, and observe the new locations and orientations of the reflected shape. Be sure to change the direction of your marked line, too. Why does the final position of your shape turn out the way it does? Before you move your original shape or your marked line, try to visualize how the location and orientation of the reflected shape will change.

Class Activity 9D: Exploring Translations with Geometer's Sketchpad

For this activity, you will need to use the drawing tools for drawing circles, line segments, lines, rays, and points in Geometer's Sketchpad. You will also need to select objects with the selection tool.

1. Draw a shape or design that is not symmetrical, such as the one shown in the next figure. Then, separately, draw a line segment. Select the two endpoints of this line segment and choose *Mark Vector* from the *Transform* menu.

2. Your next step will be to translate your shape with the direction and distance specified by your line segment. Before you continue, try to visualize the location and orientation of the translated shape.
3. Select your shape, and then choose *Translate* from the *Transform* menu. Geometer's Sketchpad shows you both the initial location of your shape and its final location after translation by the distance and direction determined by your line segment.
4. Move your original shape around on the screen, and change the length and direction of your line segment. Why does the final position of your shape turn out the way it does? Before you move your original shape or your line segment, try to visualize how the location of the translated shape will change.

Class Activity 9E: Exploring Rotations with Geometer's Sketchpad

For this activity, you will need to use the tools for drawing circles, line segments, lines, rays, and points in Geometer's Sketchpad. You will also need to select objects with the selection tool.

Exploring rotations by a fixed angle:

1. Draw a shape or design that is not symmetrical, such as the one shown in the next figure. Then, separately, draw a single point. Select that point, and then choose *Mark Center* from the *Transform* menu.

2. Your next step will be to rotate your shape about your chosen point by an angle that you will specify. Before you continue, try to visualize the location and orientation of your shape after rotation.
3. Select your shape, and choose *Rotate* from the *Transform* menu. A dialog box will appear. Select *By fixed angle*, and enter any angle you wish in the appropriate place. (If the dialog box indicates that angles are being measured in radians, then choose *Preferences* from the *Display* menu, and change the angle unit to degrees using the "pull-down" menu.) Geometer's Sketchpad shows you both the initial location of your shape and its final location after rotation about the point you marked as center, by the angle you selected.
4. Move your original shape, and move the point that you marked as center around your screen. Observe the location and orientation of your rotated shape. Why does the final position of your shape turn out the way it does? Before you move your original shape or your marked center, try to visualize how the location and orientation of the rotated shape will change.

Exploring rotations by a marked angle:

1. Draw a shape or design that is not symmetrical, such as the one shown in the next figure. Then, separately, draw a single point. Select that point, then choose *Mark Center* from the *Transform* menu.

2. Make an angle by drawing two line segments that share an endpoint. Select the three endpoints so that the second point you select is the shared endpoint. Choose *Mark Angle* from the *Transform* menu.
3. Your next step will be to rotate your shape about your marked point, by your marked angle. Before you continue, try to visualize the location and orientation of your rotated shape.
4. Select your shape, and choose *Rotate* from the *Transform* menu. Geometer's Sketchpad shows you both the initial location of your shape and its final location after rotating about your marked point, by your marked angle.
5. Move your original shape and your marked center point around the screen. Why does the final position of your shape turn out the way it does? Before you move your original shape or your marked center, try to visualize how the location and orientation of your rotated shape will change.
6. Now move one of the endpoints of your marked angle. Why does the final position of your shape move the way it does? Before you move your marked angle, try to visualize how the location and orientation of the rotated shape will change.

Class Activity 9F: Reflections, Rotations, and Translations in a Coordinate Plane

1. Draw the result of translating the shaded shapes in the next figure according to the direction and the distance given by the arrow. Explain how you know where to draw your translated shapes. It may help you to consider the coordinates of the vertices of the shapes.

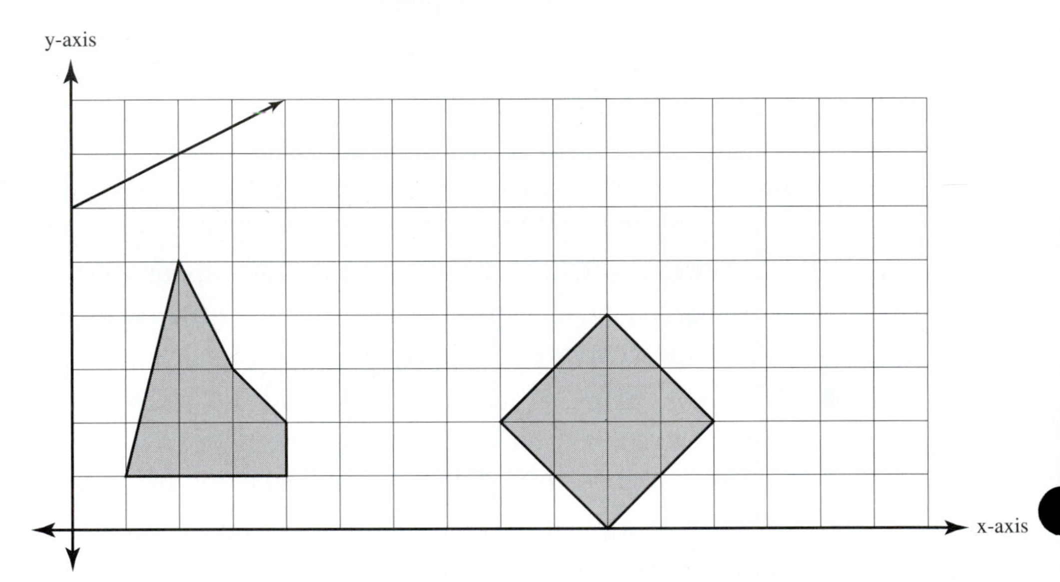

2. Draw the result of reflecting the shaded shapes in the next figure across the y-axis. Explain how you know where to draw your reflected shapes. It may help you to consider the coordinates of the vertices of the shapes.

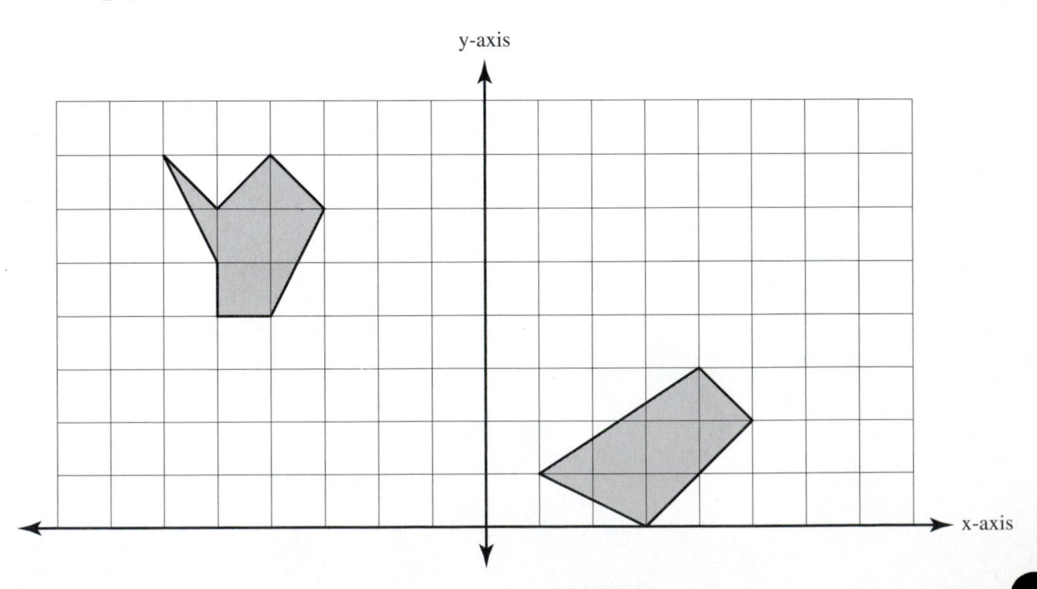

3. Draw the result of rotating the shaded shapes in the next figure by 180° around the origin, where the x- and y-axes meet. Explain how you know where to draw your rotated shapes. It may help you to consider the coordinates of the vertices of the shapes.

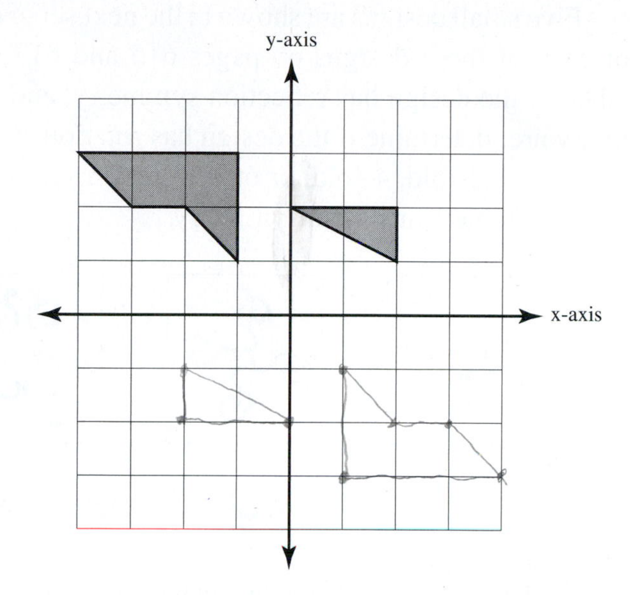

4. Draw the result of rotating the shaded shapes in the next figure by 90° counterclockwise around the origin, where the x- and y-axes meet. Explain how you know where to draw your rotated shapes. It may help you to consider the coordinates of the vertices of the shapes.

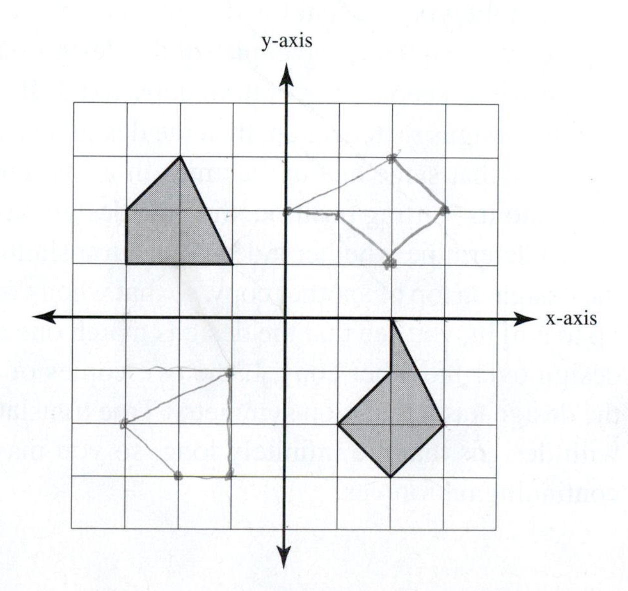

9.2 Symmetry

Class Activity 9G: Checking for Symmetry

You will need scissors, and a toothpick or paperclip, for this activity.

Five small designs are shown in the next set of diagrams. Cut out two copies of each of these designs on pages 615 and 617. For each design, determine whether the design has reflection symmetry, and if so, what the lines of symmetry are; determine if the design has rotation symmetry, and if so, whether it has 2-fold, 3-fold, 4-fold, or other rotation symmetry; and determine whether the design has translation symmetry.

To determine whether a design has reflection symmetry, see if there is a line such that, when you fold the paper along the line, the portions of the design on either side of the fold line match one another.

To determine whether a design has rotation symmetry, put one copy of the design on top of another copy, so that when you hold the two pieces of paper up to a light, you see that the designs match one another. Poke a toothpick or paperclip through the two copies of the design. Then rotate the top copy of the design while keeping the bottom copy fixed. If you can rotate less than 360° and the designs match again, then the design has rotation symmetry. If you can rotate by that same amount n times in a row and the top design then moves back into its starting position, then the design has n-fold rotation symmetry.

To determine whether a design has translation symmetry, put one copy of the design on top of another copy, so that when you hold the two pieces of paper up to a light, you see that the designs match one another. Slide one copy of the design over the other copy. If the two copies of the design match again, then the design has translation symmetry. True translation symmetry can occur only with designs that are infinitely long, so you may need to imagine the design continuing on forever.

Class Activity 9H: Frieze Patterns

You will need scissors for this activity.

Seven different frieze patterns are shown next. Cut out two copies of each of these seven frieze patterns on pages 619, 621, 623, and 625 by cutting along the gray lines. Picture each of the seven frieze patterns repeating on an infinitely long strip of paper, continuing forever, forward and backward.

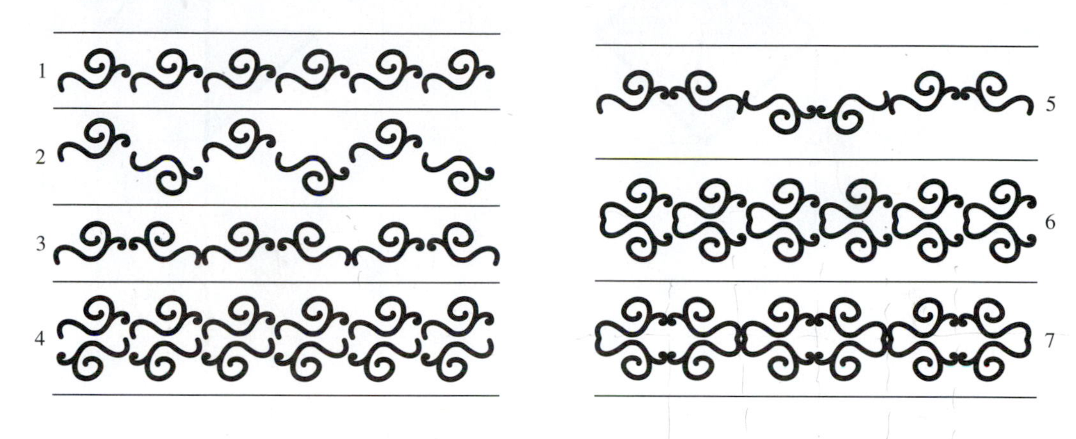

Show that each frieze pattern has translation symmetry. For each frieze pattern, determine whether the pattern has reflection symmetry and, if so, what the lines of symmetry are; determine whether the pattern has rotation symmetry and, if so, whether it has 2-fold, 3-fold, 4-fold, or other rotation symmetry; and determine whether the pattern has glide-reflection symmetry.

See the text or see Class Activity 9G to learn how to check for translation, rotation, and reflection symmetry.

To check for glide-reflection symmetry, put one copy of the pattern on top of another copy, so that when you hold the two pieces of paper up to a light, you see that the patterns match one another. Flip the top copy over, so that the two patterns are face to face. Slide one copy of the pattern over the other copy. If the two copies of the pattern can be made to match one another again, then the pattern has glide-reflection symmetry. As with translation symmetry, true glide-reflection symmetry can occur only with patterns that are infinitely long, so you will need to imagine the pattern continuing on forever.

Class Activity 9I: Traditional Quilt Designs

1. Complete the next four traditional quilt patch designs so that each one has both a horizontal and a vertical line of symmetry.

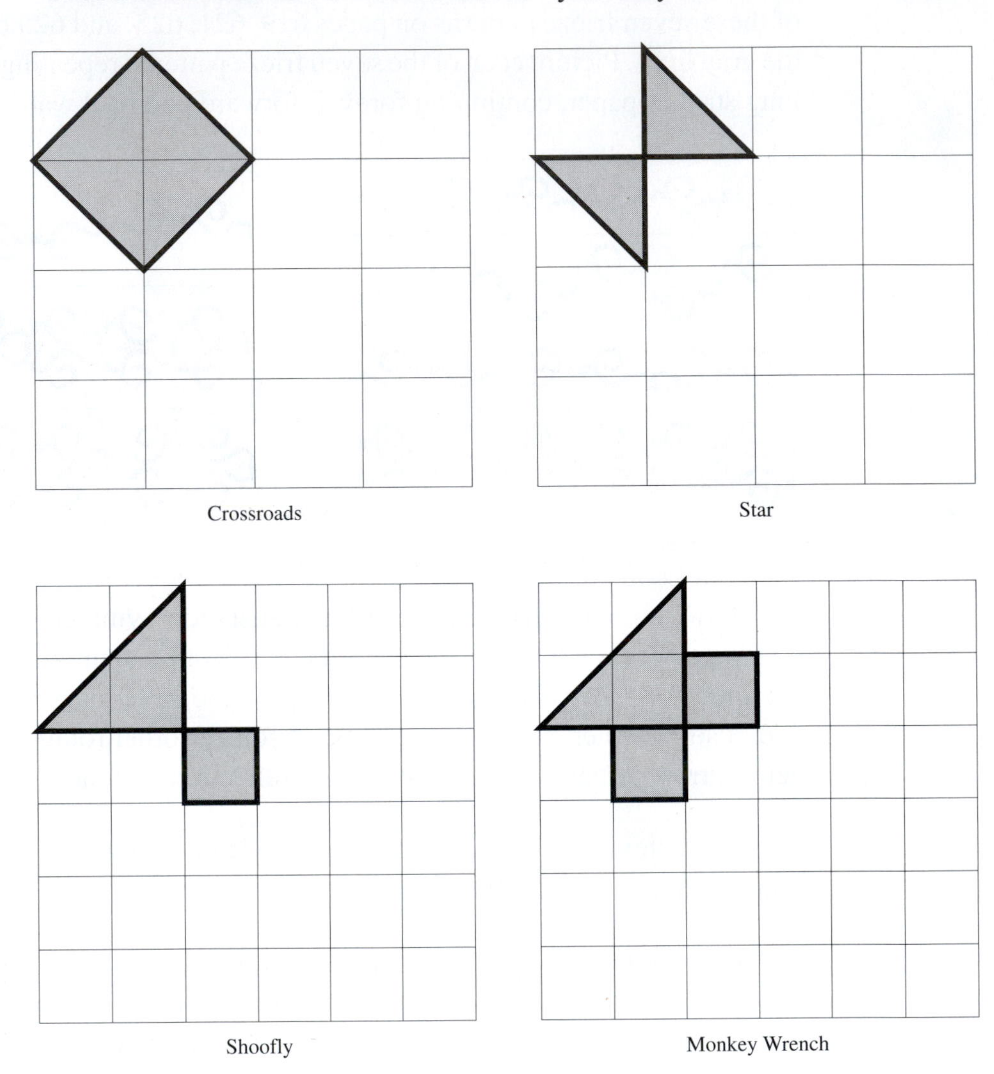

2. Complete the next four traditional quilt patch designs so that each one has 4-fold rotation symmetry.

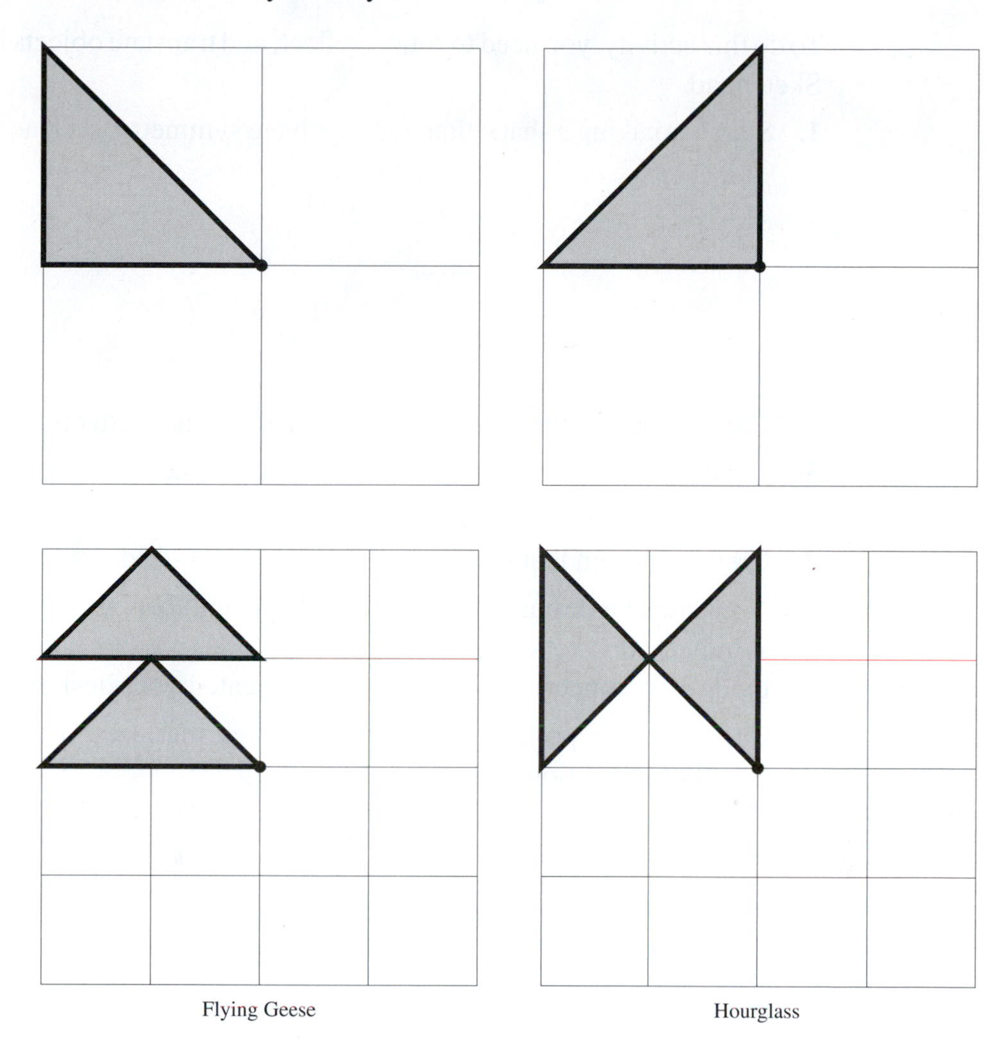

Class Activity 9J: Creating Symmetrical Designs with Geometer's Sketchpad

To do this activity, you need to rotate, reflect, and translate objects in Geometer's Sketchpad.

1. Start by making a shape that *does not* have symmetry, such as the following:

2. Create a design that uses your shape and that has reflection symmetry.
3. Create a finite portion of an infinite design that uses your shape and that has translation symmetry.
4. Create a design that uses your shape and has 6-fold rotation symmetry.
5. Now create a design that uses your shape and that has *both* 3-fold rotation symmetry *and* reflection symmetry. The design as a whole must have both types of symmetry. Describe how you created your design.
6. Create a finite portion of an infinite design that uses your shape and that has glide-reflection symmetry.

Class Activity 9K: Creating Symmetrical Designs (Alternate)

You will need graph paper for this activity.

1. On graph paper, draw a simple asymmetrical shape, such as this asymmetrical triangle.

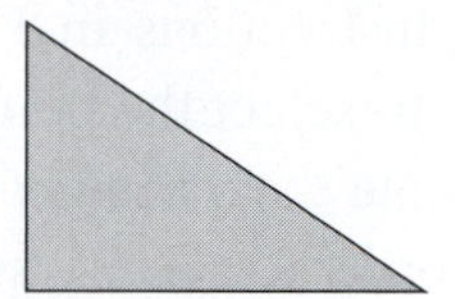

2. By drawing copies of your shape, create a finite portion of an infinite design that has translation symmetry.
3. By drawing copies of your shape, create a design that has 4-fold rotation symmetry.
4. By drawing copies of your shape, create a design that has *both* 4-fold rotation symmetry *and* reflection symmetry. The design as a whole must have both types of symmetry simultaneously. Describe how you created your design.
5. By drawing copies of your shape, create a finite portion of an infinite design that has glide-reflection symmetry.

Class Activity 9L: ✲ Creating Escher-Type Designs with Geometer's Sketchpad (for Fun)

To do this activity, you must rotate and translate objects in Geometer's Sketchpad.

Follow the instructions in Figure 9L.1 to create your own Escher-type design. Be sure to select the points, and not just the line segments, when you rotate and translate the various portions of the design. You will need some of those points in order to specify subsequent rotations or translations.

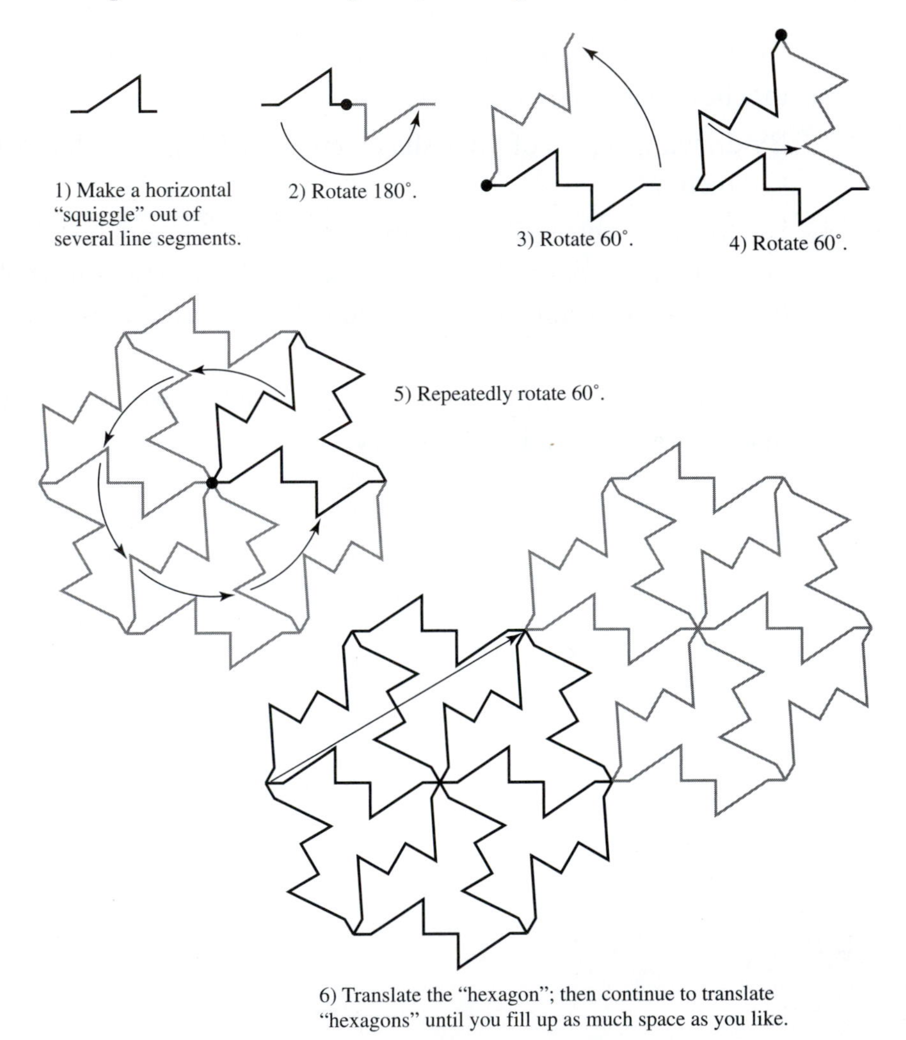

1) Make a horizontal "squiggle" out of several line segments.
2) Rotate 180°.
3) Rotate 60°.
4) Rotate 60°.
5) Repeatedly rotate 60°.
6) Translate the "hexagon"; then continue to translate "hexagons" until you fill up as much space as you like.

Figure 9L.1

How to Make an Escher-Type Design

Class Activity 9M: Analyzing Designs

1. Describe one or more ways to create the design in Figure 9M.3 by starting with the basic square design in Figure 9M.2 and using reflections, rotations, and translations.

2. Cut out the (enlarged) copies on page 627 of the basic square design of Figure 9M.2 and its reflection. Use some or all of your cut-out copies to create your own design by using rotations, reflections, and translations. Describe how you used rotations, reflections, and translations to achieve your final design.

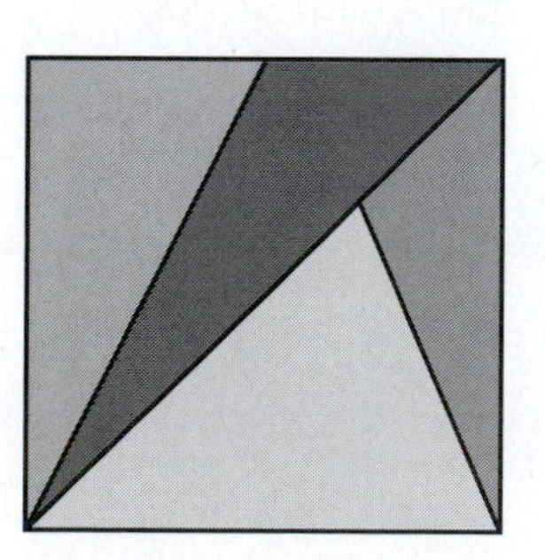

Figure 9M.2
A Square Design

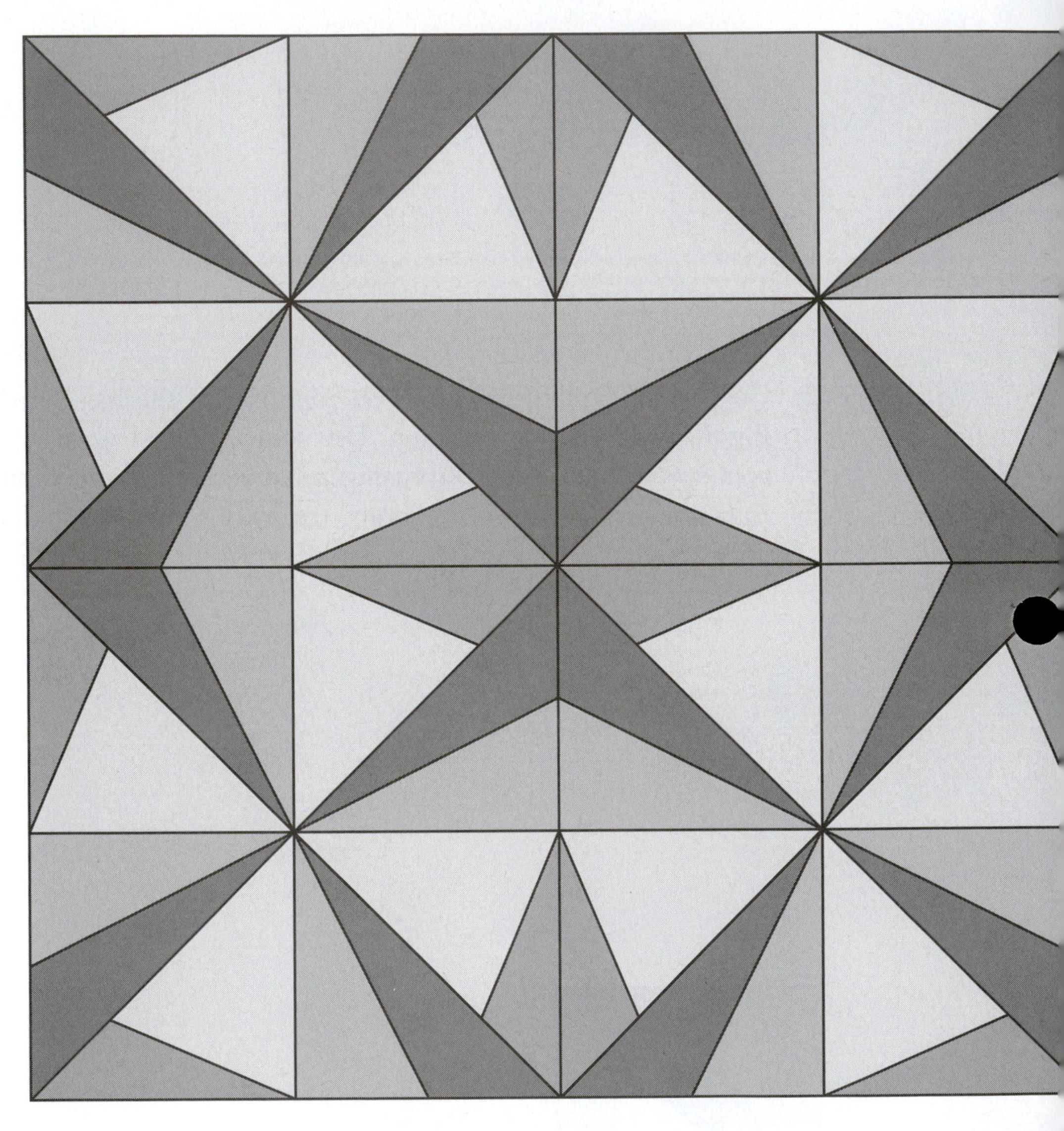

Figure 9N
A Des

9.3 Congruence

Class Activity 9N: Triangles and Quadrilaterals of Specified Side Lengths

You will need scissors, several straws, and some string for this activity.

1. Cut a 3-inch, a 4-inch, and a 5-inch piece of straw, and thread all three straw pieces onto a piece of string. Tie a knot so as to form a triangle from the three pieces of straw.
2. Now cut two 3-inch pieces of straw and two 4-inch pieces of straw, and thread all four straw pieces onto another piece of string in the following order: 3-inch, 4-inch, 3-inch, 4-inch. Tie a knot so as to form a quadrilateral from the four pieces of straw.
3. Compare your straw triangle and your straw quadrilateral. What is an obvious difference between them (other than the fact that the triangle is made of three pieces and the quadrilateral is made of four)?
4. When you made your triangle, if you had strung your three pieces of straw in a different order, would your triangle be different, or not?

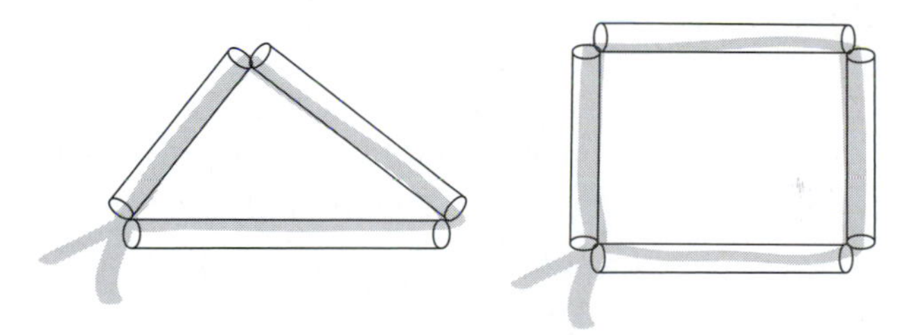

Class Activity 9O: Describing a Triangle

You will need a compass, protractor, and ruler for this activity.

1. Work with a partner. Draw a triangle on a piece of paper, but keep it hidden from your partner. Label the vertices of your triangle A, B, and C. Give your partner 3 numerical pieces of information about the angles or side lengths of your triangle *other than* the lengths of *all three* sides of your triangle. (Your partner should do the same for you.) Record the information you gave your partner.

2. Let your partner try to draw a triangle by using the information you gave her or him. Is your partner's triangle congruent to yours? Use the information your partner gave you to draw a triangle. Is this triangle congruent to your partner's triangle?

3. What if your partner gave you the information that her or his triangle has angles 20°, 70°, and 90°; would that be enough information to recreate your partner's triangle? Why, or why not?

4. What if your partner gave you the information that her or his triangle ABC has angle 20° at point A, that AB is 2 inches long, and that BC is 1 inch long; would that be enough information to recreate your partner's triangle? Why, or why not?

Class Activity 9P: Triangles with an Angle, a Side, and an Angle Specified

You will need a protractor and one or two straightedges or rulers for this activity.

1. Create a triangle that has AB (in the next figure) as one side, a 30° angle at A, and a 45° angle at B. Is there more than one way to do this? If so, compare the different triangles.

2. Create a triangle that has *AB* (in the next figure) as one side, a 100° angle at *A*, and a 30° angle at *B*. Is there more than one way to do this? If so, compare the different triangles.

A ——————————————— *B*

3. Make up a problem like the previous two, but with different angles. Can you use any pair of angles, or is there some restriction on which pairs of angles you can specify at *A* and *B* in order to form a triangle? Explain.

A ——————————————— *B*

Class Activity 9Q: ✻ Using Triangle Congruence Criteria

1. We defined a parallelogram to be a quadrilateral for which opposite sides are parallel. When we look at parallelograms, it appears that opposite sides also have the same length. Use a triangle congruence criterion and facts we studied previously about angles to explain why opposite sides of a parallelogram really must have the same length. In order to do so, consider the two triangles formed by the diagonal.

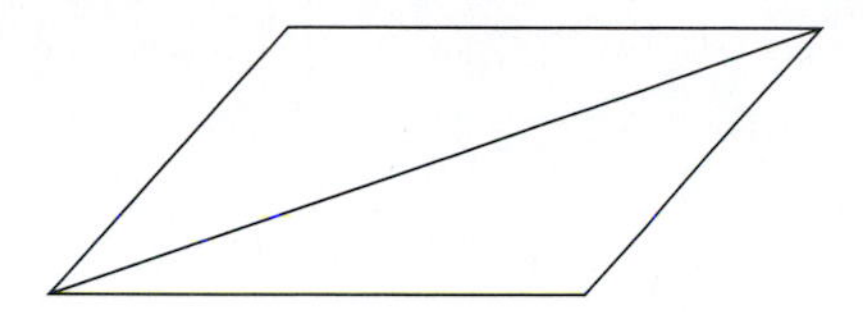

2. Here is an old-fashioned way to make a rectangular foundation for a house. Take a pair of identical pieces of wood for the length of the house and another pair of identical pieces of wood for the width of the house. Place the wood on the ground to show approximately where the foundation will go. The pieces of wood now form a quadrilateral whose opposite sides are the same length. Measure the two diagonals of the quadrilateral, and keep adjusting the quadrilateral until the two diagonals are the same length. Explain why the quadrilateral must now be a rectangle. (Remember that we defined rectangles to be quadrilaterals that have 4 right angles, so you must show that this quadrilateral has 4 right angles).

9.4 Similarity

Class Activity 9R: A First Look at Solving Scaling Problems

You have a poster that is 2 feet wide and 4 feet long. The poster has a simple design on it that you would like to scale up and draw onto a larger poster. The larger poster is to be 6 feet wide. How long should the poster be?

Find as many different ways as you can to solve this poster-scaling problem. In each case, explain your reasoning.

Class Activity 9S: Using the "Scale Factor," "Relative Sizes," and "Set up a Proportion" Methods

1. Suppose that you have a postcard with an attractive picture on it and that you would like to scale up this picture and draw it onto paper that you can cut from a roll. The roll of paper is 20 inches wide, and you can cut the paper to virtually any length. If the postcard picture is 4 inches wide and 6 inches long, then how long should you cut the 20-inch-wide paper? Assume that the 4-inch side will become 20 inches long.

Use three different methods to solve the postcard problem: the *scale factor* method, the *relative sizes* method, and the *set up a proportion* method. In each case, explain why the method makes sense. Explain the first two methods in as concrete a way as you can, as if you were teaching 5th or 6th graders who know about multiplication and division, but who do not know about setting up proportions.

Show how you can also apply the first two methods to the proportion you set up for the proportion method.

2. Decide whether the next problem is easier to solve with the scale factor method or with the relative sizes method. Explain your answer.

   A stuffed-animal company wants to produce an enlarged version of a popular stuffed bunny. The original bunny is 6 inches wide and 11 inches tall. The enlarged bunny is to be 33 inches tall. How wide should the enlarged bunny be?

3. Decide whether the next problem is easier to solve with the scale factor method or with the relative sizes method. Explain your answer.

   A toy company wants to produce a scale model of a car. The actual car is 6 feet wide and 15 feet long. The scale model of the car is to be $2\frac{1}{2}$ inches wide. How long should the scale model of the car be?

Class Activity 9T: A Common Misconception about Scaling

Johnny is working on the following problem:

The picture on a poster that is 4 feet wide and 6 feet long is to be scaled down and drawn onto a small poster that is 1 foot wide. How long should the small poster be?

Johnny solves the problem this way:

> One foot is 3 feet less than 4 feet, so the length of the small poster should also be 3 feet less than the length of the big poster. This means the small poster should be $6 - 3 = 3$ feet long.

Is Johnny's reasoning valid? Why, or why not? If not, how might you convince Johnny that his reasoning is not correct? What would be a correct way to solve the problem in that case?

Class Activity 9U: Using Scaling to Understand Astronomical Distances

Ms. Frizzle's class has been studying planets and stars. Ms. Frizzle wants to help the children get a better sense of astronomical distances by scaling down these distances. The table that follows shows the distances in kilometers from the sun to the earth, the sun to Pluto, and the sun to Alpha Centauri. Alpha Centauri is one of the closest stars to the sun.

Heavenly Body	Approximate Distance from Sun
Earth	150 million km
Pluto	5.9 billion km
Alpha Centauri	38 trillion km

If Ms. Frizzle represents the distance from the earth to the sun as 10 centimeters (about the width of a hand), then how should she represent the distance from Pluto to the sun? How should she represent the distance from Alpha Centauri to the sun? Explain your reasoning in such a way that 5th or 6th graders who understand multiplication and division, but who do not know about setting up proportions, might be able to understand.

Will Ms. Frizzle be able to show these distances in her classroom? (Recall that 100 centimeters = 1 meter, and 1 meter is about 1 yard. One thousand meters is 1 kilometer, and 1 kilometer is about $\frac{6}{10}$ of a mile, so a little over half a mile.)

Class Activity 9V: More Scaling Problems

1. A large American flag can be 5 feet tall by 9 feet 6 inches wide. Suppose you want to make a scaled-down version of the flag. If the smaller flag is to be 2 feet tall, then how long should this smaller flag be? Give your answer in feet and inches.

2. A museum wants to put a scaled copy of one of its paintings onto a 3-inch-by-5-inch card. The painting is 42 inches by 65 inches. Explain why the copy of the painting can't fill the whole card without leaving blank spaces (i.e., explain why there will have to be a border). Recommend to the museum a size for the copy of the painting that will fit on a $3'' \times 5''$ card. Show your recommendation here in the $3'' \times 5''$ rectangle.

Class Activity 9W: Measuring Distances by "Sighting"

To do this activity, you will need your own ruler, and one or more yard sticks and tape measures that can be shared by the class.

This activity will help you understand how the theory of similar triangles is used in finding distances by surveying.

1. Stand a yard stick on end on the edge of a chalkboard, or tape the yard stick vertically to the wall.

2. Stand back, away from the yard stick, in a location where you can see the yard stick. Your goal is to find your distance to the yard stick. Guess or estimate this distance before you continue.

   Your guess of how far away the yard stick is as follows:

3. Hold your ruler in front of you with an outstretched arm. Make the ruler vertical, so that it is parallel to the yard stick. Close one eye, and with your open eye, "sight" from the ruler to the yard stick. Use the ruler to determine how big the yard stick appears to be from your location.

   Record the apparent size of the yard stick here:

4. With your arm still stretched out in front of you, have a classmate measure the distance from your sighting eye to the ruler.

   Record the distance from your sighting eye to the ruler here:

5. Use the theory of similar triangles to determine your distance to the yard stick. Sketch your eye, the ruler, and the yardstick, showing the relevant similar triangles. (Your sketch does not need to be to scale.) Explain why the triangles are similar.

   Is your calculated distance close to your estimated distance? If not, which one seems to be faulty, and why?

6. Now move to a new location, and find your distance to the yard stick again with the same technique.

Class Activity 9X: Using Shadows to Determine the Height of a Tree

This activity requires several tape measures that can be shared by the class.

How could you find the height of a tree, for example, without measuring it directly? This class activity provides two ways to do this with *similar triangles*.

First Method: Go outside and find a tree whose height you will measure. Before you continue, guess or estimate the height of the tree.

This method will work only if your tree is on level ground and casts a fully visible shadow. (So you need a sunny day.)

1. Measure the length of the shadow of the tree (from the base of the tree to the shadow of the tip of the tree).

2. Measure the height of a classmate, and measure the length of that person's shadow.

3. Sketch the two similar triangles in this situation. Explain why the triangles are similar. Use your similar triangles to find the height of the tree.

   Is your calculated height fairly close to your estimated height of the tree? If not, which one do you think is faulty? Why?

Second Method: Go outside and find a tree whose height you will measure. Before you continue, guess or estimate the height of the tree.

As long as you can go outside and have tape measures that are long enough, this method will work. (If you can't go outside, you can use this method to measure the height of your room.)

1. Stand away from the tree, hold one arm out straight, close one eye, and raise your arm until you see the top of the tree aligned with your fingertips. Have a classmate measure the height from your fingertips straight down to the ground (while you are sighting the top of the tree) as well as the distance from your feet (directly below your eyes) to where the tape measure touches the ground. Also, have a classmate measure the height of your eyes. Then find your distance to the base of the tree. (You can do this either by measuring directly or by "pacing off" the distance and measuring your stride length.)

Record the following measurements:

a. Height of fingertips above the ground:

b. Distance *along the ground* from directly below your eyes to where the tape measure measuring the height of your fingertips hits the ground:

c. Height of eyes:

d. Distance to tree:

2. Now draw a diagram showing the relationships between the height of the tree, your arm, and your eyes, and showing all the various measurements you took. (Your drawing does not have to be to scale.) Find the height of the tree. (Look for similar triangles.)

Is your calculated height close to your estimated height of the tree? If not, which number seems to be faulty? Why?

10

Measurement

10.1 Fundamentals of Measurement

Class Activity 10A: The Biggest Tree in the World

Listed next are several different trees that could perhaps qualify as the biggest tree in the world. Compare these trees. Why can reasonable people differ about which tree is biggest?

Tree # 1: "General Sherman" is a giant sequoia located in Sequoia National Park in California. (See Figure 10A.1.) According to the National Park Service, General Sherman is 275 feet tall, has a circumference (at its base) of 103 feet, and has a volume of 52,500 cubic feet.

Tree # 2: "General Grant" is a giant sequoia located in Sequoia National Park in California. According to the National Park Service, General Grant is 268 feet tall, has a circumference (at its base) of 108 feet, and has a volume of 46,600 cubic feet.

Tree #3: "Mendocino Tree" is a redwood tree in Montgomery Redwoods State Reserve near Ukiah, California. It is 368 feet tall and has a diameter of 10.4 feet, which means that its circumference should be about 33 feet.

Tree #4: A Banyan tree in Calcutta, India, has a circumference of 1350 feet (meaning the circumference of the whole tree, not just the trunk) and covers three acres.

Tree #5: A tree in Santa Maria del Tule near Oaxaca, Mexico, is 130 feet tall and is described as requiring 40 people holding hands to encircle it.

Some websites have photos of these large trees. See
`www.aw-bc.com/beckmann`

Figure 10A.1
Is "General Sherman" the Biggest Tree in the World?

Class Activity 10B: What Do "6 Square Inches" and "6 Cubic Inches" Mean?

If available, square tiles that are 1-inch-by-1-inch and 1-centimeter-by-1-centimeter would be helpful for this activity. So would cubes that are 1-inch-by-1-inch-by-1-inch and 1-centimeter-by-1-centimeter-by-1-centimeter.

1. Discuss the following as clearly and concretely as you can:

a. What does it mean to say that a shape has an area of 6 square inches? Illustrate concretely with tiles, if available, or use the graph paper in Figure 10A.2.

b. What does it mean to say that a shape has an area of 12 square centimeters? Illustrate concretely with tiles, if available, or use the graph paper in Figure 10A.3.

2. If we weren't thinking carefully, we might try to illustrate 6 square inches by drawing a square that is 6 inches wide and 6 inches long. Why is it easy to think that such a square has area 6 square inches, and why is this *not correct*?

3. People sometimes say, "Area is length times width." Why is it not correct to characterize area this way?

Figure 10A
Inch Graph Paper

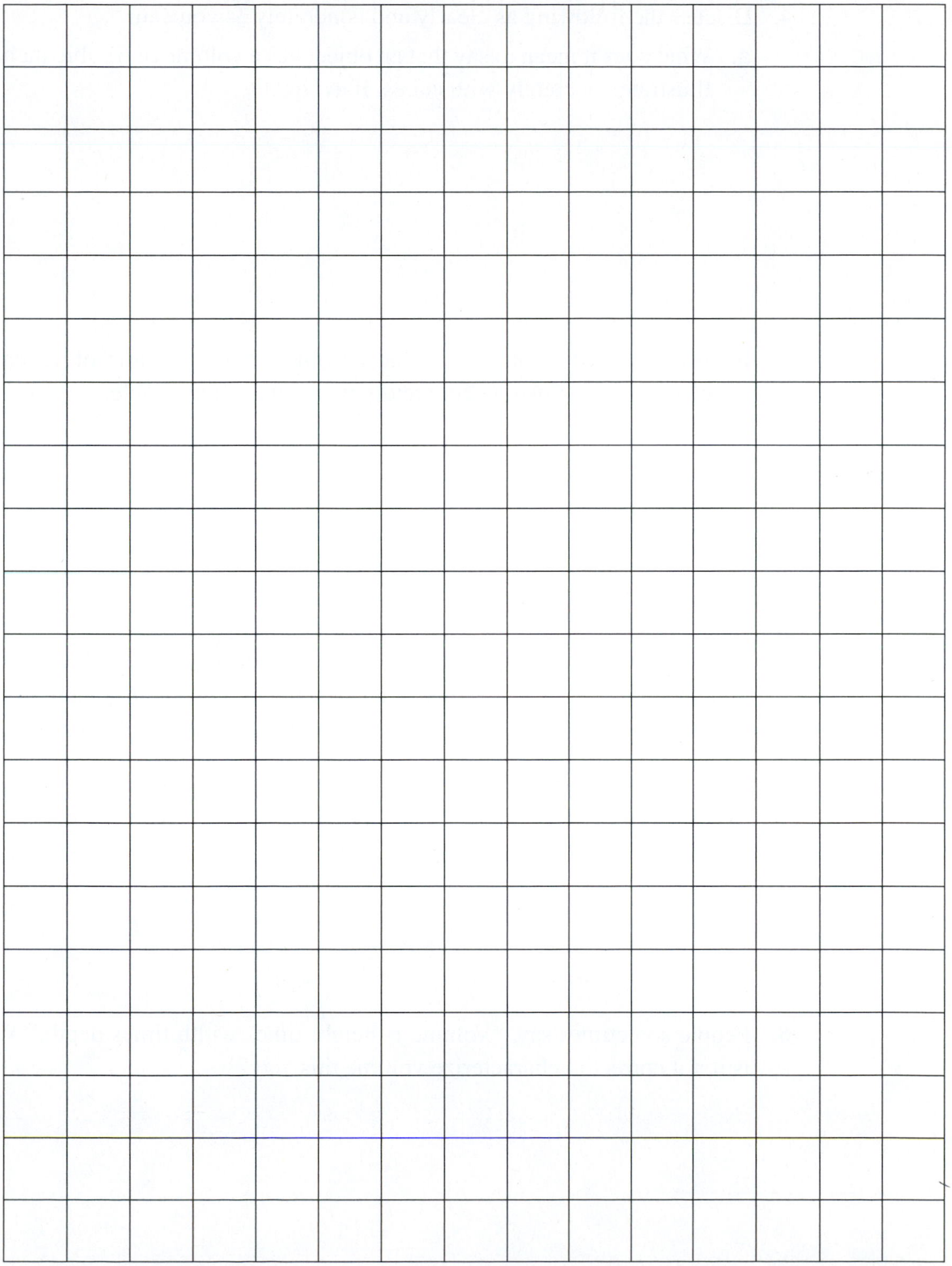

Figure 10A.3
Centimeter Graph Paper

4. Discuss the following as clearly and concretely as you can:

a. What does it mean to say that an object has a volume of 6 cubic inches? Illustrate concretely with cubes, if available.

b. What does it mean to say that an object has a volume of 12 cubic centimeters? Illustrate concretely with cubes, if available.

5. If we weren't thinking carefully, we might try to illustrate 6 cubic inches by making a cube that is 6 inches wide, 6 inches deep, and 6 inches high. Why is it easy to think that such a cube has a volume of 6 cubic inches, and why is this *not correct*?

6. People sometimes say, "Volume is height times width times depth." Why is it not correct to characterize volume this way?

Class Activity 10C: Using a Ruler

1. The most primitive way to measure the length of an object in inches is to place inch-long strips (or objects) end-to-end along the object. The number of inch-long strips it takes is the length of the object in inches. On the other hand, when we use a ruler to measure the length of an object, we read a number on a tick mark. Explain how these two processes of measuring are related.

2. Children sometimes try to measure the length of an object by placing one end of the object at the 1 marking instead of the 0 marking, as shown on the centimeter ruler in the figure. How might you help a child understand why the strip below is not 5 cm long, even though the end of the strip is at 5? Why might a child put one end of the strip at the 1 marking? Is it okay to measure by starting at 1 or at another tick mark?

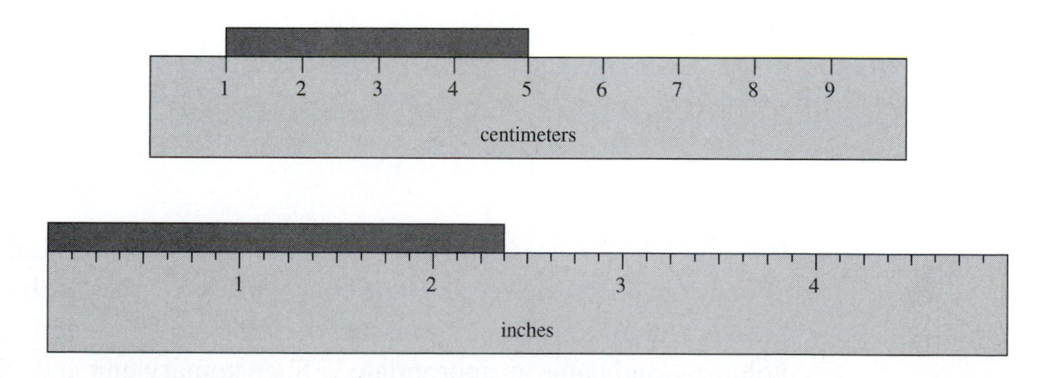

3. Some children might report that the strip measured by the inch ruler shown is 2.3 inches long. Why is this not correct? What is a correct way to report the length of the strip?

10.2 Length, Area, Volume, and Dimension

Class Activity 10D: Dimension and Size

1. In the book *Captain Underpants and the Attack of the Talking Toilets* [11, p. 30], the character Melvin describes a machine called the PATSY 2000 that he has invented:

   "The PATSY 2000 can take any one-dimensional image and create a living, breathing, three-dimensional copy of that image. For example, take this ordinary photograph of a mouse."

   Are the terms *one-dimensional* and *three-dimensional* used properly here? Discuss.

2. Imagine a lake. Describe one-dimensional, two-dimensional, and three-dimensional parts or aspects of the lake. In each case, state how you would measure the size of that part or aspect of the lake—by length, by area, or by volume—and name an appropriate U.S. customary unit and an appropriate metric unit for measuring or describing the size of that part or aspect of the lake. What are practical reasons for wanting to know the sizes of these parts or aspects of the lake?

3. Imagine a house. Describe one-dimensional, two-dimensional, and three-dimensional parts or aspects of the house. In each case, state how you would measure the size of that part or aspect of the house—by length, by area, or by volume—and name an appropriate U.S. customary unit and an appropriate metric unit for measuring or describing the size of that part or aspect of the house. What are practical reasons for wanting to know the sizes of these parts or aspects of the house?

10.3 Calculating Perimeters of Polygons, Areas of Rectangles, and Volumes of Boxes

Class Activity 10E: Explaining Why We Add to Calculate Perimeters of Polygons

If available, string will be helpful for this activity.

Imagine you are working with elementary school children. Tell the children what we mean by the term "perimeter." You may wish to use the string to help you.

If you didn't have string, how would you calculate the perimeter of the polygon in the next figure? Why does it make sense that we can calculate the perimeter by adding? The grid lines are 1 cm apart.

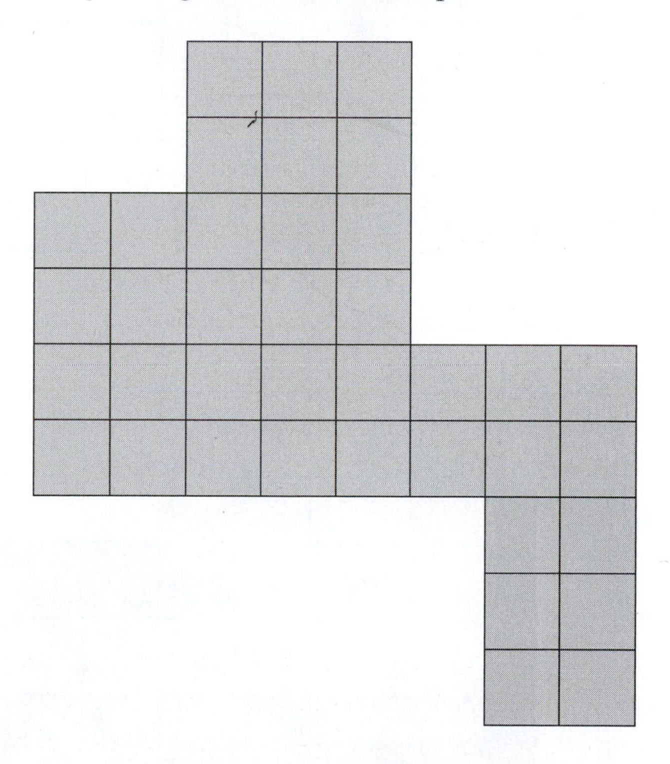

Class Activity 10F: Perimeter Misconceptions

You will need a centimeter ruler for Problem 2.

1. Johnny wants to calculate the perimeter of the shaded shape in Figure (a). Johnny's method is to shade the squares along the border of the shape, as shown in Figure (b), and to count these border squares. Therefore, Johnny says the perimeter of the shape is 24 cm. Is Johnny's method valid? If not, why not?

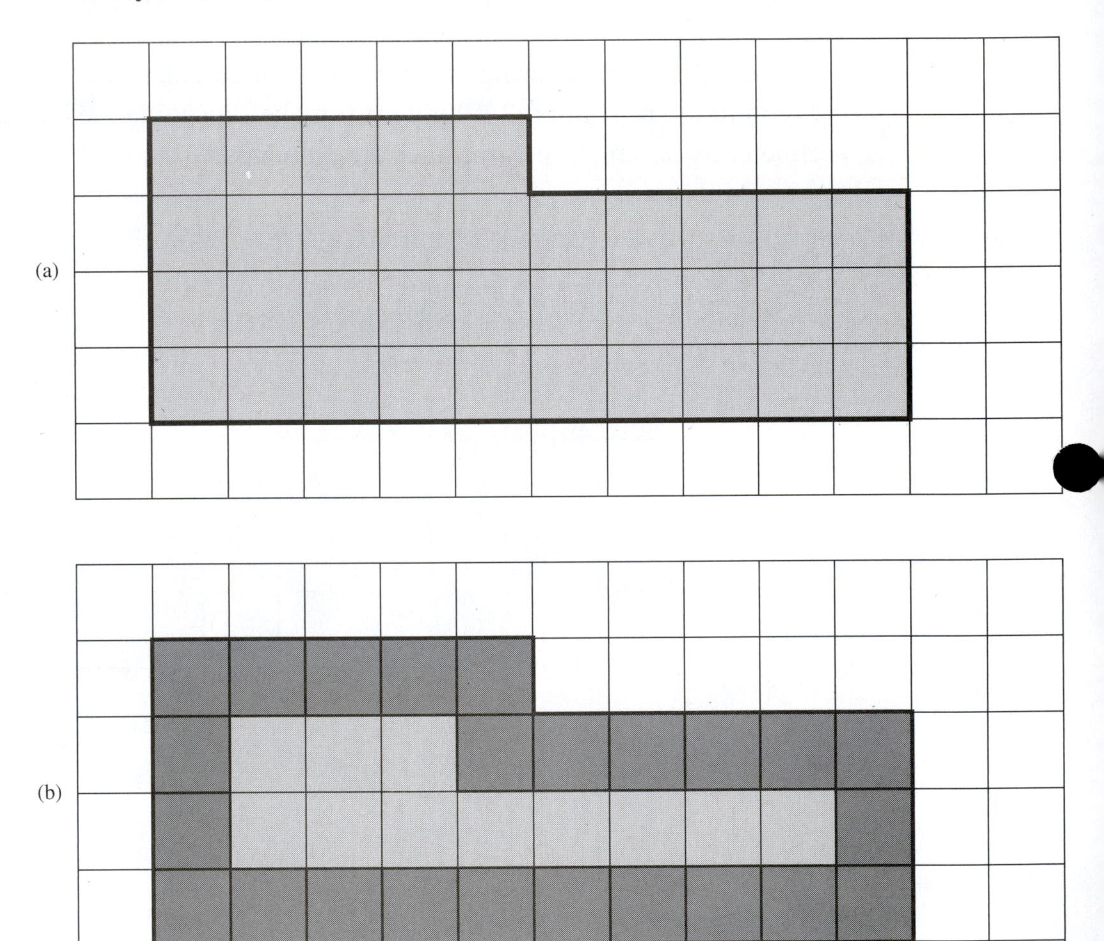

2. When Susie was asked to draw a shape with perimeter 15 cm, she drew a shape like the shaded one shown in the next figure on centimeter grid paper.

a. Carefully measure the diagonal line segment in the shaded shape with a centimeter ruler. Then explain why the shape does not have perimeter 15 cm.

b. Draw a shape that has perimeter 15 cm on the same graph paper. (The corners of your shape do not have to be located where grid lines meet.)

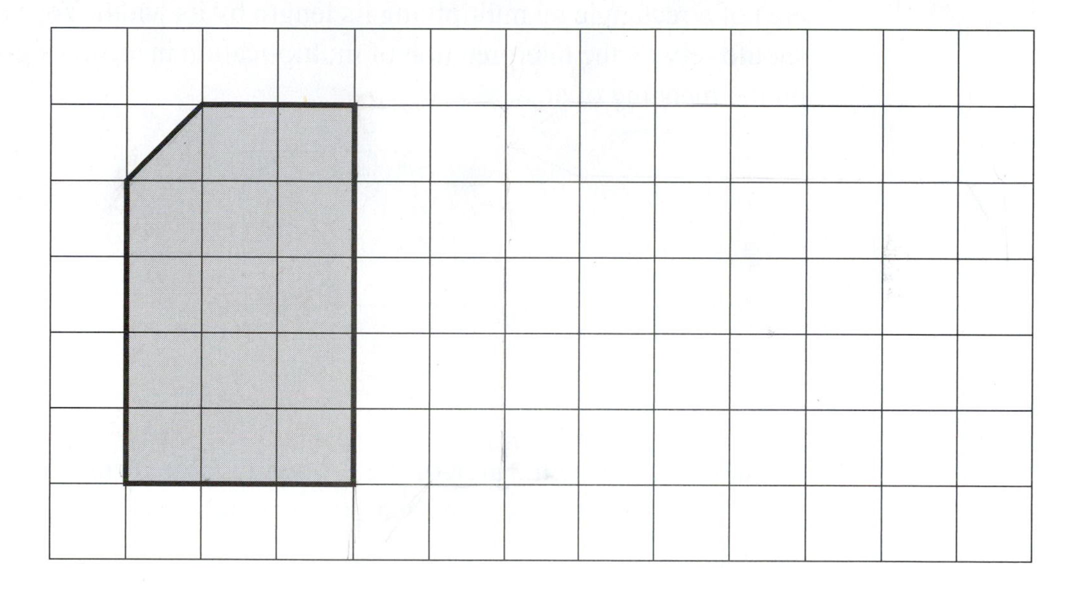

Class Activity 10G: Explaining Why We Multiply to Determine Areas of Rectangles

If available, a set of 1-inch-by-1-inch square tiles would be helpful; if they are not available, then you will need scissors for this activity.

1. If square tiles are not available, cut out the small 1-inch-by-1-inch squares on page 629. Use the rectangle on page 629 and either the paper squares or a set of 1-inch-by-1-inch tiles to help you explain why we can determine the area of a rectangle by multiplying its length by its width. Your explanation should rely on the interpretation of multiplication in terms of grouping and on the meaning of area.

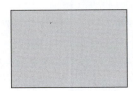

2. Explain how to see the area of the following large rectangle as consisting of $3\frac{1}{2}$ groups with $4\frac{1}{2}$ cm^2 in each group, thereby explaining why it makes sense to multiply

$$3\frac{1}{2} \times 4\frac{1}{2}$$

to determine the area of the rectangle in square centimeters:

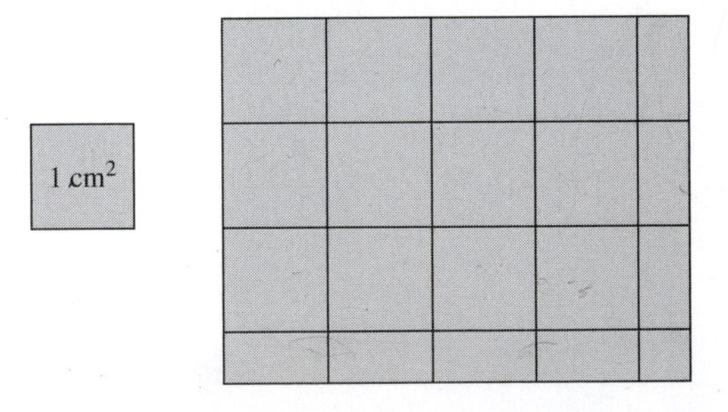

Then calculate $3\frac{1}{2} \times 4\frac{1}{2}$, and verify that this calculation has the same answer as when you determine the area of the rectangle by counting full 1-cm-by-1-cm squares and combining partial squares.

Class Activity 10H: Explaining Why We Multiply to Determine Volumes of Boxes

You will need scissors and tape for this activity. If available, a set of 1-inch-by-1-inch-by-1-inch cubes would be helpful.

Cut out the pattern on page 631, fold it, and tape it to make an open-top box. If a set of cubes is not available, then cut out the patterns for cubes on page 631, and fold and tape them to make 1-inch-by-1-inch-by-1-inch cubes. Use the box and the 1-inch-by-1-inch-by-1-inch cubes to help you explain why we can multiply the height times the length times the width of the box to determine the volume of the box. Your explanation should rely on the interpretation of multiplication in terms of grouping and on the meaning of volume.

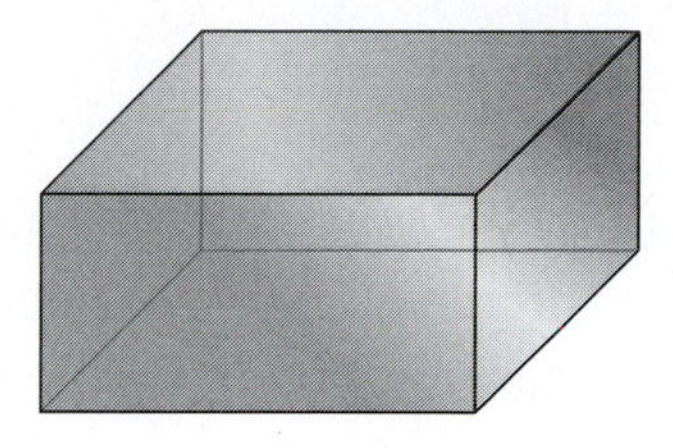

Class Activity 10I: Who Can Make the Biggest Box?

You will need several pieces of graph paper for this activity. Scissors and tape would also be useful.

For each problem in this activity, individuals or teams compete to make the "biggest" box. If scissors and tape are available, you may wish to test your patterns.

1. Use no more than one ordinary piece of graph paper to make a pattern for a tall open-top box. The box must have a base. The winner is the person (or team) whose box is tallest when oriented with the open top up.

2. Use no more than one ordinary piece of graph paper to make a pattern for a closed box that has a large surface area. The winner is the person (or team) whose box has the largest surface area.

3. Use no more than one ordinary piece of graph paper to make a pattern for a closed box that has a large volume. The winner is the person (or team) whose box has the largest volume.

4. Would the winning box in contest 2 do well in contest 3? Would the winning box in contest 3 do well in contest 2? What's the moral?

10.4 Error and Accuracy in Measurements

Class Activity 10J: Reporting and Interpreting Measurements

1. The following picture shows two rulers measuring identical shaded strips:

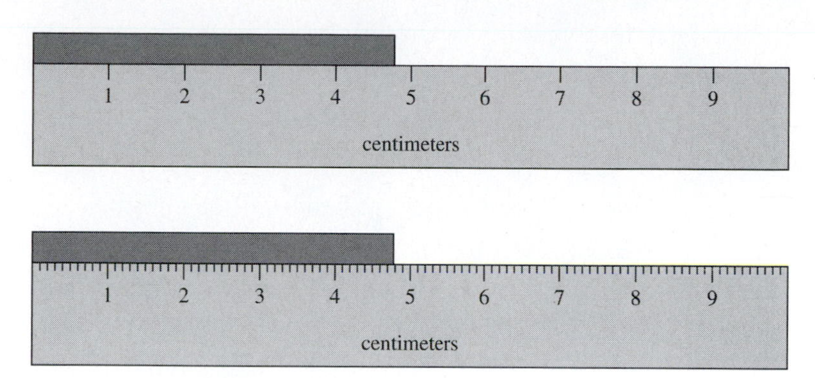

   a. Using the first ruler, how should you report the length of the strip? Explain.

   b. Using the second ruler, how should you report the length of the strip? Explain.

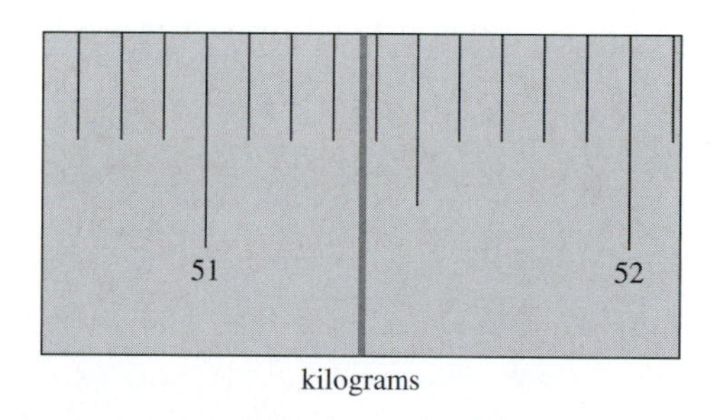

2. The long shaded line on the dial in the figure indicates the weight of a person in kilograms. Which of the following should you report as the weight of the person, and why?

   51 kg, 51.3 kg, 51.4 kg, 51.5 kg, 51.37 kg, 51.38 kg

3. As abstract numbers, 4 = 4.0, but does it mean the same thing to say that an object weighs 4 kg as to say that it weighs 4.0 kg? If not, what is the difference?

4. One source says that the average distance from the earth to the sun is 93,000,000 miles, and another source says that the average distance from the earth to the sun is 92,960,000 miles. Can both of these descriptions be correct, or must at least one of them be wrong? Explain.

10.5 Converting from One Unit of Measurement to Another

Class Activity 10K: Conversions: When Do We Multiply? When Do We Divide?

1. Julie is confused about why we *multiply* by 3 to convert 6 yards to feet. She thinks we should *divide* by 3 because feet are smaller than yards.

   a. Draw a picture to show Julie how yards and feet are related. Use your picture and the meaning of multiplication to explain to Julie why we multiply by 3 to convert 6 yards to feet.

   b. Julie knows that if she measures the length of a table in yards, she will get a smaller number than if she measures the length of the table in feet. Given this, how might you help Julie understand why we multiply by 3 to convert 6 yards to feet?

   c. Try to think of other ways to help Julie better understand conversions. What problems or questions could you pose to Julie?

2. Nate is confused about why we *divide* by 100 to convert 200 centimeters to meters. He thinks we should *multiply* by 100 because meters are bigger than centimeters. What are some ways you could help Nate better understand conversions?

Class Activity 10L: Conversion Problems

1. Shaquila is 57 inches tall. How tall is Shaquila in feet?

   Should you multiply or divide to solve this problem? Explain. Describe a number of different correct ways to write the answer to the conversion problem. Explain briefly why these different ways of writing the answer mean the same thing.

2. Carlton used identical paperclips to measure the length of a piece of wood. He found that the wood was 35 paperclips long. Next, Carlton will measure the length of the wood, using identical rods. Carlton found that 2 rods are as long as 5 paperclips. How many rods long is the wood? Explain your reasoning.

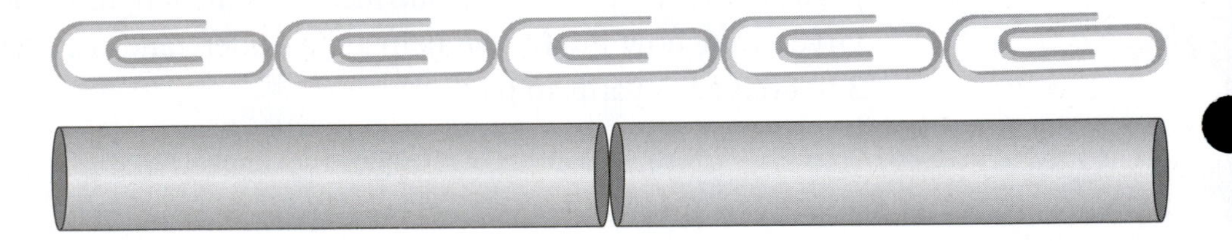

3. Suppose that the children in your class want to have a party at which they will serve punch to drink. The punch that the children want to serve is sold in half-gallon containers. If 25 people attend the party, and if each person drinks 8 fluid ounces of punch, then how many containers of punch will you need? Describe several different ways that children could correctly solve this problem. For each method of solving the problem, explain simply and clearly why the method makes sense.

Class Activity 10M: Converting Measurements with and without Dimensional Analysis

1. If a woman is 165 centimeters tall, how tall is she in feet and inches? First estimate your answer. Then use the fact that 1 inch = 2.54 cm to determine the woman's height in feet and inches in the following two different ways:

a. using logical thinking about multiplication and division

b. using dimensional analysis

2. Convert 6 feet to meters, using the fact that 1 inch = 2.54 cm. Calculate your answer in the following two different ways:

a. using logical thinking about multiplication and division

b. using dimensional analysis

Class Activity 10N: Areas of Rectangles in Square Yards and Square Feet

1. One yard is 3 feet. Does this mean that one square yard is 3 square feet? Draw a picture to show how many square feet are in a square yard. Use the meaning of multiplication to explain why you can determine the number of square feet in a square yard by multiplying.

2. A rug is 5 yards long and 4 yards wide. What is the area of the rug in square yards? What is the area of the rug in square feet? Show two different ways to solve this problem. In each case, refer to the meaning of multiplication.

3. A room has a floor area of 35 square yards. What is the floor area of the room in square feet? Explain your answer, referring to the meaning of multiplication.

Class Activity 10O: Volumes of Boxes in Cubic Yards and Cubic Feet

1. One yard is 3 feet. Does this mean that 1 cubic yard is 3 cubic feet? Describe how to use the meaning of multiplication to determine what 1 cubic yard is in terms of cubic feet.

2. A compost pile is 2 yards high, 2 yards deep, and 2 yards wide. Does this mean that the compost pile has a volume of 2 cubic yards? Describe how to use the meaning of multiplication to determine the volume of the compost pile in cubic yards.

3. Determine the volume in cubic feet of the compost pile described in the previous problem in two different ways. In each case, refer to the meaning of multiplication.

Class Activity 10P: Area and Volume Conversions: Which Are Correct and Which Are Not?

1. Analyze the calculations that follow, which are intended to convert 25 square meters to square feet. Which use legitimate methods and are correct, and which are not? Explain.

   a.

$$25 \text{ m}^2 = 25 \text{ m} \times \frac{100 \text{ cm}}{1 \text{ m}} \times \frac{1 \text{ in}}{2.54 \text{ cm}} \times \frac{1 \text{ ft}}{12 \text{ in}} = 82 \text{ ft}^2$$

   b.

$$25 \text{ m}^2 = 25 \text{ m}^2 \times \frac{100 \times 100 \text{ cm}^2}{1 \text{ m}^2} \times \frac{1 \text{ in}^2}{2.54 \times 2.54 \text{ cm}^2} \times \frac{1 \text{ ft}^2}{12 \times 12 \text{ in}^2} = 269 \text{ ft}^2$$

   c.

$$25 \text{ m} = 25 \text{ m} \times \frac{100 \text{ cm}}{1 \text{ m}} \times \frac{1 \text{ in}}{2.54 \text{ cm}} \times \frac{1 \text{ ft}}{12 \text{ in}} = 82 \text{ ft}$$

   Therefore,

$$25 \text{ m}^2 = 82^2 \text{ ft}^2 = 6727 \text{ ft}^2$$

   d. 25 square meters is the area of a square that is 5 meters wide and 5 meters long, so

$$5 \text{ m} = 5 \text{ m} \times \frac{100 \text{ cm}}{1 \text{ m}} \times \frac{1 \text{ in}}{2.54 \text{ cm}} \times \frac{1 \text{ ft}}{12 \text{ in}} = 16.404 \text{ ft}$$

   Therefore,

$$25 \text{ m}^2 = 16.404 \times 16.404 \text{ ft}^2 = 269 \text{ ft}^2$$

2. Use the fact that 1 inch = 2.54 cm to convert 27 cubic feet to cubic meters in at least two different ways.

11

More about Area and Volume

11.1 The Moving and Additivity Principles about Area

Class Activity 11A: Different Shapes with the Same Area

You will need scissors and tape for this activity.

Cut out the four 10-cm-by-15-cm rectangles on pages 639 and 641.

1. Cut one of the two rectangles on page 639 diagonally in half, creating two identical triangles, as shown in the middle of the next figure. Use the moving and additivity principles to determine the area of each triangle.

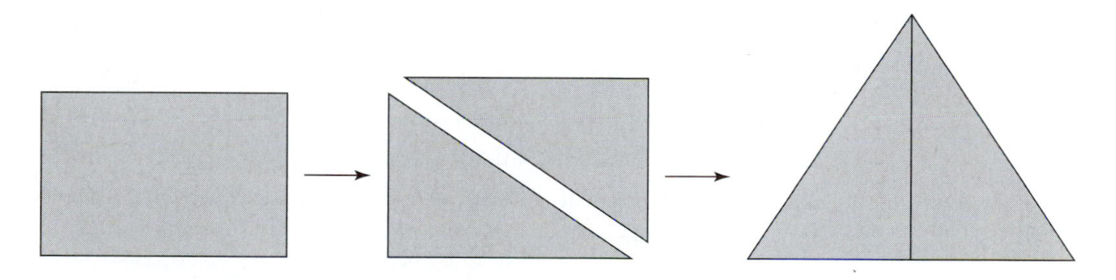

Rearrange the two triangles and tape them together, without overlapping, to create a single large triangle, as shown on the right in the previous figure. Use the moving and additivity principles to determine the area of this new, large triangle.

2. Cut the other rectangle on page 639 diagonally in half, creating two triangles. Rearrange the two triangles and tape them together, without overlapping, to create a single large triangle that is different from the triangle you created in Problem 1. Use the moving and additivity principles to determine the area of this new, large triangle.

3. Cut one of the two rectangles on page 641 diagonally in half, creating two triangles. Rearrange the two triangles and tape them together, without overlapping, to create a parallelogram that is not a rectangle. (There are two different ways you can do this.) Use the moving and additivity principles to determine the area of the parallelogram.

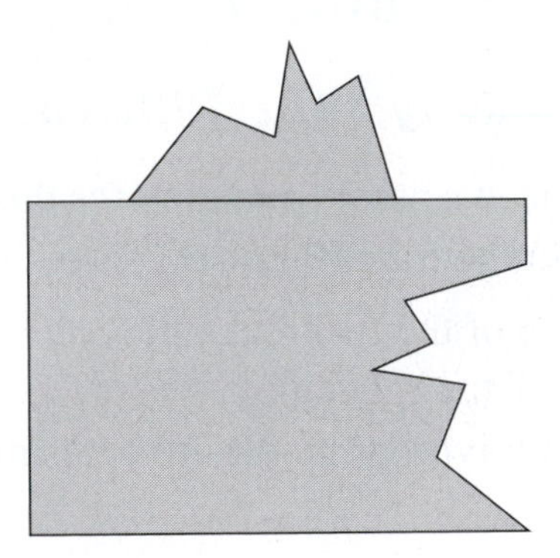

4. Cut the other rectangle on page 641 into pieces of any size or shape you like. Rearrange those pieces and tape them together, without overlapping, to create a new shape, such as the one in the figure. You can even make a shape that is not flat, if you like. Use the moving and additivity principles to determine the area of your new shape.

5. Now compare all four shapes that you created in Problems 2, 3, and 4. What do all the shapes have in common, and what is different about the shapes? Do all the shapes have the same perimeter? Do all the shapes have the same length?

Class Activity 11B: Using the Moving and Additivity Principles

You will need blank paper, scissors, and a ruler for Problem 2 of this activity.

1. The grid lines in the next figure are 1 cm apart. Use the moving and additivity principles about areas to help you determine the area of each of the three triangles shown (in square centimeters). Explain why your answers are correct. *Do not* use a formula for areas of triangles.

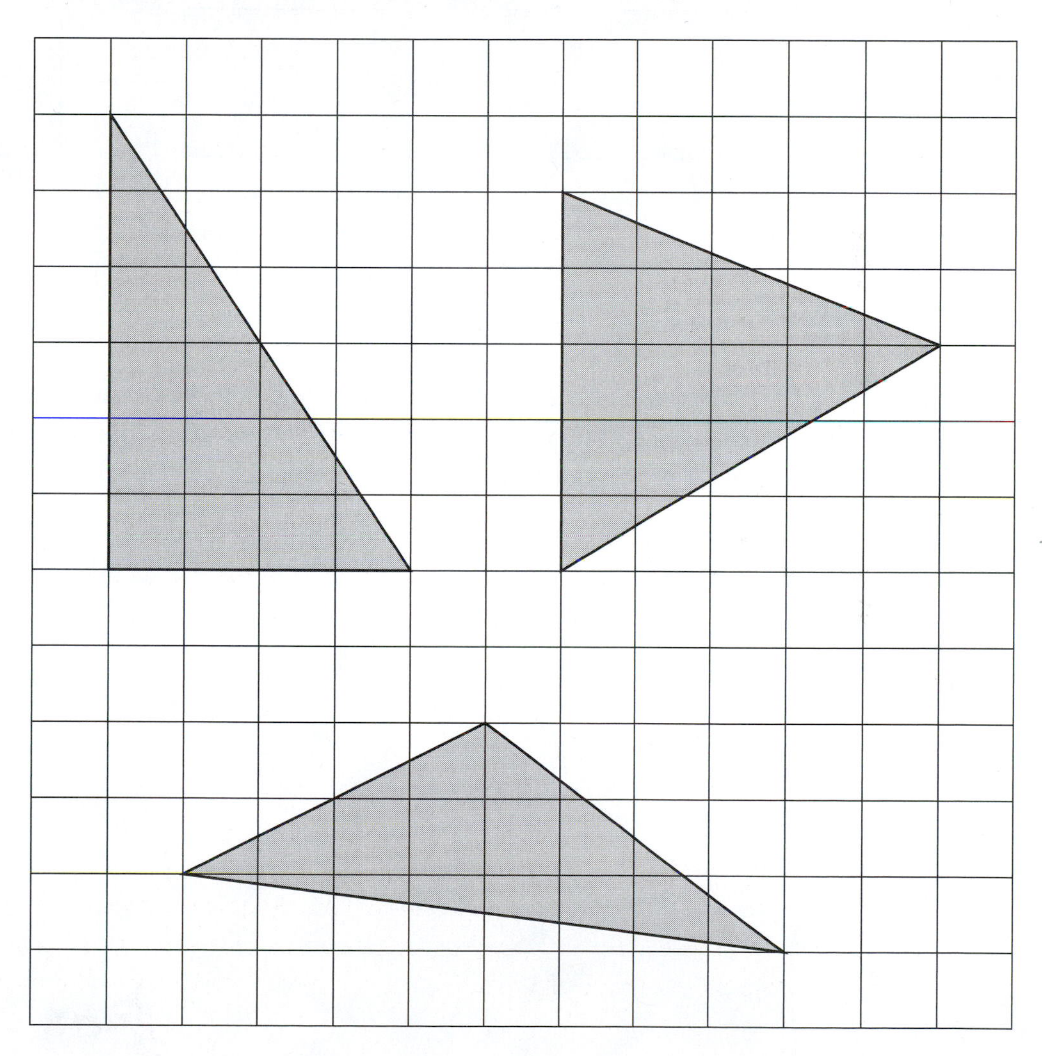

2. Take a piece of paper and fold it into fourths by folding it in half and then in half again, as shown in the next diagram. While the paper is still folded, cut off the "open" corner—so that you cut off four triangles, as indicated on the right. When you unfold the paper, you will have either an octagon, a hexagon, or a rhombus, depending on how you cut. (Can you see how to make each of these shapes?)

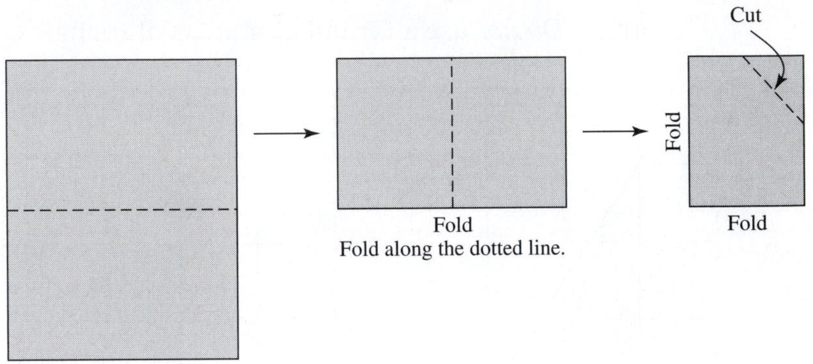

Determine the area of your octagon, hexagon, or rhombus. Use only the moving and additivity principles, actual measurements, and the formula for areas of rectangles to determine the area. Do not use any other formulas.

Class Activity 11C: Using the Moving and Additivity Principles to Determine Surface Area

You will need blank paper, scissors, and tape for this activity.

1. Take a standard 8.5-in-by-11-in piece of paper, roll it up, and tape it, without overlapping the paper, to make a cylinder without bases, as shown in the next figure. What is the surface area of the cylinder? Why?

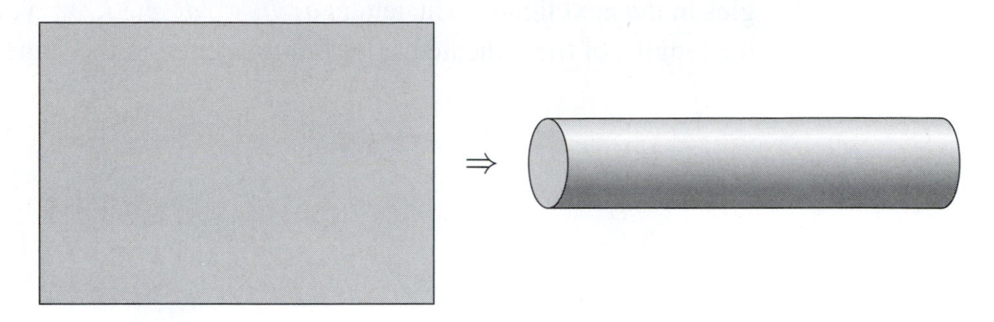

2. Determine the surface area of the shape you will make when you follow the next instructions, explaining your reasoning. Predict the shape you will make:
   Cut out the rectangle on page 635. Cut along the curve, then attach the side labeled A to the side labeled B (without overlaps). Attach the side labeled C to the side labeled D (without overlaps).

3. The eight triangles on the centimeter graph paper on page 637 can be cut out and taped together, without overlapping, to form an octahedron. Use the moving and additivity principles to determine the surface area of the octahedron. Do not use any formula for the area of a triangle.

11.2 Using the Moving and Additivity Principles to Prove the Pythagorean Theorem

Class Activity 11D: Using the Pythagorean Theorem

You will need a calculator for Problem 2 of this activity.

1. What, if anything, does the Pythagorean theorem say about the three triangles in the next figure? The letters a, b, c, d, e, f, x, y, and z represent the lengths of the indicated sides (all measured in the same units).

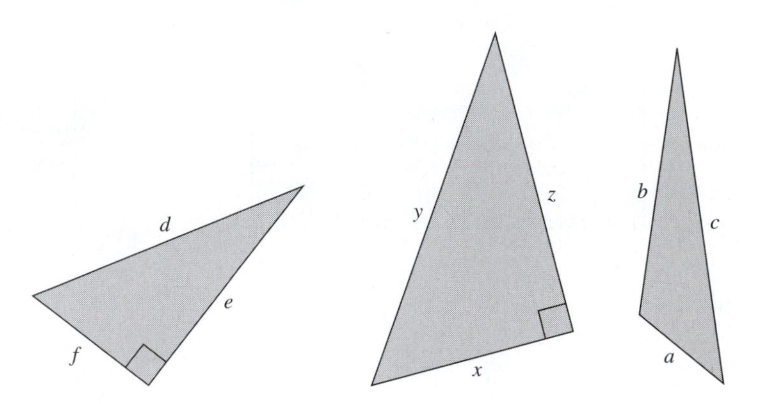

2. Tyler and Sarah measured the distance between two trees and found it to be 18 feet. But when they measured this distance, Tyler held up his end of the tape measure 8 inches higher than Sarah's end. If Sarah and Tyler had measured the distance between the trees by holding the tape measure horizontally, would they get an appreciably different result? Explain.

Class Activity 11E: Can We Prove the Pythagorean Theorem by Checking Examples?

You will need a ruler with $\frac{1}{4}$-inch markings (or finer) for this activity.

1. For each of the right triangles, measure the lengths of the sides and check that the Pythagorean theorem really is valid in each case. In other words, check that the sum of the squares of the lengths of the two shorter sides really is equal to the square of the length of the hypotenuse.

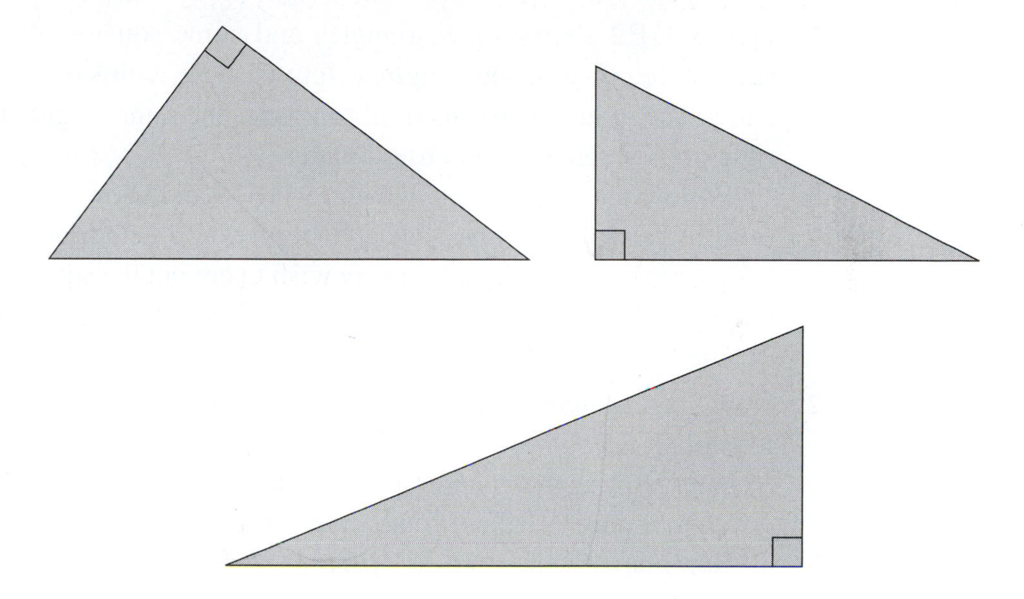

2. Would it be possible to prove that the Pythagorean theorem is true by checking many right triangles, continuing what you did in Problem 1? Explain.

Class Activity 11F: A Proof of the Pythagorean Theorem

Start with any arbitrary right triangle. Let a and b be the lengths of the short sides, and let c be the length of the hypotenuse. (See Figure 11F.1). To prove the Pythagorean theorem, we must explain why $a^2 + b^2 = c^2$ is true. Even though we will be working with copies of the *particular* triangle shown in Figure 11F.1, our explanation for why the Pythagorean theorem is true will be *general* because it would work in the same way for *any* right triangle.

1. Figure 11F.2 shows some triangles and some squares. All the triangles shown are copies of our original right triangle with sides of lengths a, b, and c. The square shapes have all four sides the same length and four angles right angles, making them true squares.

   Follow the instructions in Figure 11F.2, and show two different ways of filling a square that has sides of length $a + b$ with triangles and squares without gaps or overlaps. You may wish to cut out the squares and triangles on page 643 and use these to help you figure out how to fill in the squares of side length $a + b$.

2. Now use Problem 1 and the moving and additivity principles about areas to explain why $a^2 + b^2 = c^2$. There are several different ways to do this.

   Hint: Notice that *both* of the two ways of filling a square of side length $a + b$ in Figure 11F.2 use four copies of the original right triangle.

   Summarize your proof of Pythagoras's theorem.

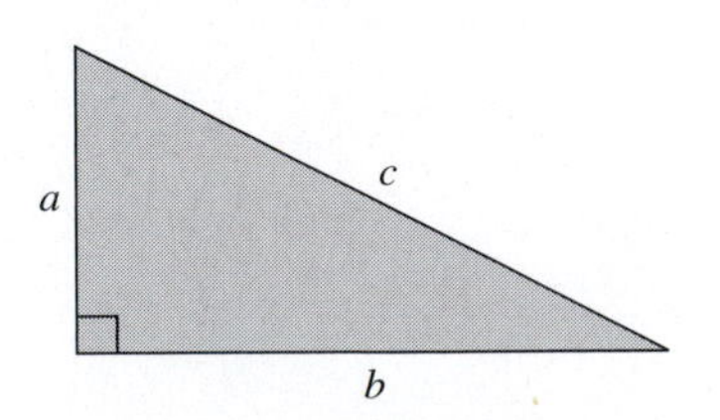

Figure 11F.1
A Right Triangle

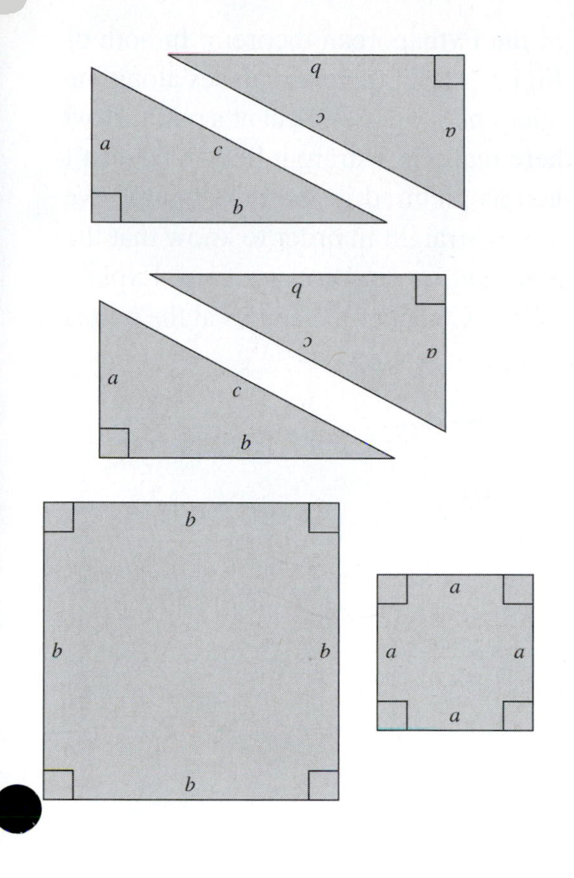

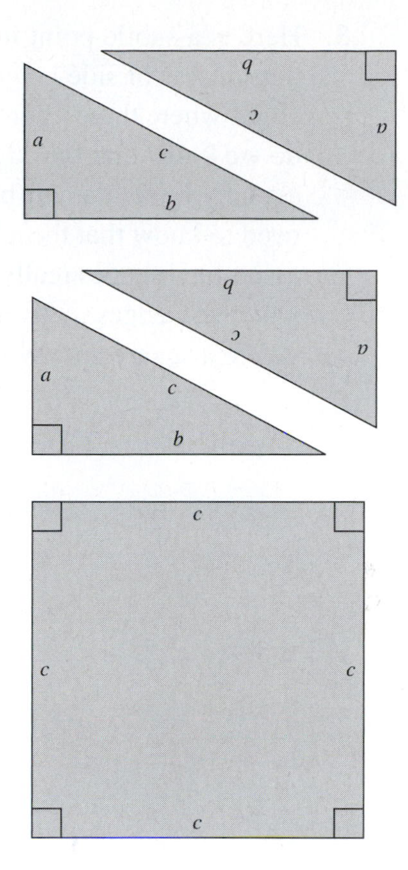

Draw the triangles and squares above into the square below so that they fill it without gaps or overlaps.

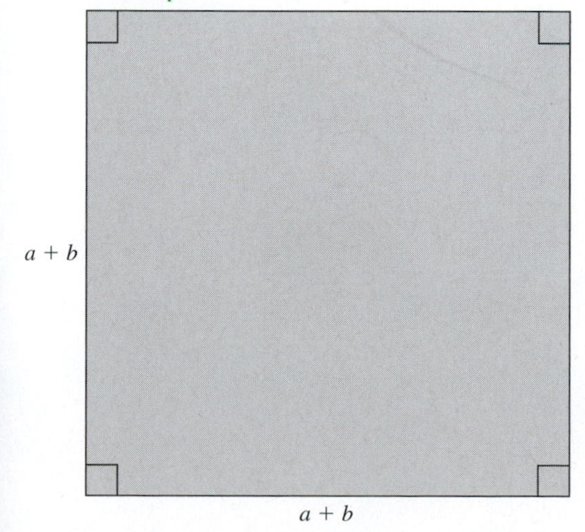

Draw the triangles and square above into the square below so that they fill it without gaps or overlaps.

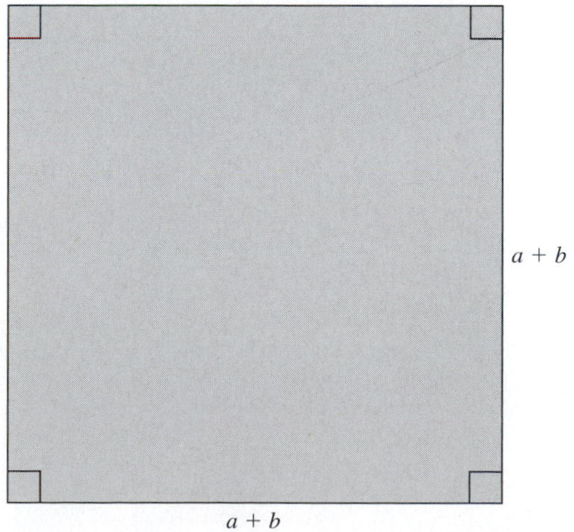

Figure 11F.2

Two Ways to Fill a Square of Side Length $a + b$

3. Here is a subtle point in the proof of the Pythagorean theorem: In both of the squares of side length $a + b$ in Figure 11F.2, there are places along the edges where three figures (two triangles and a square) meet at a point. How do we know that the edge formed there really is a straight line and doesn't actually have a small bend in it, such as pictured in the next figure? We need to know that the edge there really is straight in order to know that the assembled shapes really do create large *squares*, and not *octagons*. Explain why these edges really are straight. (*Hint*: Consider the angles at the points where a square and two triangles meet.)

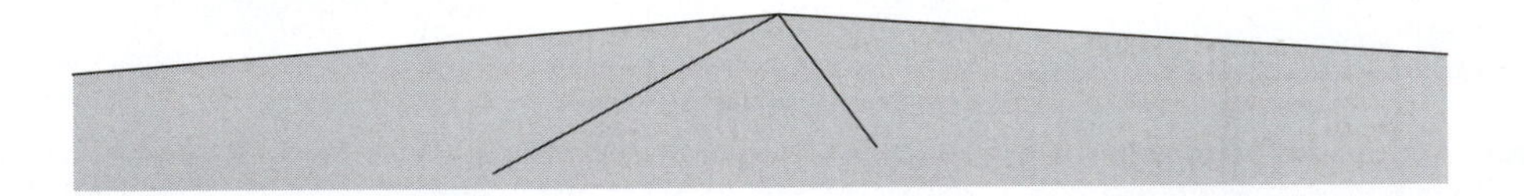

11.3 Areas of Triangles

Class Activity 11G: Choosing the Base and Height of Triangles

Show the three different ways to choose the base and height of the triangles shown in the next figure. Once you have chosen a base, the right angle formed by the corner of a piece of paper may help you determine where to draw the height, which must be perpendicular to the base (or an extension of the base).

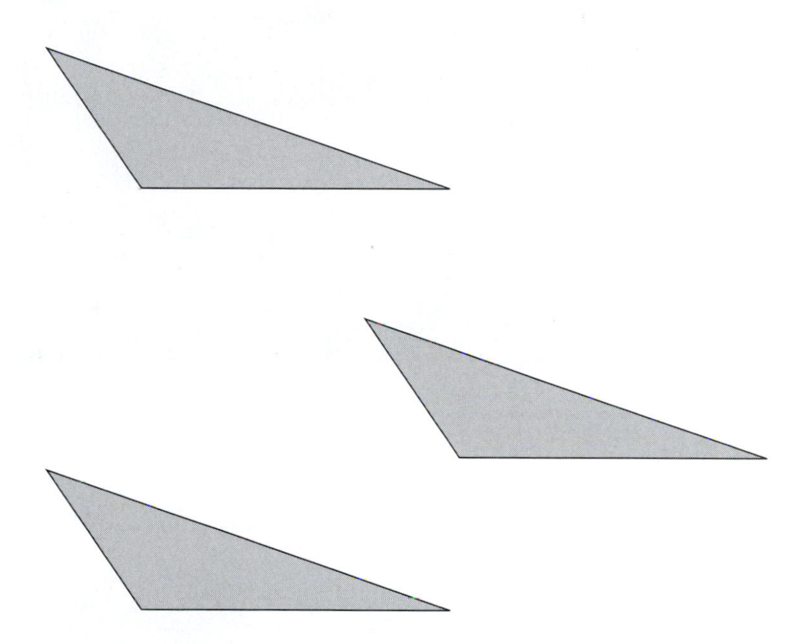

Class Activity 11H: Explaining Why the Area Formula for Triangles Is Valid

1. Explain why each of the next triangles has area $\frac{1}{2}(b \times h)$ square units for the given choices of b and h.

   Depending on your explanation, the formula $\frac{1}{2}(b \times h)$, the formula $(\frac{1}{2}b) \times h$, or the formula $b \times (\frac{1}{2}h)$ may be most suitable for describing the area of the triangle. Why is it valid to describe the area with any one of these three formulas?

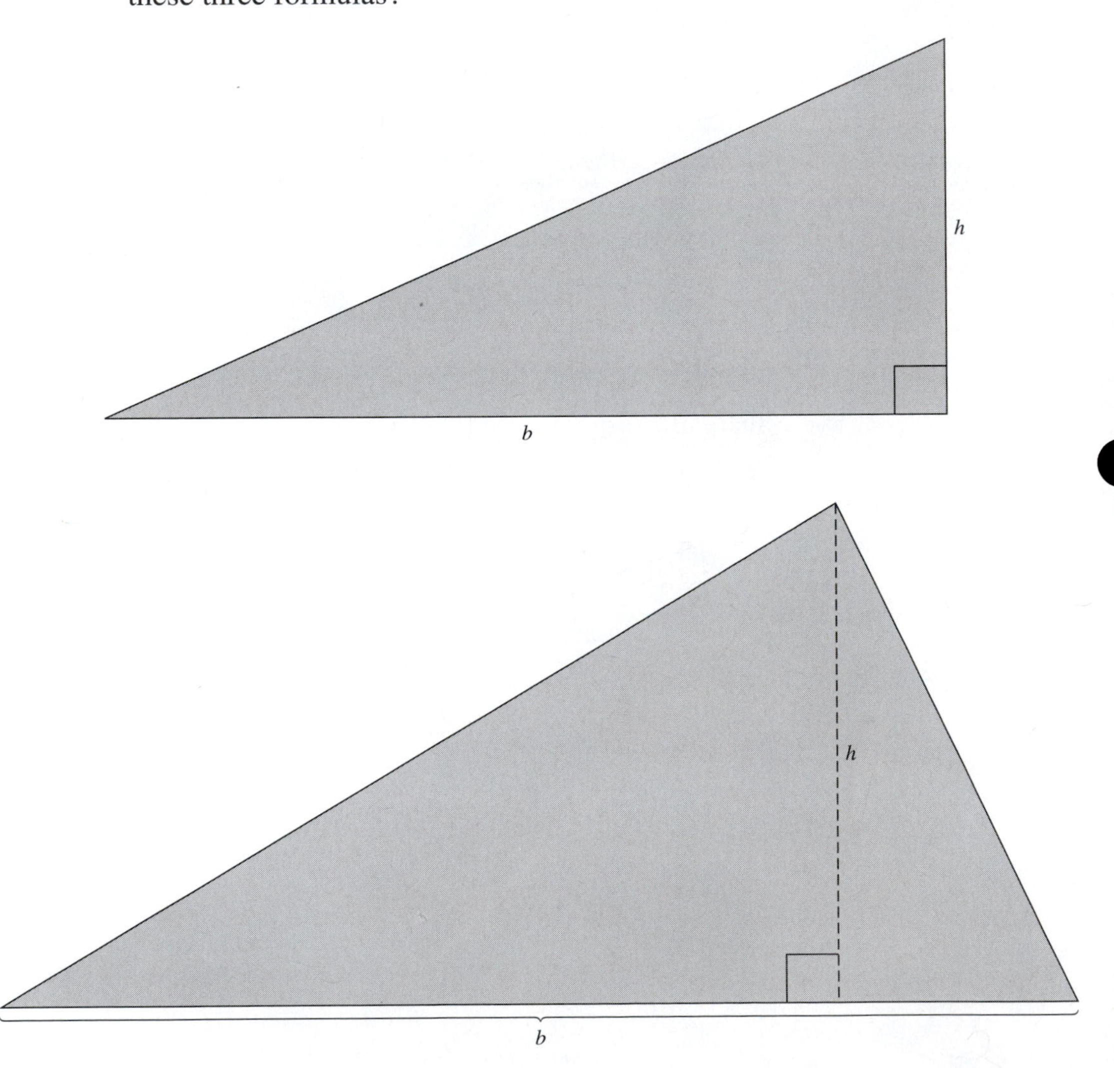

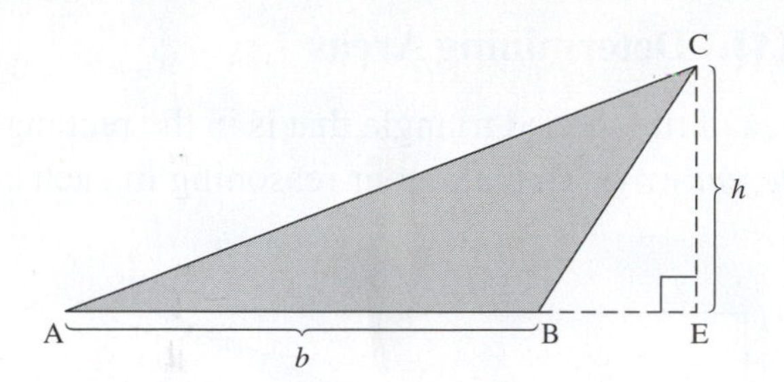

Figure 11H.1

A Triangle

2. The area of the triangle in Figure 11H.1 is still $\frac{1}{2}(b \times h)$ square units for the given choices of b and h. What is wrong with the following reasoning that claims to show that the area of the triangle ABC is $\frac{1}{2}(b \times h)$ square units?

   Draw a rectangle around the triangle ABC, as shown in Figure 11H.2. The area of this rectangle is $b \times h$ square units. The line AC cuts the rectangle in half, so the area of the triangle ABC is half of $b \times h$ square units—in other words, $\frac{1}{2}(b \times h)$ square units.

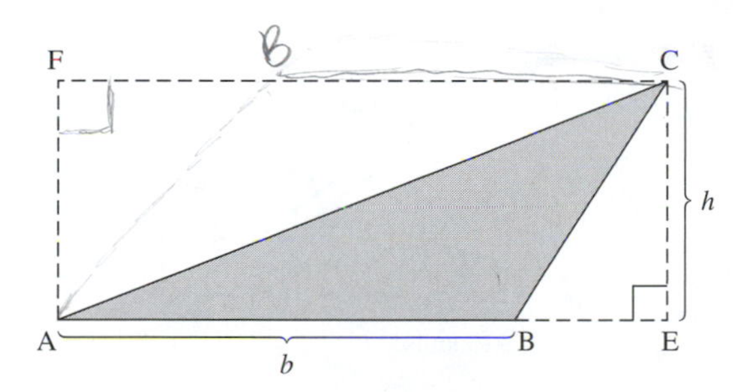

Figure 11H.2

Can You Find the Area This Way?

3. What is a valid way to explain why the triangle in Problem 2 has area $\frac{1}{2}(b \times h)$ square units for the given choice of b and h?

Class Activity 11I: Determining Areas

1. Determine the area of the shaded triangle that is in the rectangle in the next figure in *two different ways*. Explain your reasoning in each case.

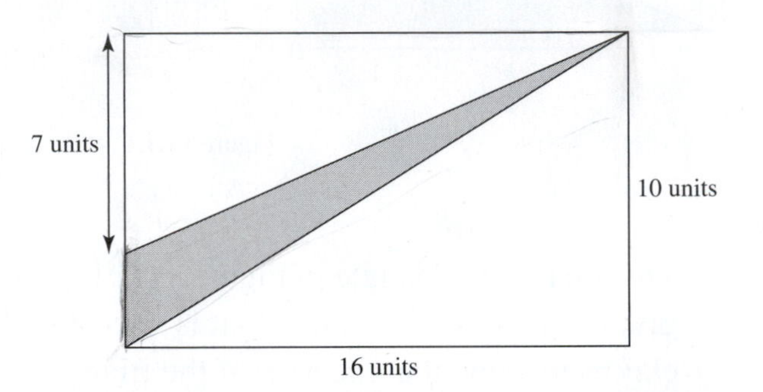

2. Determine the area of the shaded triangle in the next figure in *two different ways*. Explain your reasoning in each case.

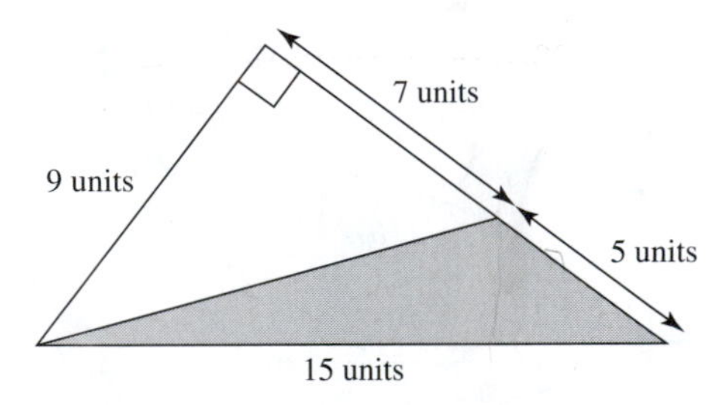

3. Determine the areas of the shaded shapes in the next pair of figures. The figure on the left consists of a 3-unit by 3-unit square and a 5-unit by 5-unit square. Explain your reasoning in each case.

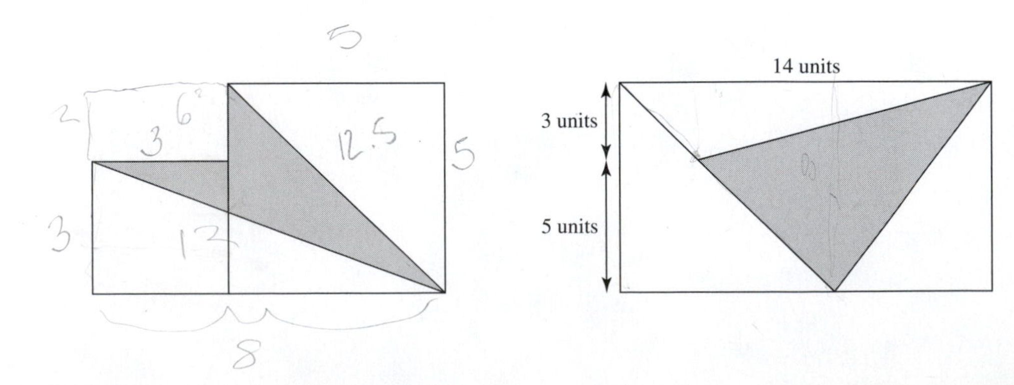

11.4 Areas of Parallelograms

Class Activity 11J: Do Side Lengths Determine the Area of a Parallelogram?

1. A rectangle is a special kind of parallelogram. We can calculate the area of a rectangle in terms of the lengths of the sides of the rectangle. Does this mean we can therefore calculate the area of a parallelogram in terms of the lengths of the sides of the parallelogram? Investigate this question as follows:

   The three parallelograms in Figure 11J.1 (the first of which is also a rectangle) all have two sides that are 3 cm long and two sides that are 7 cm long. The small squares in the grid are 1 cm by 1 cm.

   a. Use the moving and additivity principles to determine the areas of the three parallelograms in Figure 11J.1.

   b. Can there be a formula for areas of parallelograms that is only in terms of the lengths of the sides? Explain why or why not.

2. There is no formula for areas of parallelograms that is expressed solely in terms of the lengths of the *sides* of the parallelogram. However, there *is* a formula for areas of parallelograms. Try to find a formula *in terms of lengths of parts of the parallelogram.* Use the following parallelogram to help you explain your formula:

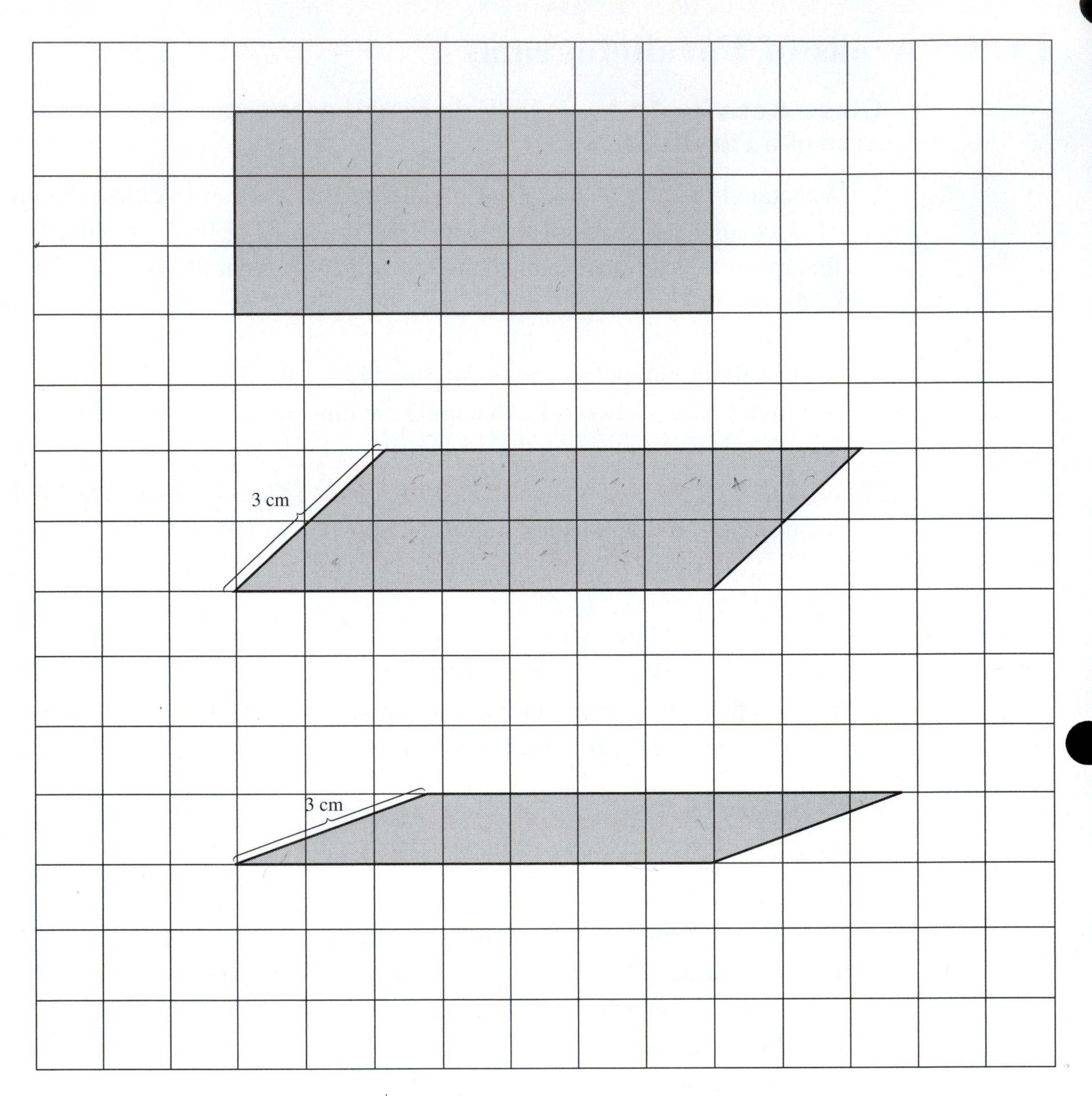

Figure 11J.1
Three Parallelograms

Class Activity 11K: Explaining Why the Area Formula for Parallelograms Is Valid

1. Show how to subdivide and recombine the next parallelogram to form a b by h rectangle, thereby explaining why the area of the parallelogram is $b \times h$.

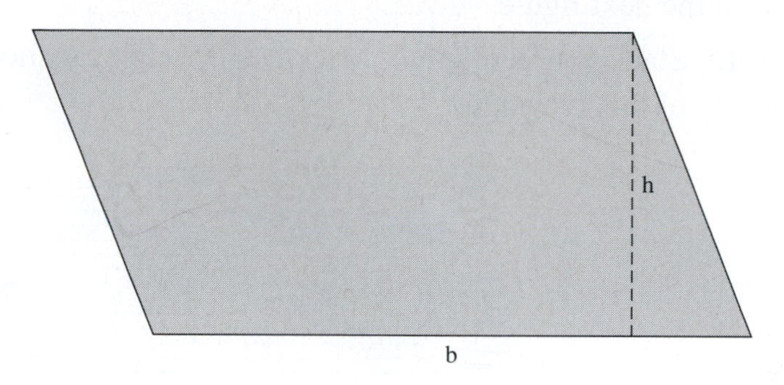

2. Show how to subdivide the shaded parallelogram and recombine it to fill the b by h rectangle, thereby explaining why the area of the parallelogram is $b \times h$. It may help you to cut apart the enlarged Figure A.26 on page 645.

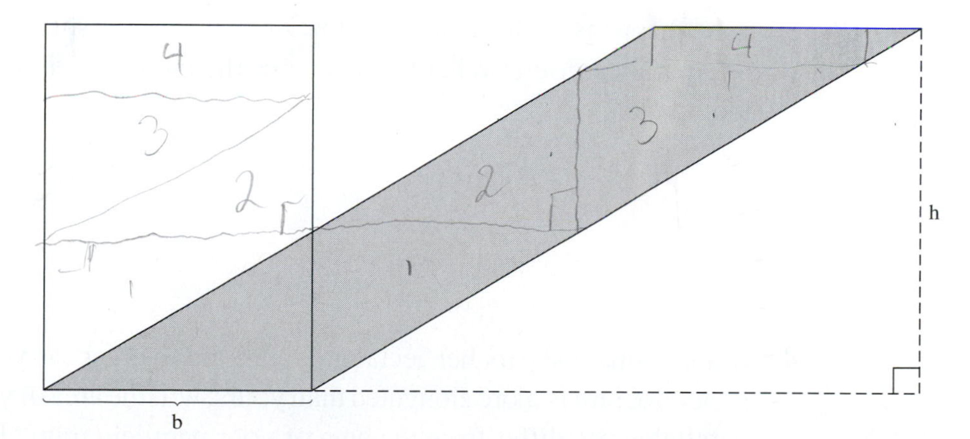

11.5 * Cavalieri's Principle about Shearing and Area

Class Activity 11L: Shearing a Toothpick Rectangle to Make a Parallelogram

For this activity you will need enough toothpicks to form a rectangle, as shown in the next figure.

1. Stack your toothpicks to form a rectangle, as shown, and determine the area of this rectangle.

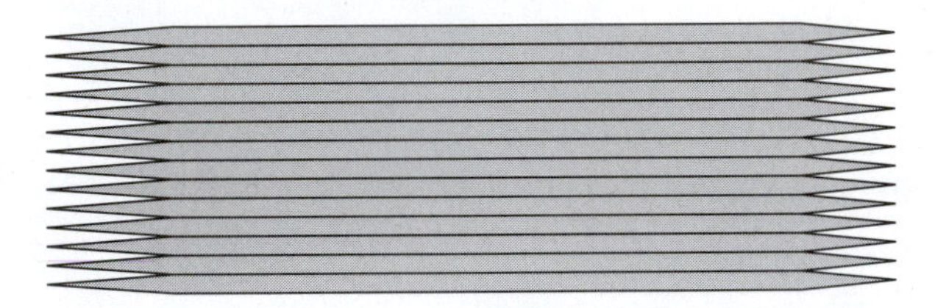

2. Now push on the ends of the toothpicks so as to shear your rectangle, forming a parallelogram as shown in the following figure:

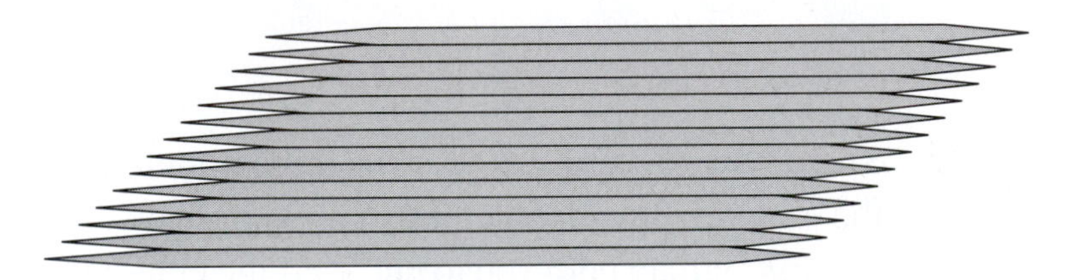

3. What happens to the area of the rectangle when it is sheared into the shape of a parallelogram? What is the area of the parallelogram formed by your toothpicks?

4. If a classmate shears her rectangle more than you shear yours, so that her parallelogram is more elongated than yours, will the area of your classmate's parallelogram differ from the area of your parallelogram? Explain.

Class Activity 11M: Is This Shearing?

The next figure shows a rectangle made of toothpicks being sheared into a parallelogram. During the process of shearing, which of the following change, and which remain the same?

- the area
- the lengths of the sides
- the height of the stack of toothpicks

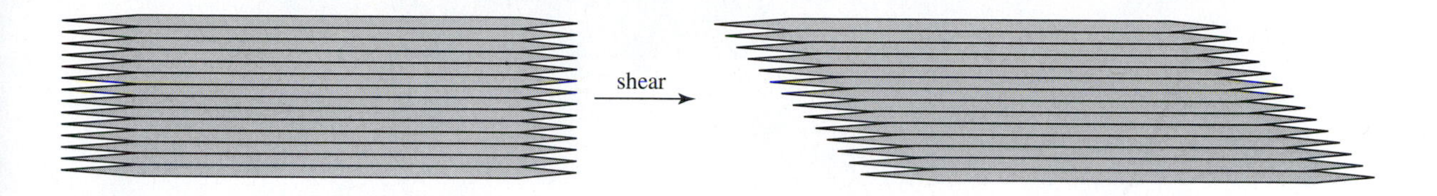

The next figure shows a rectangle made with pieces of drinking straws tied with a string being "squashed" into a parallelogram. During the process of "squashing," which of the following change and which remain the same?

- the area
- the lengths of the sides
- the "vertical height" of the straw figure

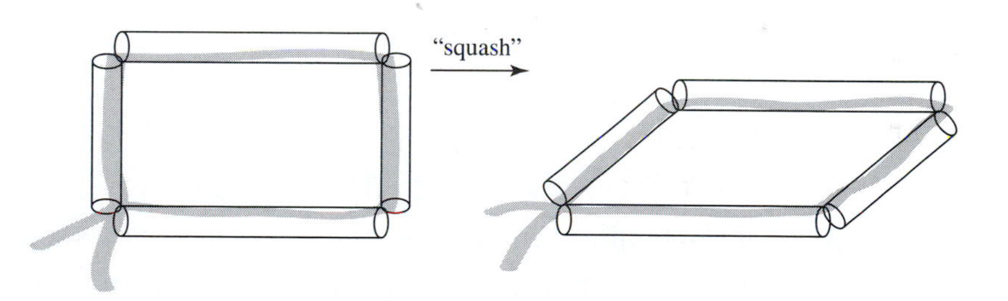

Is the process of "squashing" the same as shearing? Why or why not?

Class Activity 11N: Shearing Parallelograms

1. Imagine shearing parallelogram ABCD so that side AB remains fixed. Draw the path that the line segment CD could move along during shearing. How is this path related to the line segment AB?

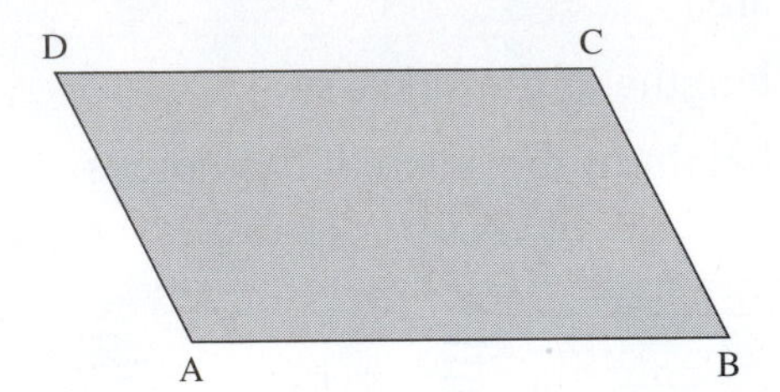

2. Determine the area of the next parallelogram by shearing it into a rectangle. Explain your reasoning. Adjacent dots are 1 cm apart.

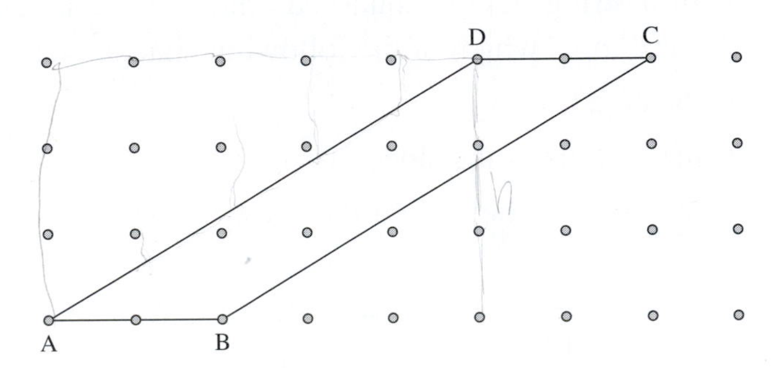

3. Determine the area of the next parallelogram by shearing it into a rectangle. Explain your reasoning. Adjacent dots are 1 cm apart. Note: Shearing does not have to be horizontal.

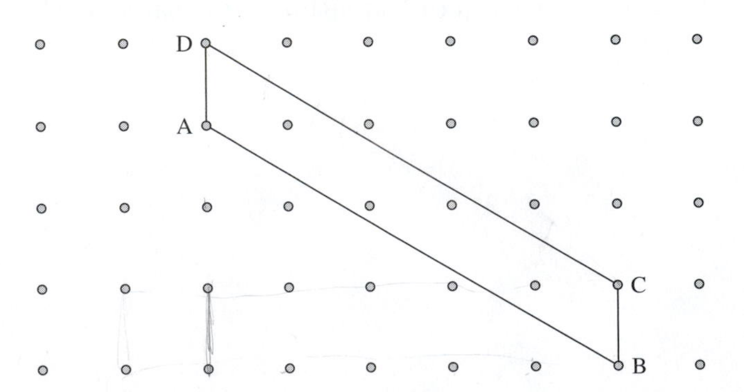

Class Activity 11O: Shearing Triangles

You will need scissors for this activity.

1. Cut out the large triangle on page 647, and cut it into strips along the dotted lines. On a flat surface, put the paper strips back together to make the original triangle. Now give your paper strips a push from the side, simulating shearing, as indicated in the next figure. When you shear your paper triangle you can form new shapes that are basically triangles, except that the edges are jagged. (If your paper strips were infinitesimally thin, then the edges would not be jagged.)

   By shearing your paper triangle, form various different (jagged) triangles, including a right triangle.

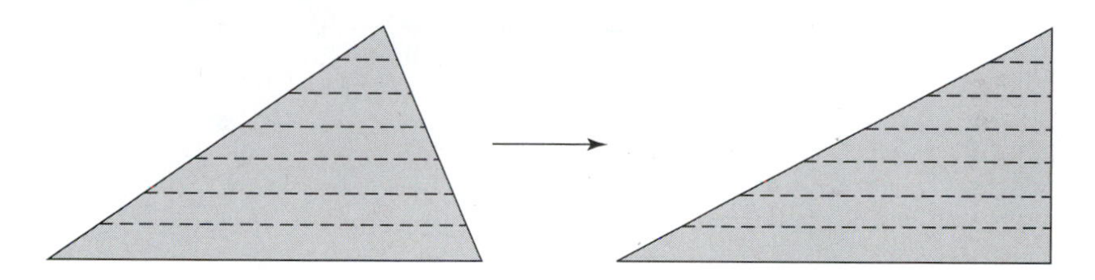

   Determine which of the following change during shearing and which remain the same:

   - the area
   - the lengths of the sides
   - the height of the "stack" of paper strips

2. Imagine shearing the triangle ABC so that side AB remains fixed. Draw all the locations to which point C could move during shearing. Describe the nature of these locations, in particular; how are they related to the line segment AB?

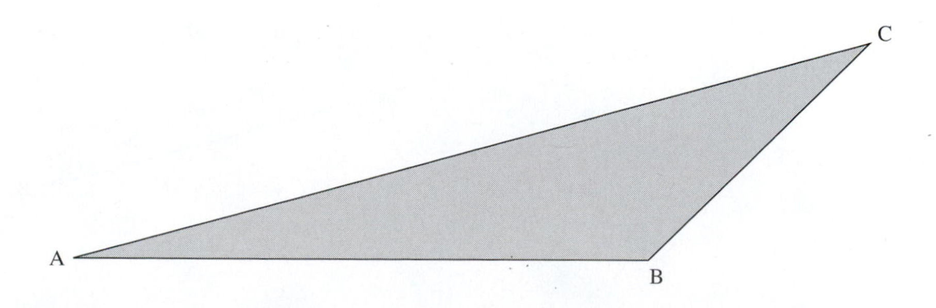

3. Determine the area of the next triangle by first shearing the triangle into a right triangle and then using the moving and additivity principles to determine the area of the right triangle. Explain why you can determine the area of the original triangle by this method. Adjacent dots are 1 cm apart.

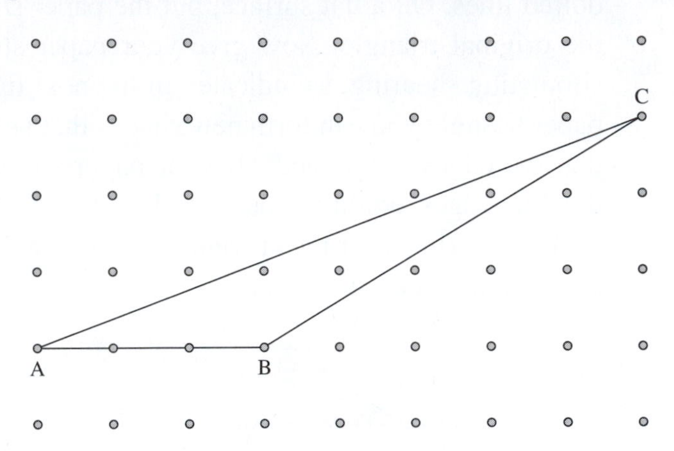

4. Determine the area of the next triangle by first shearing the triangle into a right triangle and then using the moving and additivity principles to determine the area of the right triangle. Explain why you can determine the area of the original triangle by this method. Note: Shearing does not have to be horizontal.

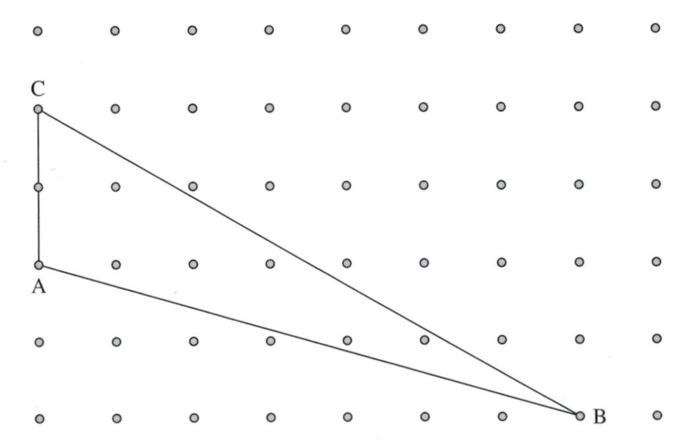

11.6 Areas of Circles and the Number Pi

Class Activity 11P: How Big Is the Number π?

Suppose that the only thing you know about the number π is that it is the circumference of a circle of diameter 1. Explain how you can use the next picture to give information about the size of the number π. (Based on the picture, you should be able to say that π lies between two numbers. Which two numbers?)

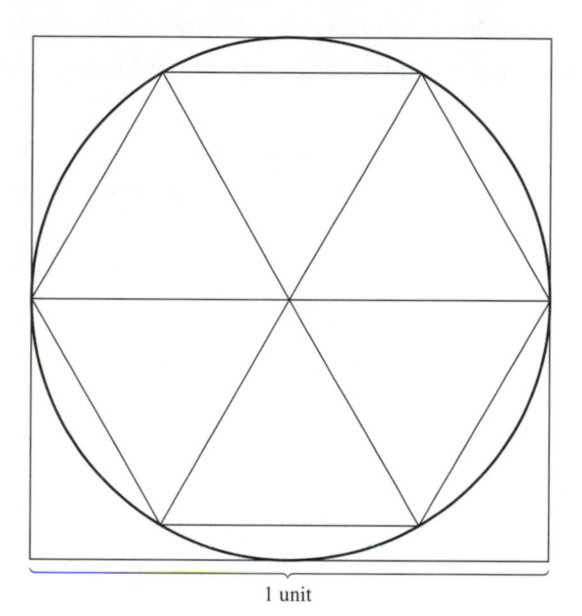

Class Activity 11Q: Over- and Underestimates for the Area of a Circle

Scissors might be helpful for this activity.

Use the next picture to explain why the area of a circle that has radius r units is less than $4r^2$ square units and greater than $2r^2$ square units. In other words, explain why

$$2r^2 \text{ units}^2 < \text{area of circle} < 4r^2 \text{ units}^2$$

In order to explain why the area of the circle is greater than $2r^2$ square units, it may help you to trace, cut out, and rearrange the triangles inside the circle.

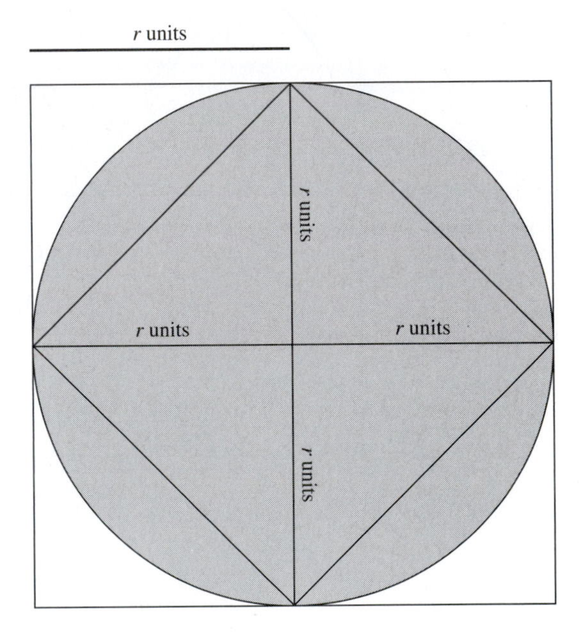

Class Activity 11R: Why the Area Formula for Circles Makes Sense

You will need scissors for this activity.

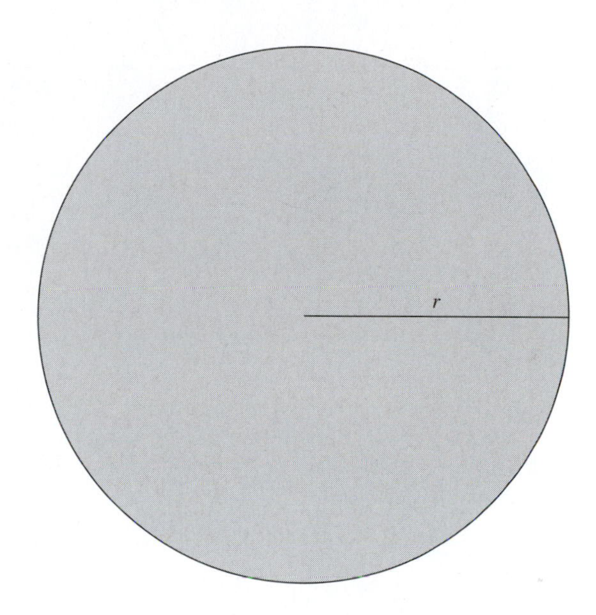

1. Trace the given circle, and cut your traced circle into 8 pie pieces. Arrange the 8 pie pieces as shown at the top of Figure 11R.1.
2. Now cut your traced circle into 16 pie pieces. Arrange the 16 pie pieces as shown in the middle of Figure 11R.1.
3. Imagine cutting a circle into more and more smaller and smaller pie pieces, and rearrange them as in Figure 11R.1.
   What shape would your rearranged circle become more and more like?
   What would the lengths of the sides of this shape be?
   What would the area of this shape be?
4. Using your answers to Problem 3, explain why it makes sense that a circle of radius r units has area πr^2 square units, given that the circumference of a circle of radius r is $2\pi r$.

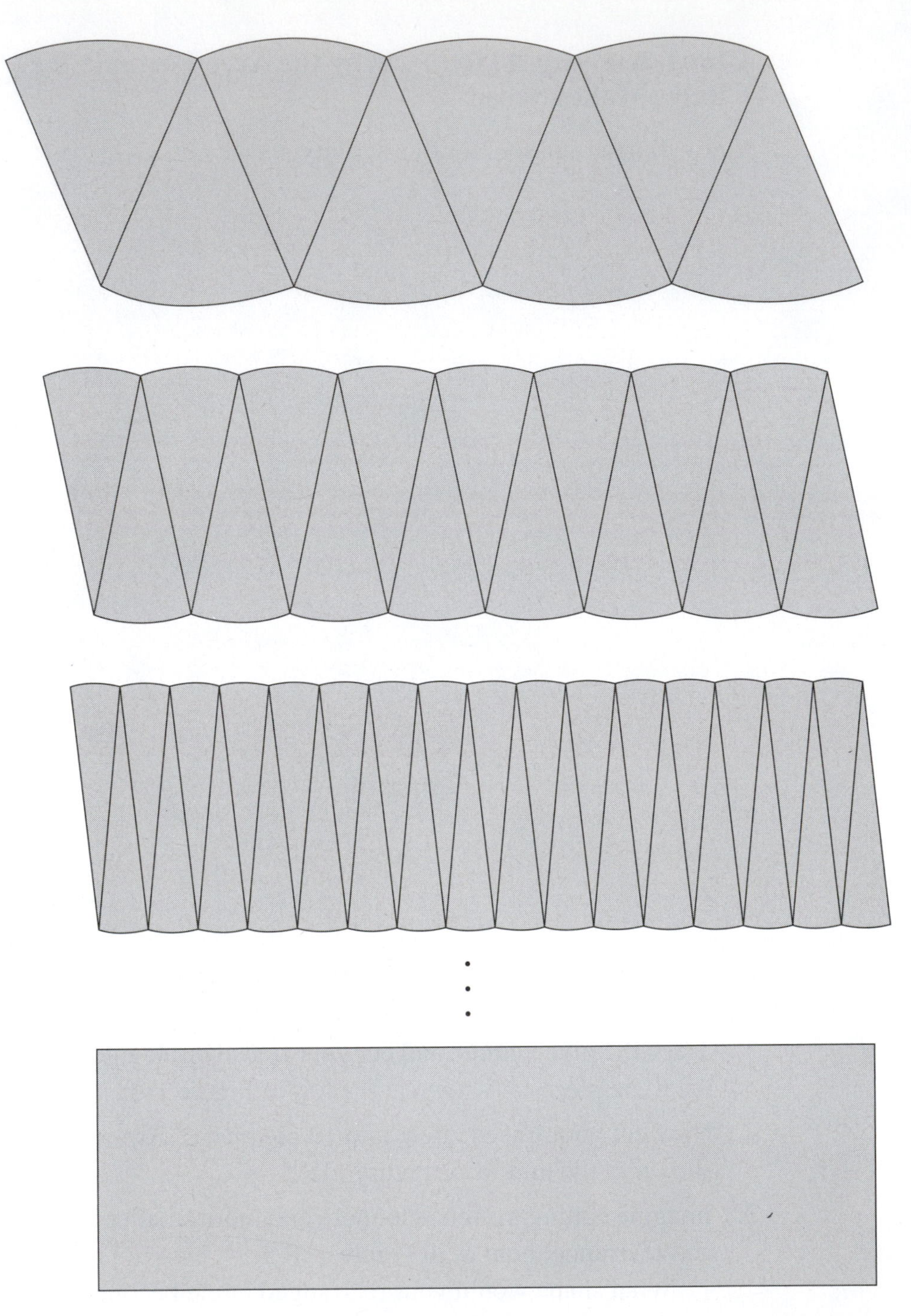

Figure 11R.1
Rearranging a Circle

Class Activity 11S: Using the Circle Circumference and Area Formulas to Find Areas and Surface Areas

You will need paper, a ruler, a compass, a protractor, and scissors for Problems 3, 4, and 5.

1. A company wants to manufacture tin cans that are 3 inches in diameter and 4 inches tall. Describe the shape and dimensions of the paper label the company will need to wrap around the curved portion of each can. Explain your reasoning.

2. A reflecting pool will be made in the shape of the shaded region shown in the next figure. The curves shown are $\frac{1}{4}$ circles. What is the area of the surface of the pool? Explain your reasoning.

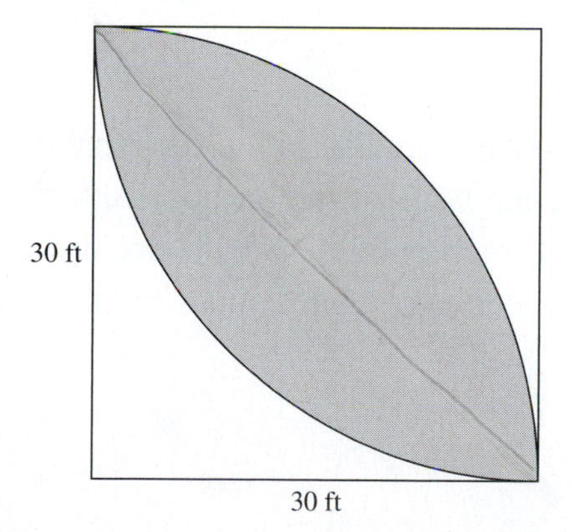

3. On a piece of paper, draw half of a circle that has radius 5 inches. (*Suggestion*: Put the point of your compass in the middle of the long edge of the piece of paper.) Cut out the half-circle, and attach the two radii on the straight edge (the diameter) in order to form a cone without a base. Calculate the radius of the circle that will form a base for your cone. Explain your reasoning. Then draw this circle, and verify that it does form the base of the cone.

4. Make a pattern for a cone such that the base is a circle of radius 2 inches and the cone without the base is made from a half-circle. Determine the total surface area of your cone. Explain your reasoning.

5. Make a pattern for a cone such that the base is a circle of radius 2 inches and the cone without the base is made from part of a circle of radius 6 inches. What fraction of the 6-inch circle will you need to use? Determine the total surface area of your cone. Explain your reasoning.

11.7 Approximating Areas of Irregular Shapes

Class Activity 11T: Determining the Area of an Irregular Shape

1. Think about several different ways that you might determine (at least approximately) the surface area of Lake Lalovely shown on the map in Figure 11T.1. Suppose that you have the following items on hand:

   - many 1-inch-by-1-inch paper or plastic squares
   - graph paper (adjacent lines separated by 1/4 inch)
   - a scale for weighing (such as one used to determine postage)
   - string
   - modeling dough or clay

   Which of these items could help you to determine approximately the surface area of the lake. How?

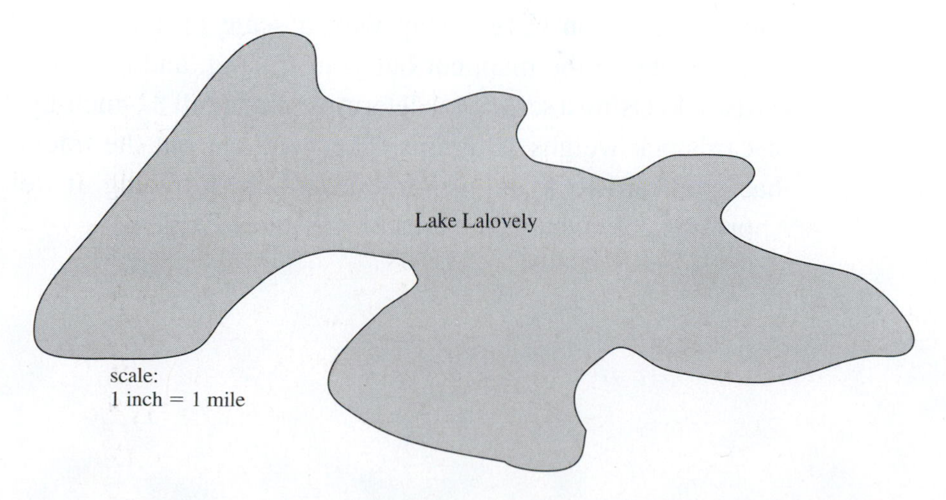

Figure 11T.1
Lake Lalovely

2. Suppose that you have a map with a scale of 1 inch = 100 miles. You trace a state on the map onto $\frac{1}{4}$-inch graph paper. (The grid lines are spaced $\frac{1}{4}$ inch apart.) You count that the state takes up about 100 squares of graph paper. Approximately what is the area of the state? Explain.

3. Suppose that you have a map with a scale of 1 inch = 15 miles. You cover a county on the map with a $\frac{1}{8}$-inch-thick layer of modeling dough. Then you re-form this piece of modeling dough into a $\frac{1}{8}$-inch-thick rectangle. The rectangle is $2\frac{1}{2}$ inches by $3\frac{3}{4}$ inches. Approximately what is the area of the county? Explain.

4. Suppose that you have a map with a scale of 1 inch = 50 miles. You trace a state on the map, cut out your tracing, and draw this tracing onto cardstock. Using a scale, you determine that a full $8\frac{1}{2}$-inch-by-11-inch sheet of cardstock weighs 10 grams. Then you cut out the tracing of the state that is on cardstock and weigh this card-stock tracing. It weighs 6 grams. Approximately what is the area of the state? Explain.

11.8 Relating the Perimeter and Area of a Shape

Class Activity 11U: How Are Perimeter and Area Related?

If available, string would be useful for this activity.

1. Suppose surveyors have determined that a forest on flat terrain has a perimeter of 200 miles. What is the area of the forest? Can you answer this question without additional information, or is more than one answer possible? Draw pictures, experiment, discuss, and debate this question for a few minutes. If available, use a loop of string to help you think about the question.

2. Now suppose that you learn that the forest is in the shape of a rectangle. Can you determine the area now, or is more than one answer possible? Figure 11U.1 shows centimeter graph paper. Draw at least 5 different rectangles of perimeter 20 cm on this graph paper. Include some rectangles that have sides whose lengths *aren't whole numbers* of centimeters.

3. Determine the areas of the rectangles you just drew. In the next table, list the areas of your rectangles in decreasing order. Below each area, draw a small sketch showing the approximate shape of the corresponding rectangle.

   How are the larger area rectangles qualitatively different in shape from the smaller area ones?

area						
shape						

4. Show how two people can use a loop of string and 4 fingers to represent all rectangles of a certain fixed perimeter.

Now consider all the rectangles of perimeter 20 cm, *including those whose side lengths aren't whole numbers of centimeters*. What are all the theoretical possibilities for the areas of rectangles of perimeter 20 cm? What is the largest possible area, and what is the smallest possible area (if there is one)?

5. Suppose that a forest on flat terrain is shaped like a rectangle and has perimeter 200 miles. What can you say about the forest's area? Give the most informative answer that you can.

6. Now suppose that a forest on flat terrain has perimeter 200 miles, but there is no information on the shape of the forest. Now what can you say about the forest's area? Use a loop of string to help you think about this question.

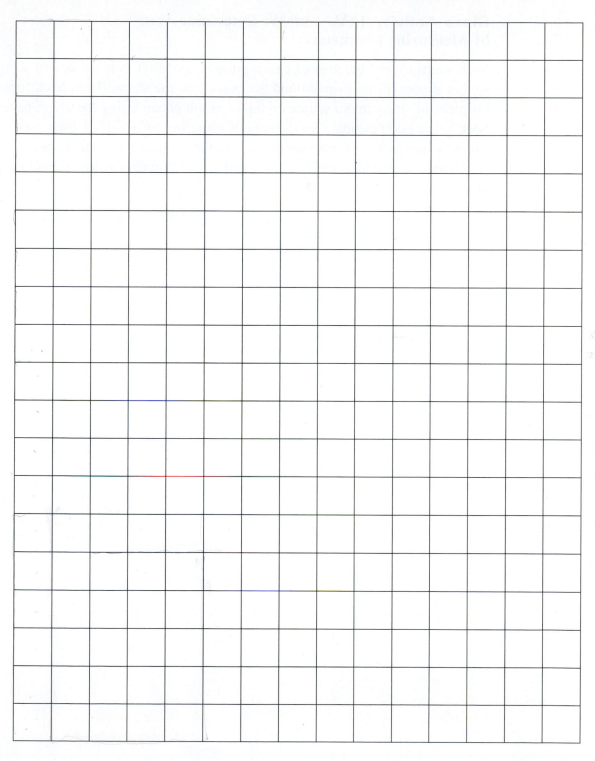

Figure 11U.1
Centimeter Graph Paper

Class Activity 11V: Can We Determine Area by Measuring Perimeter?

Nick wants to find the area of an irregular shape. Nick cuts a piece of string so that it goes all the way around the outside of the shape. Then Nick forms his piece of string into a square on top of graph paper. Using the graph paper, Nick gets a good estimate for the area of his string square. Nick then uses the square's area as his estimate for the area of the original irregular shape.

Discuss whether or not Nick's method is a legitimate way to estimate areas of irregular shapes.

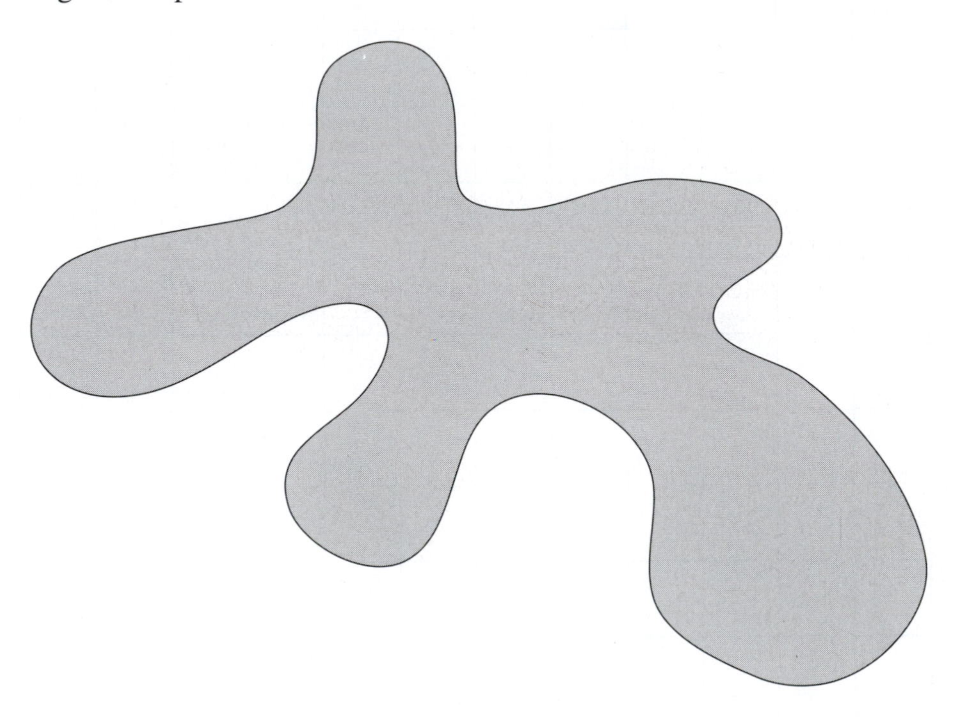

11.9 Principles for Determining Volumes

Class Activity 11W: Using the Moving and Additivity Principles to Determine Volumes

1. Using only the volume formula for *rectangular* prisms and the moving and additivity principles about volumes, determine a formula for the volume of the triangular prism in the next figure. You may wish to cut out the patterns on page 649 to make two models of the triangular prism.

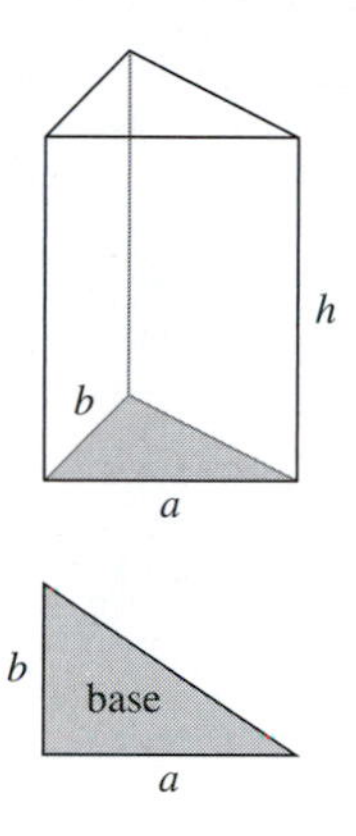

2. Using only the volume formula for *rectangular* prisms and the moving and additivity principles about volumes, determine the volume of the *parallelogram-base* prism on the left in the next figure in terms of a, b, and h. You may wish to cut out the patterns on pages 651 and 653 to make models of the three prisms shown.

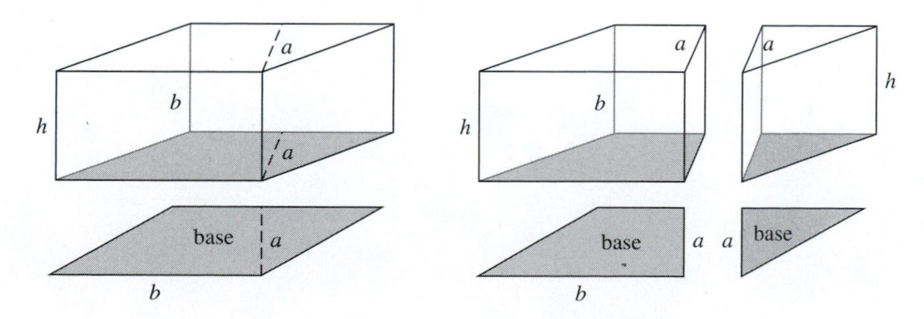

3. Explain how to use the idea of Class Activity 11R to "rearrange" a right cylinder to form a rectangular prism, as indicated in the next figure. Use this idea to determine the volume of the cylinder in terms of r and h.

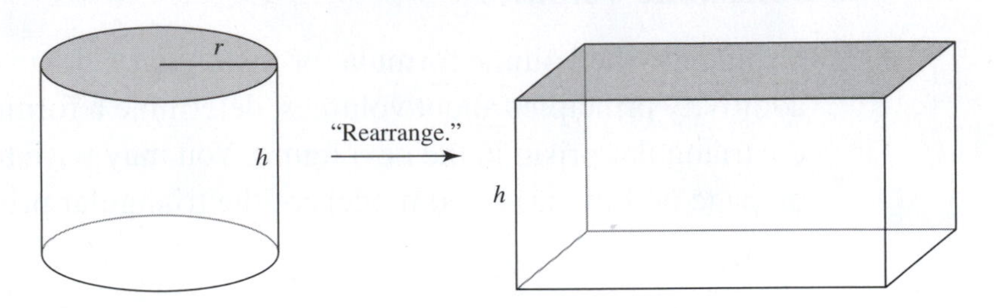

Class Activity 11X: Determining Volumes by Submersing in Water

For parts of this activity you will need a milliliter measuring cup, water, a centimeter ruler, and a good-sized lump of clay that still fits comfortably in your measuring cup. Use synthetic clay that does not dissolve in water. (Modeling dough dissolves in water.)

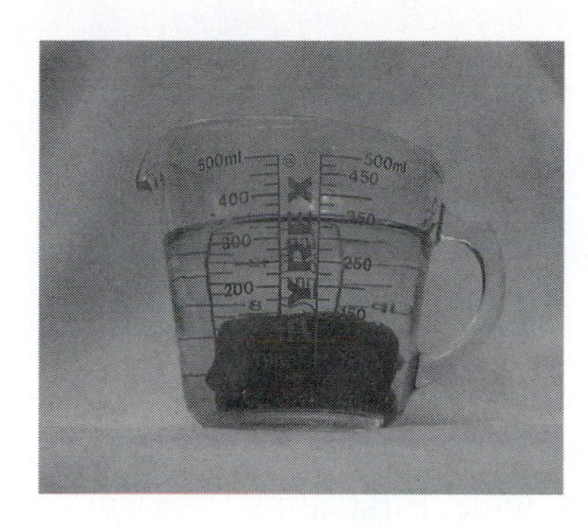

1. Form your lump of clay into a rectangular prism, and determine the volume of the clay in cubic centimeters by measuring the height, width, and length of the rectangular prism.

2. Pour water into your measuring cup, and note the volume of water in milliliters. Predict what will happen to the water level when you submerse your clay prism in the water. Then test your prediction.

3. What if you subdivided your clay into several pieces, or formed the clay into a different shape? Would the clay still displace the same amount of water when it is submersed?

4. A container can hold 2 liters. Initially, the container is $\frac{1}{2}$ full of water. When an object is placed in the container, the object sinks to the bottom and the container becomes $\frac{2}{3}$ full. What is the volume of the object in cubic centimeters? Explain.

5. A tank in the shape of a rectangular prism is 50 cm tall, 80 cm long, and 30 cm wide. First, some rocks were placed at the bottom of the tank. Then 80 liters of water were poured into the tank. At that point, the tank was $\frac{3}{4}$ full. What is the total volume of the rocks in cubic meters? Explain.

Class Activity 11Y: Floating Versus Sinking: Archimedes's Principle

For this activity you will need a milliliter measuring cup, water, and modeling clay that does not dissolve in water. In Problem 3 you will need a scale for weighing. In Problem 4 you will need a small paper cup, and several coins or other small, heavy objects that fit inside the cup.

1. Pour water into your measuring cup and note the volume of water in milliliters. Form your clay into a "boat" that will float. By how much does the water level rise when you float your clay boat? Does this increase in water level tell you the volume of the clay?

2. Predict what will happen to the water level when you sink your clay boat: Will the water level go up, or will the water level go down? Sink your boat, and see if your prediction was correct.

   Did the increase in water level in Problem 1 tell you the volume of the clay, or not? Explain.

3. Exactly how much water does a floating object displace? Do the following to find out:

   a. Weigh your clay boat from Problem 1.

   b. Weigh an amount of water that weighs as much as your clay boat.

   c. Float your clay boat in the measuring cup, and record the water level.

   d. Remove your clay boat from the water, and pour the water you measured in part (b) into the measuring cup. Compare the water level now to the water level in part (c). If you did everything correctly, these two water levels should be the same.

   This experiment illustrates **Archimedes's principle**, that a floating object displaces an amount of water that weighs as much as the object.

4. Determine approximately how much a quarter weighs in grams by using Archimedes's principle. Observe how much the water level rises when you put several quarters in a small cup floating in water in a measuring cup.

5. Ships usually have markings on their sides to show how low or high the ship is floating in the water. Besides showing how low or high the ship is floating, what other information could the captain of the ship get by observing which mark is at the water line?

6. The Great Salt Lake in Utah and the Dead Sea in Israel and Jordan are lakes that have so much salt in them that people float in the water. (See Figure 11Y.1.) Use Archimedes's principle to explain why people float in these lakes. What is it about this salty water that causes people to float?

Figure 11Y.1
Floating in the Dead Sea

11.10 Volumes of Prisms, Cylinders, Pyramids, and Cones

Class Activity 11Z: Why the Volume Formula for Prisms and Cylinders Makes Sense

Cubic inch blocks would be helpful for this activity.

1. Explain why the

$$(\text{height}) \times (\text{area of base})$$

formula gives the correct volume for right prisms and cylinders. Do so by imagining that the prism or cylinder is built with 1-unit-by-1-unit-by-1-unit clay cubes (or other cubes which could be cut into pieces, if necessary). Think about building the shape in layers.

- How many layers would you need?
- How many cubes would be in each layer?

If cubic inch blocks are available, it may help you to build prisms on the bases shown here, as instructed. Use these models to help you explain why the volume formula is valid.

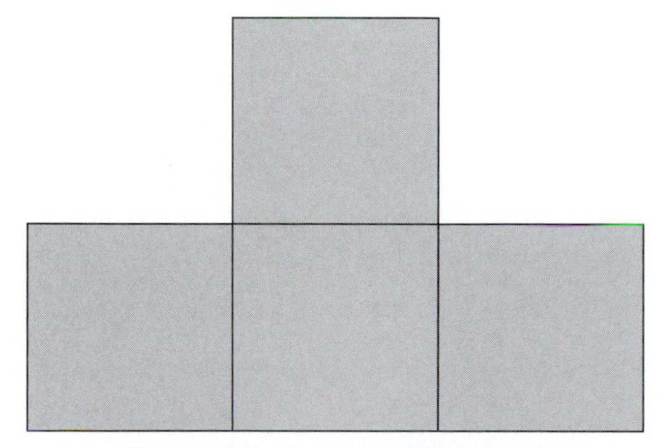

Build a 3 inch tall prism that has this base.

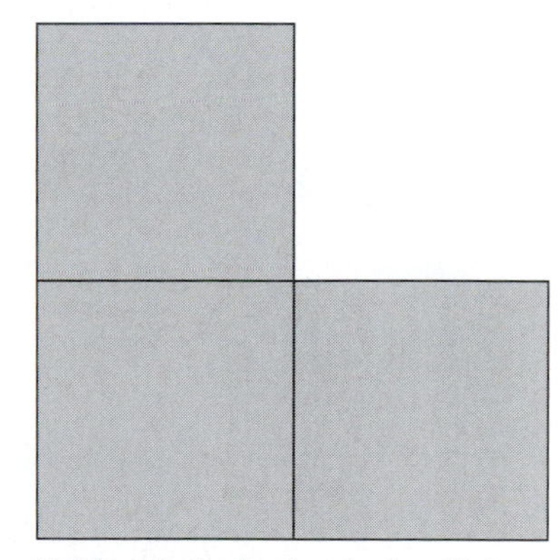

Build a 4 inch tall prism that has this base.

2. Use the result of Problem 1 and Cavalieri's principle to explain why the

$$(\text{height}) \times (\text{area of base})$$

gives the correct volume for an *oblique* prism or cylinder. Explain why the height should be measured perpendicular to the bases, and not "on the slant."

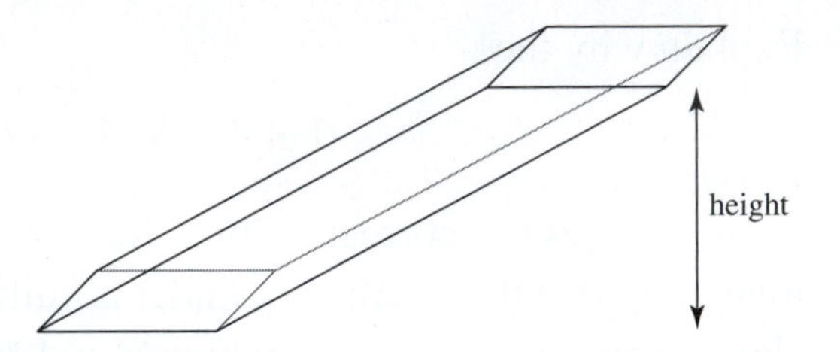

Class Activity 11AA: Filling Boxes and Jars

1. How many plastic 2-cm-by-2-cm-by-2-cm cubes can be stacked neatly in an 8-cm-by-10-cm-by-12-cm box? Explain.

2. How many plastic 2-cm-by-2-cm-by-2-cm cubes can be stacked neatly in an 8-cm-by-9-cm-by-12-cm box? Explain.

3. A BerryBombs cereal box is 10 inches tall, $7\frac{1}{2}$ inches wide, and $2\frac{1}{2}$ inches deep. Give the length, width, and height of a cardboard box that could hold exactly 12 BerryBombs cereal boxes, leaving no extra space. Now find the dimensions of another box that could hold 12 BerryBombs cereal boxes with no extra space.

4. A 9-inch-by-9-inch-by-12-inch jar in the shape of a rectangular prism is completely filled with gum balls of diameter $\frac{3}{4}$ inches. Estimate the number of gum balls in the jar, explaining your reasoning.

5. A jar in the shape of a 9-inch tall cylinder with a circular base of diameter 6 inches is completely filled with gum balls of diameter $\frac{3}{4}$ inches. Estimate the number of gum balls in the jar, explaining your reasoning.

Class Activity 11BB: Comparing the Volume of a Pyramid with the Volume of a Rectangular Prism

You will need scissors, tape, and dry beans or rice for this activity.

1. Cut out, fold, and tape the patterns on pages 657 and 659 to make an open rectangular prism and an open pyramid with a square base.

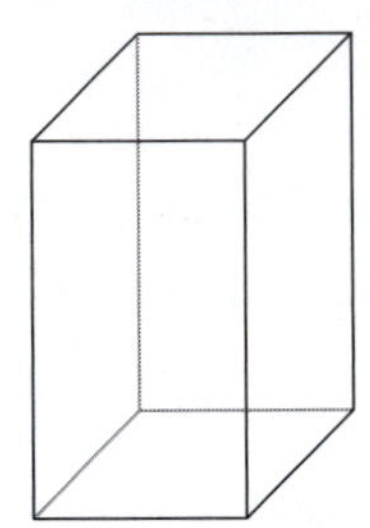

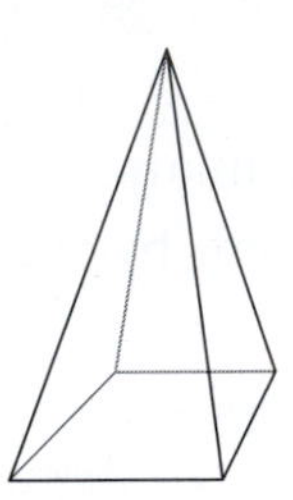

2. Verify that the prism and the pyramid have bases of the same area and have equal heights.

3. Just by looking at your shapes, make a guess: How do you think the volume of the pyramid compares with the volume of the prism?

4. Now fill the pyramid with beans, and pour the beans into the prism. Keep filling and pouring until the prism is full. Based on your results, fill in the blanks in the equations that follow:

volume of prism = ___ × volume of pyramid

volume of pyramid = ___ × volume of prism

Class Activity 11CC: The $\frac{1}{3}$ in the Volume Formula for Pyramids and Cones

You will need scissors and tape for this activity.

According to the volume formula, a right pyramid that is 1 unit high and has a 1-unit-by-1-unit square base has volume

$$\frac{1}{3} \times 1 \times (1 \times 1) \text{ unit}^3 = \frac{1}{3} \text{ unit}^3$$

Now pretend that you *don't yet* know the volume formula for pyramids and cones. This class activity will help you use fundamental principles about volumes to explain where the $\frac{1}{3}$ comes from.

1. Cut out three of the four patterns on pages 661 and 663. (The fourth is a spare.) Fold these patterns along the *undashed* line segments, and glue or tape them to make three *oblique* pyramids. Make sure the dashed lines appear on the outside of each oblique pyramid.

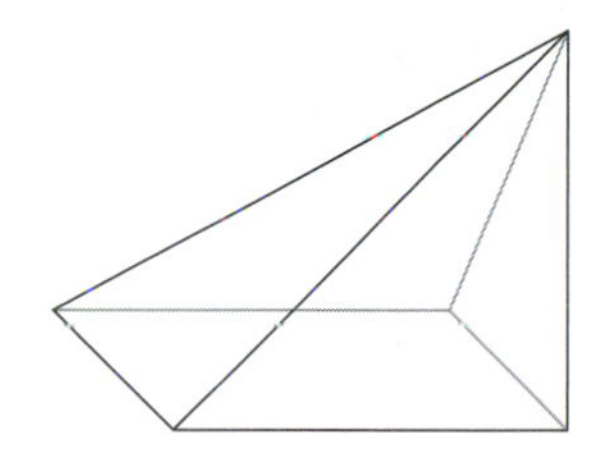

2. Fit the three oblique pyramids together to make a familiar shape. What shape is it? What is the volume of the shape formed from the three oblique pyramids? Therefore, what is the volume of one of the oblique pyramids?
3. Use your answers to Problems 2 and 3, and Cavalieri's principle, to explain why a *right* pyramid that is 1 unit high and has a 1-unit-by-1-unit square base has volume $\frac{1}{3}$ cubic units. (The dashed lines on the model oblique pyramids are meant to help you see how to shear the oblique pyramid. Imagine that the oblique pyramids are made out of a stack of small pieces of paper, where each dashed line going around the oblique pyramid represents a piece of paper.)

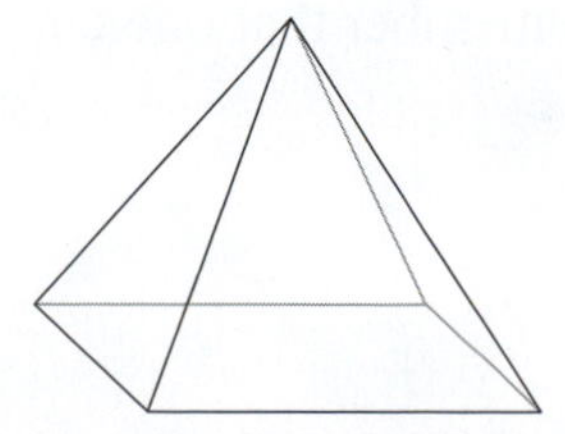

Class Activity 11DD: Using Volume Formulas with Real Objects

For this activity, you will need several drink containers that show how much they hold in milliliters or liters. You will also need either a tape measure or scissors and tape.

1. If you don't already have a centimeter tape measure, cut out Figure A.31 on page 655, and tape it together to make a centimeter tape measure.

2. Use a centimeter tape measure to measure the lengths of various parts of your drink containers. Use these measurements to calculate the approximate volume of liquid in your (full) drink containers in cubic centimeters.

3. Compare your calculated volumes of the drink containers to the capacities (in milliliters or liters) shown on the containers. In each case, do the two agree? Remember that one cubic centimeter holds one milliliter of liquid.

Class Activity 11EE: Volume and Surface Area Contests

You will need several blank pieces of paper, a ruler, and a protractor for this activity. Scissors and tape would also be useful.

Work in a group (which is ideally an entire class). For each problem in this activity, group members compete to make the object with the largest volume and the object with the largest surface area. In each case, be prepared to explain your calculations to others. If scissors and tape are available, you may wish to test your patterns.

1. Use no more than one ordinary piece of paper to make a cylinder with no bases. You may cut the paper apart to make your cylinder. Determine the surface area and the volume of your cylinder.

   Which person in the group has the cylinder of largest volume?

   Which person in the group has the cylinder of largest surface area?

2. Use no more than one ordinary piece of paper to make a pattern for a cylinder with two bases (a top and bottom). The pattern may consist of several separate pieces. Determine the surface area and the volume of the cylinder your pattern will make.

   Which person in the group has the cylinder of largest volume?

   Which person in the group has the cylinder of largest surface area? Is this the same cylinder as the one of largest volume?

3. Use no more than one ordinary piece of paper to make a pattern for a cone without a base. Determine the surface area (without the base) and the volume of the cone your pattern will make.

   Which person in the group has the cone of largest volume?

   Which person in the group has the cone of largest surface area (not including the base)? Is this the same cone as the one of largest volume?

Class Activity 11FF: Volume Problems

1. A fish tank in the shape of a rectangular prism is 1 m long and 30 cm wide. The tank was $\frac{1}{2}$ full. Then 30 liters of water were poured in, and the tank became $\frac{2}{3}$ full. How tall is the tank? Explain your reasoning.

2. A cone without a base is made from a half-circle of radius 10 cm. Determine the volume of the cone. Explain your reasoning.

3. A cup has a circular opening and a circular base. A cross section of the cup and the dimensions of the cup are shown below. Determine the volume of the cup. Explain your reasoning.

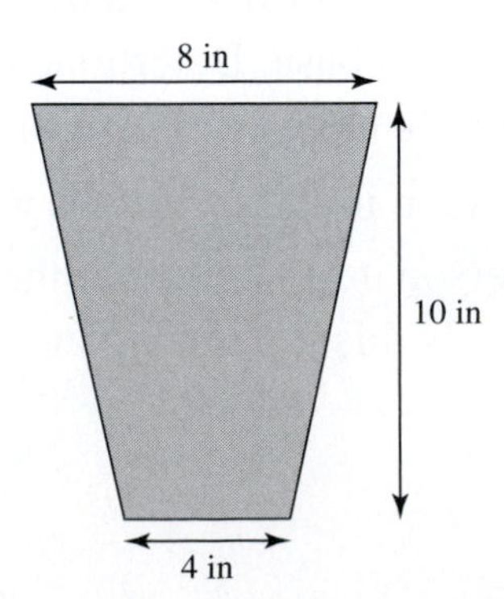

Class Activity 11GG: * The Volume of a Rhombic Dodecahedron

1. Cut out the patterns on page 665, and use them to make 6 pyramids.

2. Cut out the pattern on the bottom of page 667, and use it to make a cube.

3. Let's say the cube is 1 unit wide, 1 unit deep, and 1 unit high. Determine the volumes of the pyramids. You may put shapes together to do so.

4. Cut out the two patterns at the top of page 667, and put them together to make a closed shape. The shape will have 12 faces, each of which is a rhombus. The shape is called a **rhombic dodecahedron**.

5. Determine the volume of the rhombic dodecahedron. You may put shapes together to do so.

6. One interesting property of rhombic dodecahedra is that they can be stacked and fit together to fill up space without any gaps. If you have a number of rhombic dodecahedra, try this out. Cubes can also be stacked and fit together to fill up space without any gaps, but ordinary dodecahedra (with pentagon faces) cannot.

11.11 Areas, Volumes, and Scaling

Class Activity 11HH: Areas and Volumes of Similar Boxes

You will need scissors and tape for this activity.

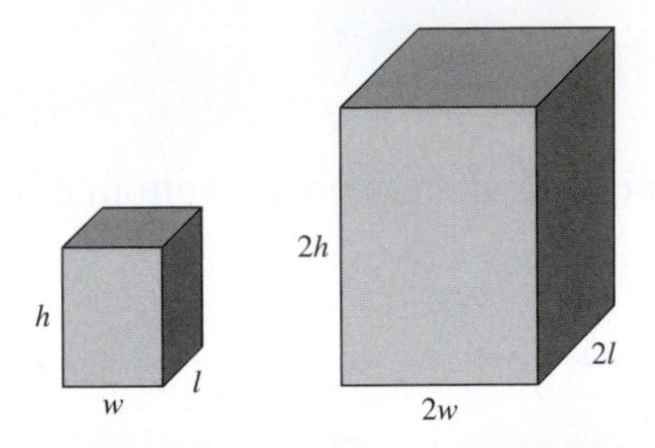

1. Cut out the patterns in Figures A.38 and A.39 on pages 669 and 671. Fold these patterns to make two boxes (rectangular prisms), but leave them untaped, so that you can still unfold them. One box will have width w, length l, and height h; and one box will have width $2w$, length $2l$, and height $2h$. So the big box is twice as wide, twice as long, and twice as high as the small box.
2. By working with the unfolded patterns of the two boxes, determine how the surface area of the big box compares with the surface area of the small box. Is the surface area of the big box twice as large, three times as large, etc., as the surface area of the small box? Look and think carefully—the answer may not be what you first think it is. Explain clearly why your answer is correct.

3. Now tape up the small box, and tape up most of the large box. Leave an opening so that you can put the small box inside it. Determine how the volume of the big box compares with the volume of the small box. Is the volume of the big box twice as large, three times as large, etc., as the volume of the small box? Once again, think carefully, and explain why your answer is correct.

4. What if there were an even bigger box whose width, length, and height were each three times the respective width, length, and height of the small box? How would the surface area of the bigger box compare with the surface area of the small box? Answer this question by thinking about how to make a pattern for the bigger box and determining how the pattern of the bigger box would compare with the pattern of the small box.

   How would the volume of the bigger box compare with the volume of the small box?

5. Now imagine a variety of bigger boxes. Fill in the first two empty columns in the next table with your previous results. Continue to fill in the remaining empty columns by extrapolating from your results.

Size of big box compared with small box					
length, width, depth	**2 times**	**3 times**	**5 times**	**2.7 times**	k **times**
surface area					
volume					

Class Activity 11II: Areas and Volumes of Similar Cylinders

You will need scissors and tape for this activity.

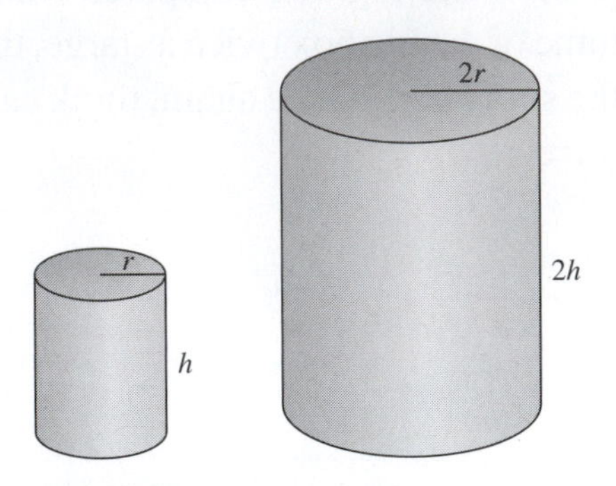

1. Cut out the patterns for cylinders, Figures A.40 and A.41 on pages 673 and 675. Use the patterns to make two cylinders, but don't glue or tape them because you will need to unfold them. The small cylinder has radius r and height h. The big cylinder has radius $2r$ and height $2h$, so the big cylinder has twice the radius and is twice as high as the small one.
2. How does the surface area of the large cylinder compare with the surface area of the small cylinder? Is it twice as large, three times as large, etc.? Use your patterns to get a feel for what the answer should be. Then give a clear and thorough explanation of your answer. To give a thorough explanation, you will have to do more than just physically compare the cylinders' patterns.

3. How does the volume of the large cylinder compare with the volume of the small cylinder? Is it twice as large, three times as large, etc.? Tape up your patterns to form cylinders, and use the cylinders to get a feel for what the answer should be. Then give a clear and thorough explanation of your answer. To give a thorough explanation, you will have to do more than just physically compare the cylinders.

4. Now imagine a variety of bigger cylinders. Fill in the first empty column in the next table with your previous results. Continue to fill in the remaining empty columns by extrapolating from your results. Compare this table with the table in Problem 5 of the previous Class Activity.

Size of big cylinder compared with small cylinder					
length, width, depth	**2 times**	**3 times**	**5 times**	**2.7 times**	k **times**
surface area					
volume					

Class Activity 11JJ: Determining Areas and Volumes of Scaled Objects

1. Compare your tables for Problem 5 of Class Activity 11HH and Problem 4 of Class Activity 11II. Extrapolate from these results to answer the following questions: If someone made a Goodyear blimp that was 1.5 times as wide, 1.5 times as long, and 1.5 times as high as the current one, how much material would it take to make the larger blimp compared with the current blimp? How much more gas would it take to fill this bigger Goodyear blimp, compared with the current one (at the same pressure)?

2. An adult alligator can be 15 feet long and weigh 475 pounds. Suppose that some excavated dinosaur bones indicate that the dinosaur was 30 feet long and was shaped roughly like an alligator. How much would you expect the dinosaur to have weighed?

Class Activity 11KK: A Scaling Proof of the Pythagorean Theorem

This activity will help you use similar shapes to prove the Pythagorean theorem.

Remember that the Pythagorean theorem says that for any right triangle with short sides of length a and b, and hypotenuse of length c,

$$a^2 + b^2 = c^2$$

1. Given any right triangle, such as the next triangle on the left, drop the perpendicular to the hypotenuse, as shown on the right.

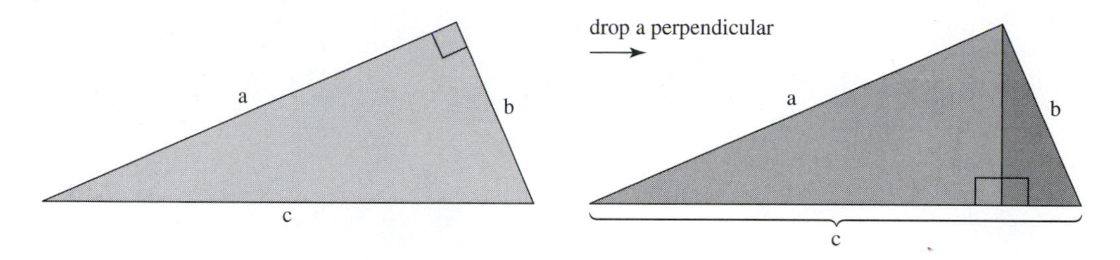

Use angles to explain why the two smaller right triangles on the right are similar to the original right triangle. (Do not use any actual measurements of angles, because the proof must be general—it must work for *any* initial right triangle.)

2. Now view each of the three right triangles from Problem 1 (the original triangle and the two smaller ones) as taking up a percentage of the area of the square formed on its hypotenuse, as in the next figure. Why must each triangle take up the same percentage of its square?

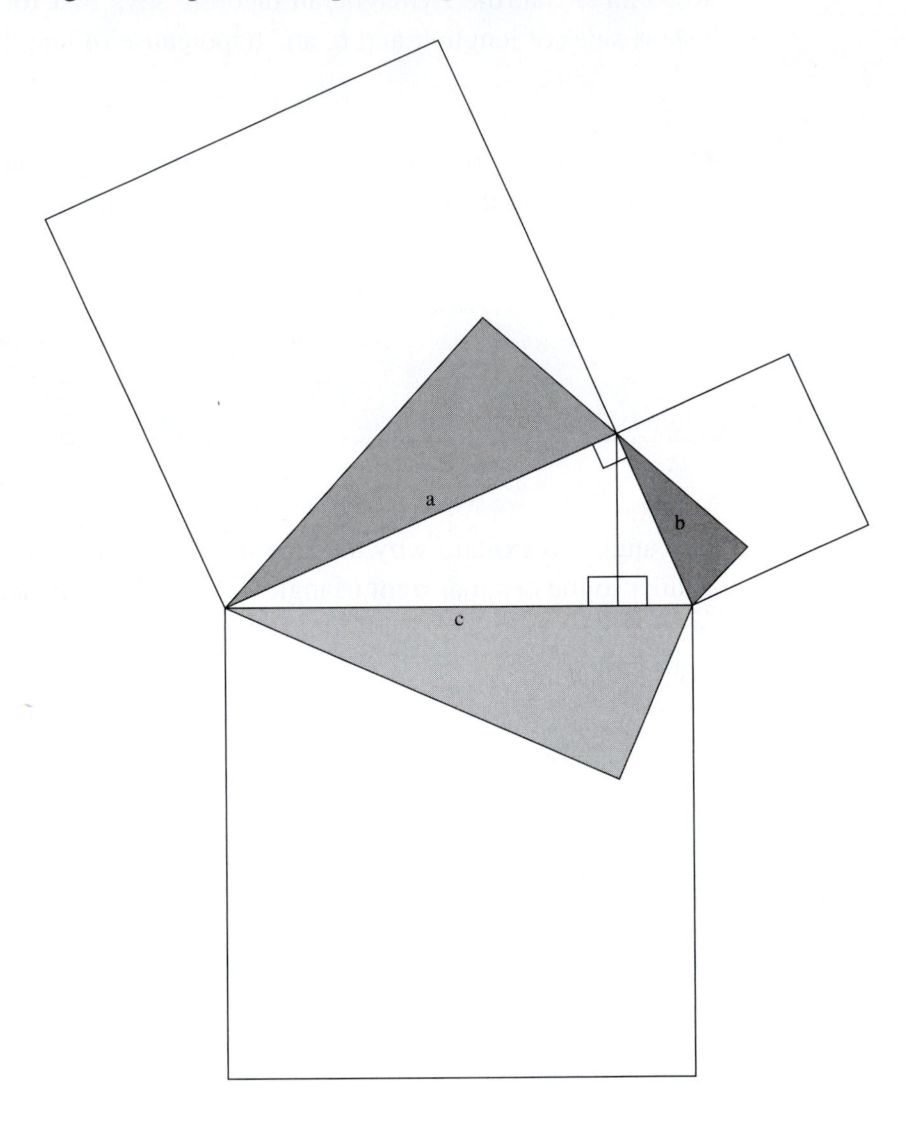

3. Let $P\%$ be the percentage of Problem 2. Express the areas of the three triangles in terms of $P\%$, and then explain why

$$P\% \cdot a^2 + P\% \cdot b^2 = P\% \cdot c^2$$

4. Use Problem 3 to explain why

$$a^2 + b^2 = c^2$$

thus proving the Pythagorean theorem.

12

Number Theory

12.1 Factors and Multiples

Class Activity 12A: Factors, Multiples, and Rectangles

1. Elsie has 24 square tiles that she wants to arrange in the shape of a rectangle in such a way that the rectangle is completely filled with tiles. How many different rectangles can Elsie make?

   Write one or more statements about factors or multiples that are related to this problem.

2. What if Elsie has only 13 tiles? Now how many rectangles can she make?

3. If Elsie has more than 24 square tiles, will she necessarily be able to make more rectangles than she could in Problem 1? Try some experiments.

4. John has many 1-inch-by-1-inch square tiles. John will use his tiles to make rectangles (in such a way that each rectangle is filled with tiles), each of which has a side that is 4 inches long. What are the areas of the rectangles that John can make?

   Write one or more statements about factors or multiples that are related to this problem.

Class Activity 12B: Problems about Factors and Multiples

1. For each of the situations listed, write a problem about the items or the scenario. In each case, solving your problem should require finding the factors or multiples of a number. Solve each problem.

   a. At the store, pencils come in packages of 12.

   b. Amy has a collection of 45 marbles.

   c. The children in a class are clapping, snapping, and stomping to a steady beat in the following pattern: clap, clap, snap, stomp, clap, clap, snap, stomp, ...

2. For each of the situations listed, describe an activity involving the items. In each case, your activity should require finding the factors or multiples of a number.

   a. Beads in several different colors and string with which to make necklaces

   b. Graph paper and markers (optional: scissors)

   c. A drum and a tambourine

3. Consider the following problem:

   A rectangular garden has an area of 64 square feet. What could its length and width be?

   Why can you not solve the problem just by finding all the factors of 64? What must you add to the statement of the problem so that it *can* be solved by finding the factors of 64?

Class Activity 12C: Finding All Factors

1. Tyrese is looking for all the factors of 156. So far, Tyrese has divided 156 by all the counting numbers from 1 to 13, listing those numbers that divide 156 and listing the corresponding quotients. Here is Tyrese's work so far:

1, 156	$1 \times 156 = 156$
2, 78	$2 \times 78 = 156$
3, 52	$3 \times 52 = 156$
4, 39	$4 \times 39 = 156$
6, 26	$6 \times 26 = 156$
12, 13	$12 \times 13 = 156$
13, 12	$13 \times 12 = 156$

Should Tyrese keep checking to see if numbers larger than 13 divide 156, or can Tyrese stop dividing at this point? If so, why? What are all the factors of 156?

2. Find all the factors of 198 in an efficient way.

Class Activity 12D: Do Factors Always Come in Pairs?

Carmina noticed that factors always seem to come in pairs. For example,

$48 = 1 \times 48$, 1 and 48 are a pair of factors of 48.
$48 = 2 \times 24$, 2 and 24 are a pair of factors of 48.
$48 = 3 \times 16$, 3 and 16 are a pair of factors of 48.
$48 = 4 \times 12$, 4 and 12 are a pair of factors of 48.
$48 = 6 \times 8$, 6 and 8 are a pair of factors of 48.

The number 48 has 10 factors that come in 5 pairs. Carmina wants to know if every counting number always has an even number of factors. Investigate Carmina's question carefully. When does a counting number have an even number of factors, and when does it not?

12.2 Greatest Common Factor and Least Common Multiple

Class Activity 12E: Finding Commonality

1. Follow the instructions on the next page.
2. You have 24 pencils, 30 stickers, and plenty of goodie bags.
   a. List all the ways you could distribute the pencils to goodie bags so that each goodie bag has the same number of pencils and there are no pencils left over.

   b. List all the ways that you could distribute the stickers to goodie bags so that each goodie bag has the same number of stickers and there are no stickers left over.

   c. Now suppose you want to use the same set of goodie bags for both the pencils (as in Part (a)) and the stickers (as in Part (b)). List all your options. What is the largest number of goodie bags you could use?

   d. Describe the options in Part (c) in mathematical terms. Describe the largest number of goodie bags in Part (c) in mathematical terms.

Shade the squares above the multiples of 4 on strip A, and shade the multiples of 6 on strip B. Predict when the shaded squares on both strips will line up.

A

1 2 3 4 5 6 7 8 9 10 11 12 13 14 15 16 17 18 19 20 21 22 23 24 25 26 27 28 29 30 31 32 33 34 35 36 37 38 39 40

B

1 2 3 4 5 6 7 8 9 10 11 12 13 14 15 16 17 18 19 20 21 22 23 24 25 26 27 28 29 30 31 32 33 34 35 36 37 38 39 40

At which numbers do the shaded squares in both strips line up? Describe these locations in mathematical terms.
What is the first place the shaded squares line up? Describe this location in mathematical terms.

Shade the squares above the multiples of 4 on strip C, and shade the multiples of 10 on strip D. Predict when the shaded squares on both strips will line up.

C

1 2 3 4 5 6 7 8 9 10 11 12 13 14 15 16 17 18 19 20 21 22 23 24 25 26 27 28 29 30 31 32 33 34 35 36 37 38 39 40

D

1 2 3 4 5 6 7 8 9 10 11 12 13 14 15 16 17 18 19 20 21 22 23 24 25 26 27 28 29 30 31 32 33 34 35 36 37 38 39 40

At which numbers do the shaded squares in both strips line up? Describe these locations in mathematical terms.
What is the first place the shaded squares line up? Describe this location in mathematical terms.

Class Activity 12F: The "Slide Method"

1. Examine the initial and final steps of a "slide," which was used to find the GCF and LCM of 900 and 360. Try to determine how it was made. Then make another "slide" to find the GCF and LCM of 900 and 360.

A Slide

initially:

	900, 360

final:

10	900, 360
2	90, 36
3	45, 18
3	15, 6
	5, 2

$$\text{GCF} = 10 \times 2 \times 3 \times 3 = 180$$
$$\text{LCM} = 10 \times 2 \times 3 \times 3 \times 5 \times 2 = 1800$$

2. Use the slide method to find the GCF and LCM of 1080 and 1200 and to find the GCF and LCM of 675 and 1125.

3. Why does the slide method work?

Class Activity 12G: Problems Involving Greatest Common Factors and Least Common Multiples

1. Pencils come in packages of 18; erasers that fit on top of these pencils come in packages of 24. What is the smallest number of pencils and erasers that you can buy so that each pencil can be matched with an eraser? How many packages of pencils will you need and how many packages of erasers? (Assume that you must buy whole packages—you can't buy partial packages.)

   Solve the problem, explaining your solution. Does this problem involve the GCF or the LCM? Explain.

2. Ko has a bag with 45 red candies and another bag with 30 green candies. Ko wants to make snack bags so that each snack bag contains the same number of red candies and each bag contains the same number of green candies (which may be different from the number of red candies), and so that all his candies will be used up in the snack bags. What is the largest number of snack bags that Ko can make this way? How many red candies and how many green candies will go in each snack bag?

   Solve the problem, explaining your solution. Does this problem involve the GCF or the LCM? Explain.

3. Sam has many 8-inch sticks that he is placing end to end to make a line of sticks. Becky has many 12-inch sticks that she is placing end to end to make a line of sticks. If Sam and Becky want their lines of sticks to be equally long, how long could they be? What is the shortest such length?

   Solve the problem, explaining your solution. Does this problem involve the GCF or the LCM? Explain.

4. Explain how a least common multiple is relevant to the following situation: A class is clapping and snapping to a steady beat. Half of the class uses the pattern

snap, snap, clap, snap, snap, clap, . . .

The other half of the class uses the pattern

snap, clap, snap, clap, snap, clap, . . .

5. Mary will make a small 8 inch by 12 inch rectangular quilt for a doll house out of identical square patches. Each square patch must have side lengths that are a whole number of inches and no partial squares are allowed in the quilt. Other than using 1 inch by 1 inch squares, what size squares can Mary use to make her quilt? Show Mary's other options below. What are the largest squares that Mary can use? How is this question related to GCFs or LCMs?

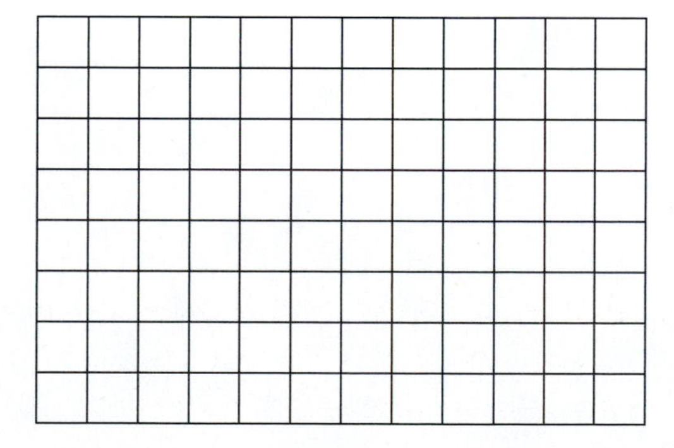

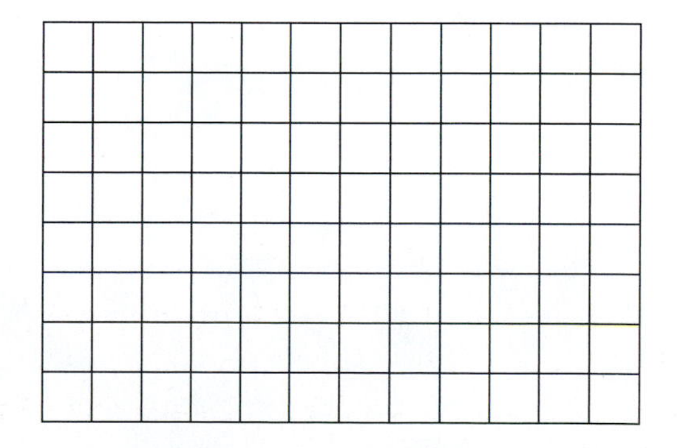

6. Two gears are meshed, as shown in the next figure, with the stars on each gear aligned. The large gear has 36 teeth, and the small gear has 15 teeth. Each gear rotates around a pin through its center. How many revolutions will the large gear have to make and how many revolutions will the small gear make in order for the stars to be aligned again? How is this question related to GCFs or LCMs?

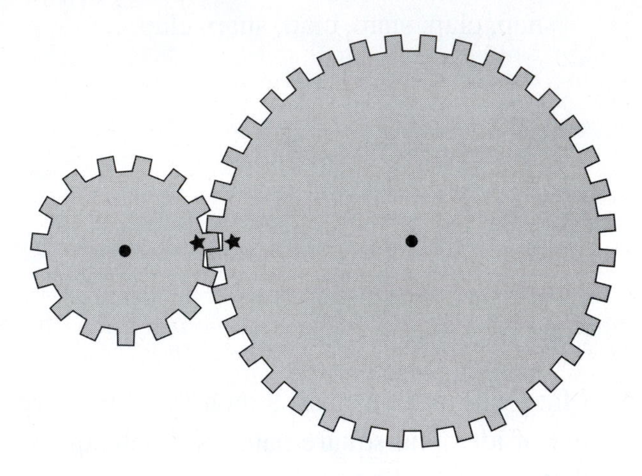

Class Activity 12H: Flower Designs

Examine the flower designs on the next page. Each flower design was created by starting with a number of dots in a circle. Then a fixed "jump number," N, is chosen. Starting at one dot, petals are formed by connecting each subsequent Nth dot until returning to the starting dot, at which point the flower design closes up.

1. Examine the first four "flower" designs in Figure 12H.1. Try to find relationships between the number of dots at the center of the flower, the "jump number," which describes how the petals were made (e.g., by connecting every 8th dot or every 15th dot), and the number of petals in the flower design. These numbers are listed in the following table:

design	number of dots	"Jump Number" a petal connects every ___th dot	number of petals
design 1	36	8	9
design 2	36	15	12
design 3	36	16	9
design 4	30	14	15
design 5	30	4	
design 6	30	12	

2. Predict the number of petals that the 5th and 6th flower designs will have. Then complete the designs to see if your prediction was correct. (To complete the designs, you might find it easiest to count dots forward and then draw the petal backwards.)

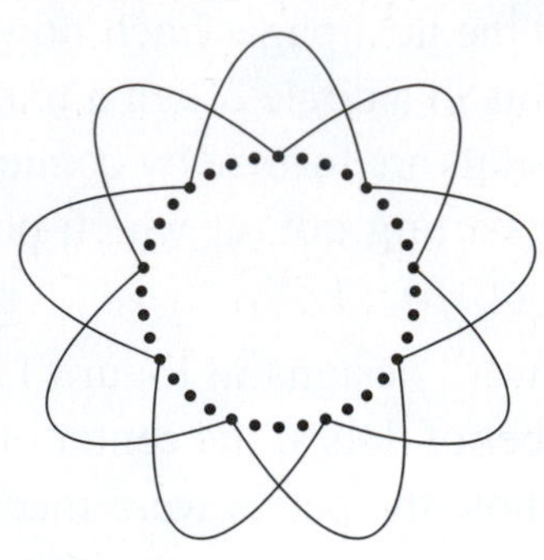

36 dots
jump #8, a petal connects
every 8th dot, 9 petals

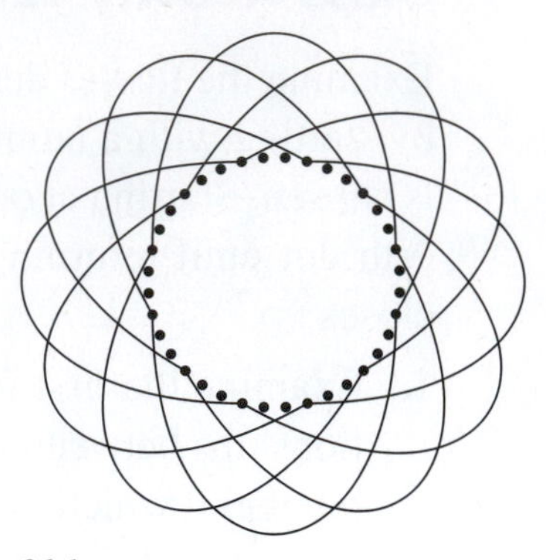

36 dots
jump #15, a petal connects
every 15th dot, 12 petals

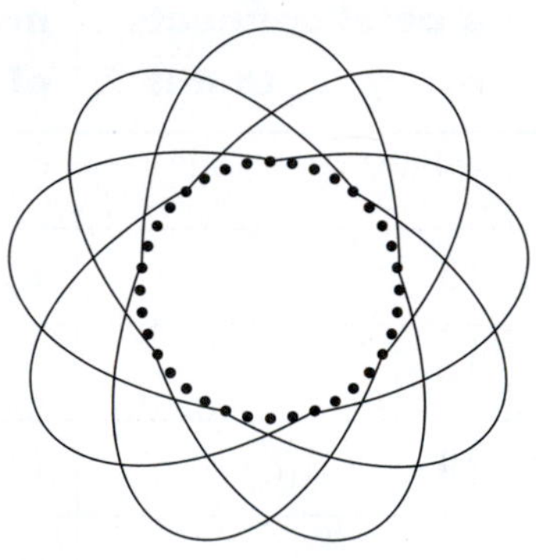

36 dots
jump #16, a petal connects
every 16th dot, 9 petals

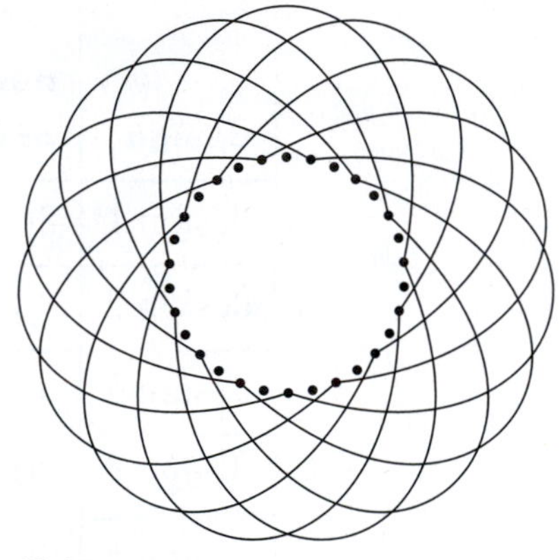

30 dots
jump #14, a petal connects
every 14th dot, 15 petals

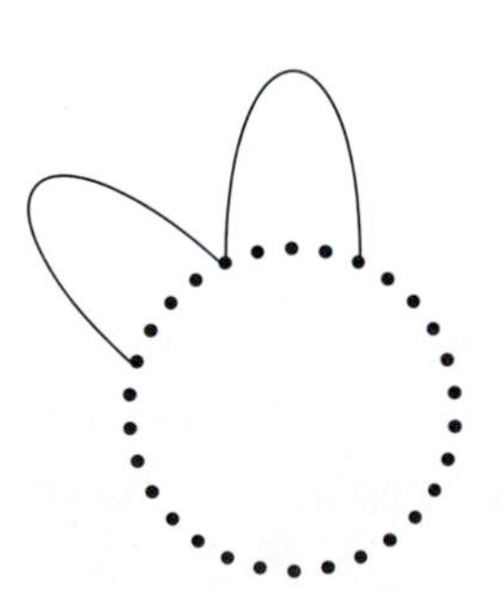

30 dots
jump #4, a petal connects
every 4th dot, ____ petals

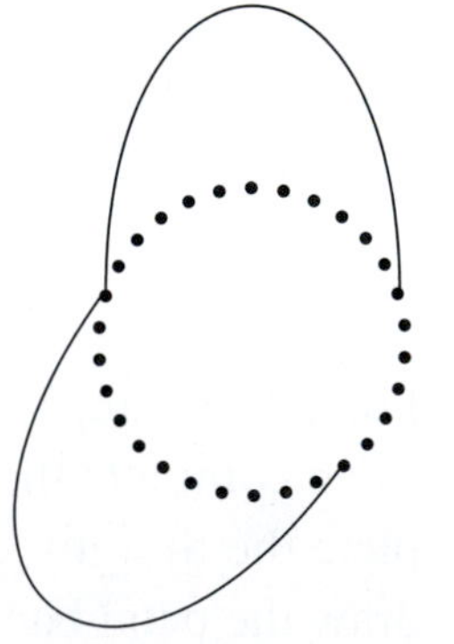

30 dots
jump #12, a petal connects
every 12th dot, ____ petals

Figure 12H.1
Flower Designs

Class Activity 12I: Relationships between the GCF and the LCM and Explaining the Flower Designs

1. By experimenting with a number of pairs of counting numbers A and B, discover one or more relationships among A, B, GCF, and LCM, where GCF stands for the greatest common factor of A and B and LCM stands for the least common multiple of A and B.

For example,

- If $A = 6$ and $B = 8$, then GCF $= 2$ and LCM $= 24$.
- If $A = 8$ and $B = 12$, then GCF $= 4$ and LCM $= 24$.
- If $A = 3$ and $B = 5$, then GCF $= 1$ and LCM $= 15$.

In these and other examples, how are A, B, GCF, and LCM related?

2. Refer to Class Activity 12H on Flower Designs.

a. Describe the number of petals in flower designs in terms of the "jump number" and the LCM of the number of dots in the design and the jump number. By considering how the flower designs were made, explain why this relationship must hold for every flower design.

b. Describe the number of petals in flower designs in terms of the number of dots in the design and the GCF of the number of dots in the design and the jump number. Use Problem 1 and Part (a) to explain why this relationship must hold for every flower design.

Class Activity 12J: Using GCFs and LCMs with Fractions

1. Show how to determine the simplest form of

$$\frac{36}{90}$$

by first finding the GCF of 36 and 90.

2. Show how to determine the simplest form of

$$\frac{36}{90}$$

without finding the GCF of 36 and 90 first.

3. Compare your work in Problems 1 and 2. Which method did you prefer? Why?

4. Show how to add

$$\frac{1}{6}+\frac{1}{10}$$

in two ways: using the least common denominator and using the denominator that is the product of the two denominators. Which method did you prefer? Why?

5. If we use the least common denominator to add two fractions, is the resulting sum necessarily in simplest form?

12.3 Prime Numbers

Class Activity 12K: The Sieve of Eratosthenes

1. Use the Sieve of Eratosthenes to find all the prime numbers up to 120. Start by circling 2 and crossing off every 2nd number after 2. Then circle 3 and cross off every 3rd number. (Cross off a number even if it already has been crossed off.) Continue in this manner, going back to the beginning of the list, circling the next number N that hasn't been crossed off, and then crossing off every Nth number until every number in the list is either circled or crossed off. The numbers that are circled at the end are the prime numbers from 2 to 120.

	2	3	4	5	6	7	8	9	10
11	12	13	14	15	16	17	18	19	20
21	22	23	24	25	26	27	28	29	30
31	32	33	34	35	36	37	38	39	40
41	42	43	44	45	46	47	48	49	50
51	52	53	54	55	56	57	58	59	60
61	62	63	64	65	66	67	68	69	70
71	72	73	74	75	76	77	78	79	80
81	82	83	84	85	86	87	88	89	90
91	92	93	94	95	96	97	98	99	100
101	102	103	104	105	106	107	108	109	110
111	112	113	114	115	116	117	118	119	120

2. Explain why the circled numbers must be prime numbers and why the numbers that are crossed off are not prime numbers.

Class Activity 12L: The Trial Division Method for Determining whether a Number Is Prime

You will need the list of prime numbers from Class Activity 12K on the Sieve of Eratosthenes for this activity.

1. Using the trial division method and your list of primes from the Sieve of Eratosthenes, determine whether the three numbers listed are prime numbers. Record the results of your trial divisions below the number. (The first few are done for you.) You will need these results for Problem 3.

239	323	4001
$239 \div 2 = 119.5$	$323 \div 2 = 161.5$	$4001 \div 2 = 2000.5$
$239 \div 3 = 79.67\ldots$	$323 \div 3 = 107.67\ldots$	$4001 \div 3 = 1333.67\ldots$

2. In the trial division method, you determine only whether your number is divisible by *prime* numbers. Why is this legitimate? Why don't you also have to find out if your number is divisible by other numbers such as 4, 6, 8, 9, 10, etc.?

3. How do you know when to stop with the trial division method? To help you answer this, look at the list of divisions you did in Problem 1. As you go down each list, what happens to the divisor and the quotient? If your number *was* divisible by some whole number, at what point would that whole number be known?

12.4 Even and Odd

Class Activity 12M: Why Can We Check the Ones Digit to Determine whether a Number Is Even or Odd?

Remember that a counting number is called "even" if that number of objects can be divided into groups of 2 with none left over:

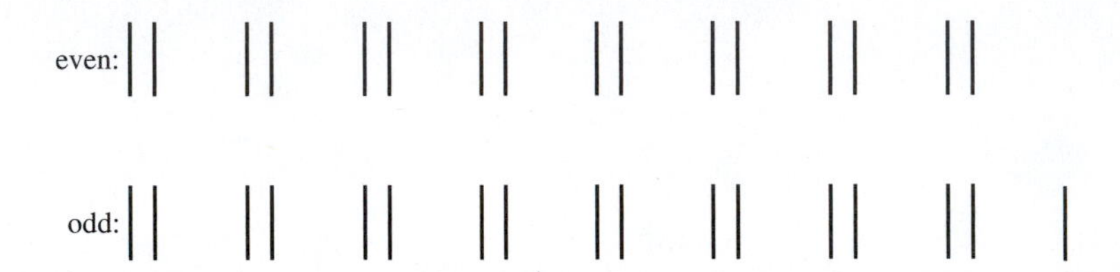

Why is it valid to determine whether a number of objects can be divided into groups of 2 with none left over by checking the ones digit of the number? We will investigate this question in this class activity.

1. Recall that we can represent the expanded form of a whole number with bundled toothpicks (or other bundled or grouped objects) as in Figure 12M.1. Using Figure 12M.1, explain why 134 toothpicks can be divided into groups of 2 with none left over, but there will be 1 toothpick left over when 357 toothpicks are divided into groups of 2. Explain why we have to consider only the ones digit to determine if there will be a toothpick left over by describing what happens with each bundle of 100 toothpicks and each bundle of 10 toothpicks when we divide the toothpicks into groups of 2.

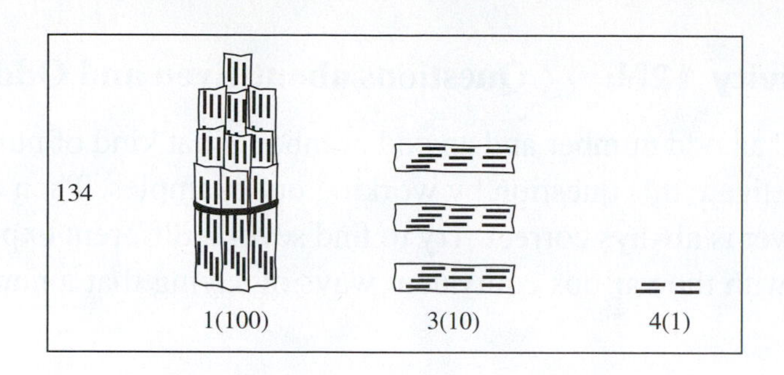

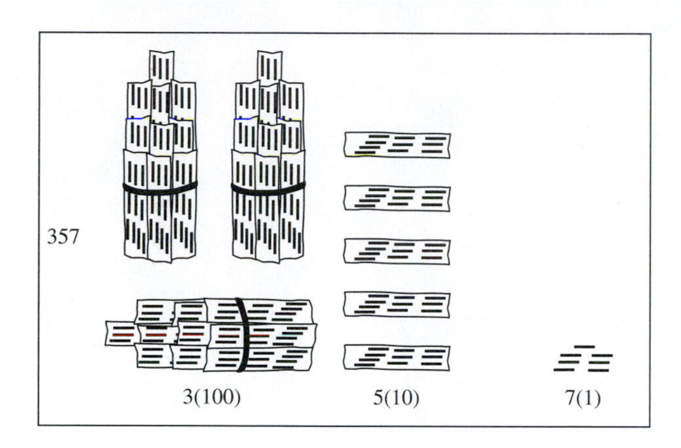

Figure 12M.1
Representing 134 and 357 with Bundled Toothpicks

2. Working more generally, let ABC be a three-digit whole number with A hundreds, B tens, and C ones. Use the idea of representing ABC with bundled toothpicks to help you explain why ABC is divisible by 2 exactly when C is either 0, 2, 4, 6, or 8. What can we say about each of the B bundles of 10 toothpicks and each of the A bundles of 100 toothpicks when we divide the toothpicks into groups of 2?

Class Activity 12N: Questions about Even and Odd Numbers

1. If you add an odd number and an odd number, what kind of number do you get? Investigate this question by working out examples. Then explain why your answer is always correct. Try to find several different explanations by working with the various equivalent ways of saying that a number is even or odd.

2. If you multiply an even number and an odd number, what kind of number do you get? Investigate this question by working out examples. Then explain why your answer is always correct. Try to find several different explanations by working with the various equivalent ways of saying that a number is even or odd.

Class Activity 12O: Extending the Definitions of Even and Odd

We have defined even and odd only for counting numbers. What if we wanted to extend the definition of even and odd to other numbers?

1. If we extend the definitions of even and odd to all the integers, then what should 0 be, even or odd? What should -5 be, even or odd? Explain.

2. Give definitions of even and odd that apply to all integers, not just to the counting numbers.

3. Would it make sense to extend the definitions of even and odd to the rational numbers (fractions)? Why or why not?

12.5 Divisibility Tests

Class Activity 12P: The Divisibility Test for 3

1. Is it possible to tell if a whole number is divisible by 3 just by checking its last digit? Investigate this question by considering a number of examples. State your conclusion.

2. The divisibility test for 3 says that you can determine whether a counting number is divisible by 3 by adding its digits and checking to see if this sum is divisible by 3. If the sum is divisible by 3, then the original number is, too; if the sum is not divisible by 3, then the original number is not either.

   For each of the numbers listed, see if the divisibility test for 3 really does accurately predict which numbers are divisible by 3. That is, for each number, check its divisibility by 3 by using long division or a calculator, and then add the digits of the number and see if that sum is divisible by 3. The two conclusions should agree.

   Example: 1437 is divisible by 3 because $1437 \div 3 = 479$, with no remainder.

   $1 + 4 + 3 + 7 = 15$ also indicates that 1437 is divisible by 3, because $15 \div 3 = 5$, with no remainder.

   2570

   123

   14,928

   7213

   555, 555

   11,111

3. Explain why the divisibility test for 3 is valid for 3-digit counting numbers. In other words, explain why you can determine whether or not a 3-digit counting number, ABC, is divisible by 3 by adding its digits, $A + B + C$, and determining if this sum is divisible by 3.

In order to develop your explanation, consider the following:

a. A counting number is divisible by 3 exactly when that many objects can be divided into groups of 3 with none left over.

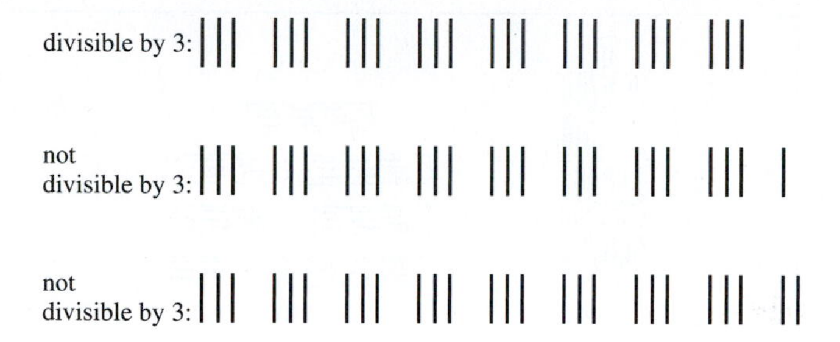

b. Think about representing 3-digit numbers with bundled toothpicks. For example, we can represent the number 247 by using 2 bundles of 100 toothpicks, 4 bundles of 10 toothpicks, and 7 individual toothpicks, as shown in Figure 12P.1. The number ABC is represented by A bundles of 100 toothpicks, B bundles of 10 toothpicks, and C individual toothpicks.

c. Consider starting to divide bundled toothpicks into groups of 3 by dividing *each individual bundle* of 10 and *each individual bundle* of 100 into groups of 3. Determine how many toothpicks will be left over if you divide the bundles this way and if you do no further dividing into groups of 3.

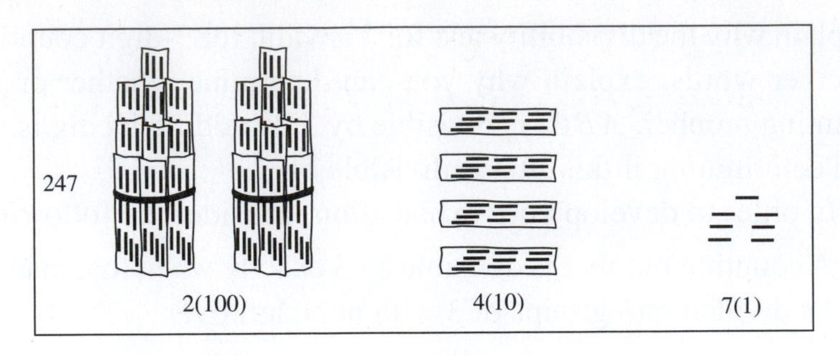

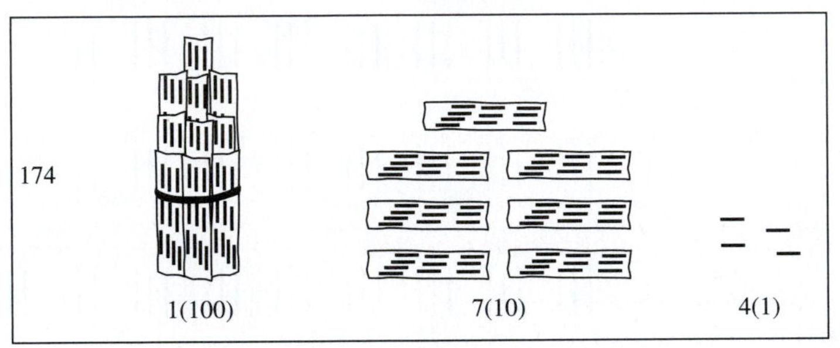

Figure 12P.1

Representing 247 and 174 with Bundled Toothpicks

4. Relate your explanation in Problem 3 to the following equations:

$$\begin{aligned} ABC &= A \cdot 100 + B \cdot 10 + C \\ &= (A \cdot 99 + B \cdot 9) + (A + B + C) \\ &= (A \cdot 33 + B \cdot 3) \cdot 3 + (A + B + C) \end{aligned}$$

12.6 Rational and Irrational Numbers

Class Activity 12Q: Decimal Representations of Fractions

1. The decimal representations of the following fractions are shown to 16 decimal places, with no rounding:

$\frac{1}{12} = 0.0833333333333333$	$\frac{1}{4} = 0.25$
$\frac{1}{11} = 0.0909090909090909$	$\frac{2}{5} = 0.4$
$\frac{113}{33} = 3.4242424242424242$	$\frac{37}{8} = 4.625$
$\frac{491}{550} = 0.8927272727272727$	$\frac{17}{50} = 0.34$
$\frac{14}{37} = 0.3783783783783783$	$\frac{1}{125} = 0.008$
$\frac{35}{101} = 0.3465346534653465$	$\frac{9}{20} = 0.45$
$\frac{1}{41} = 0.0243902439024390$	$\frac{19}{32} = 0.59375$
$\frac{5}{7} = 0.7142857142857142$	$\frac{1}{3200} = 0.0003125$
$\frac{1}{14} = 0.0714285714285714$	$\frac{1}{64{,}000} = 0.000015625$
$\frac{1}{21} = 0.0476190476190476$	$\frac{1}{625} = 0.0016$

In what way are the decimal representations in the first column similar? In what way are the decimal representations of the fractions in the second column similar?

2. Complete the next set of calculations, using longhand division (not a calculator!) to find the decimal representations of $\frac{2}{55}$ and $\frac{1}{101}$. At each step in the long-division process, write down the remainder you obtain.

$$\begin{array}{r|l} 55\overline{)2.0000000} \; (0.0) & \text{remainder } 2 \\ -\,0 & \\ \hline 20 & \text{remainder } 20 \\ \_\_\_ & \text{remainder } \_\_\_ \\ \_\_\_ & \text{remainder } \_\_\_ \\ \_\_\_ & \text{remainder } \_\_\_ \\ \_\_\_ & \text{remainder } \_\_\_ \\ \_\_\_ & \text{remainder } \_\_\_ \\ \_\_\_ & \text{remainder } \_\_\_ \end{array} \qquad \begin{array}{r|l} 101\overline{)1.0000000} \; (0.0) & \text{remainder } 1 \\ -\,0 & \\ \hline 10 & \text{remainder } 10 \\ \_\_\_ & \text{remainder } \_\_\_ \\ \_\_\_ & \text{remainder } \_\_\_ \\ \_\_\_ & \text{remainder } \_\_\_ \\ \_\_\_ & \text{remainder } \_\_\_ \\ \_\_\_ & \text{remainder } \_\_\_ \\ \_\_\_ & \text{remainder } \_\_\_ \end{array}$$

3. Use longhand division to calculate the decimal representations of $\frac{3}{8}$ and $\frac{1}{7}$. As in Problem 2, write down the remainder you get at each step.

4. Answer the following questions for the decimal representations of the fractions in Problems 2 and 3:

- What happened to the decimal representation of the fraction when you got a remainder of 0?

- What happened to the decimal representation of the fraction when you got a remainder that you had before?

5. Without actually carrying it out, imagine doing longhand division to find the decimal representation of $\frac{7}{31}$.

a. What remainders could you possibly get in the longhand division process when finding $7 \div 31$? For example, could you possibly get a remainder of 45 or 73 or 32? How many different remainders are theoretically possible?

b. If you were doing longhand division to find $7 \div 31$ and you got a remainder of 0 somewhere along the way, what would that tell you about the decimal representation of $\frac{7}{31}$?

c. If you were doing longhand division to find $7 \div 31$ and you got a remainder you'd gotten before, what would then happen in the decimal representation of $\frac{7}{31}$?

d. Now use your answer to Parts (a), (b), and (c) to explain why the decimal representation of $\frac{7}{31}$ must either terminate or eventually repeat after at most 30 decimal places.

6. In general, suppose that $\frac{A}{B}$ is a proper fraction, where A and B are whole numbers. Explain why the decimal representation of $\frac{A}{B}$ must either terminate or begin to repeat after at most $B - 1$ decimal places.

7. Could the number

$$0.1010010001000010000010000001\ldots$$

where the decimal representation continues forever with the pattern of more and more 0s in between 1s, be the decimal representation of a fraction of whole numbers? Explain your answer.

Class Activity 12R: Writing Terminating and Repeating Decimals as Fractions

1. By using denominators that are suitable powers of 10, show how to write the following terminating decimals as fractions:

$$0.137 = \qquad 0.25567 =$$

$$13.89 = \qquad 329.2 =$$

2. Write the following fractions as decimals, and observe the pattern:

$\frac{1}{9} =$

$\frac{1}{99} =$

$\frac{1}{999} =$

$\frac{1}{9999} =$

$\frac{1}{99{,}999} =$

3. Using the decimal representations of $\frac{1}{9}$, $\frac{1}{99}$, . . . , that you found in Problem 2, show how to write the following decimals as fractions:

$$0.\overline{2} = 0.222222\ldots = \qquad 0.\overline{08} = 0.080808\ldots =$$

$$0.\overline{003} = 0.003003\ldots = \qquad 0.\overline{52} = 0.525252\ldots =$$

$$0.\overline{1234} = \qquad 0.\overline{123456} =$$

4. Use the fact that

$$0.\overline{49} = \frac{49}{99}$$

to write the next four repeating decimals as fractions. *Hint*: Shift the decimal point by dividing by suitable powers of 10.

$0.0\overline{49} =$ $\qquad$ $0.00\overline{49} =$

$0.000\overline{49} =$ $\qquad$ $0.0000\overline{49} =$

5. Use the results of Problem 4, together with the fact that

$$0.3\overline{49} = 0.3 + 0.0\overline{49} \qquad 0.12\overline{49} = 0.12 + 0.00\overline{49}$$

and other similar facts, to determine how to write the following repeating decimals as fractions:

$7.3\overline{49} =$ $\qquad$ $0.12\overline{49} =$

$1.2\overline{49} =$ $\qquad$ $0.111\overline{49} =$

Class Activity 12S: What Is 0.9999 . . .?

1. Use the fact that $\frac{1}{9} = 0.\overline{1} = 0.111111111\ldots$ to determine the decimal representations of the following fractions:

$\frac{2}{9} =$ $\qquad$ $\frac{6}{9} =$

$\frac{3}{9} =$ $\qquad$ $\frac{7}{9} =$

$\frac{4}{9} =$ $\qquad$ $\frac{8}{9} =$

$\frac{5}{9} =$ $\qquad$ $\frac{9}{9} =$

What can you conclude about $0.\overline{9}$?

2. Add longhand:

$$\begin{array}{r} 0.9999999999\ldots \\ +\,0.1111111111\ldots \\ \hline \end{array}$$

Note that the nines and ones repeat forever.
Now subtract longhand:

$$\begin{array}{r} -\ \ .1111111111\ldots \\ \hline \end{array}$$

Look back at what you just did: Starting with $0.\overline{9}$, you added and then subtracted $0.\overline{1}$. What does this tell you about $0.\overline{9}$?

3. Let N stand for the number $0.\overline{9} = 0.999999\ldots$:

$$N = 0.999999999\ldots$$

Write the decimal representation of $10N$:

$$10N =$$

Now subtract N from $10N$ in two ways—in terms of N and as decimal numbers:

in terms of N :	as decimals :
$10N$	
$-N$	-0.999999999

What can you conclude about $0.\overline{9}$?

4. Given that the number 1 has two different decimal representations, namely, 1 and $0.\overline{9}$, find different decimal representations of the following numbers:

$$17 =$$

$$23.42 =$$

$$139.8 =$$

Class Activity 12T: The Square Root of 2

1. If the sides of a square are 1 unit long, then how long is the diagonal of the square?

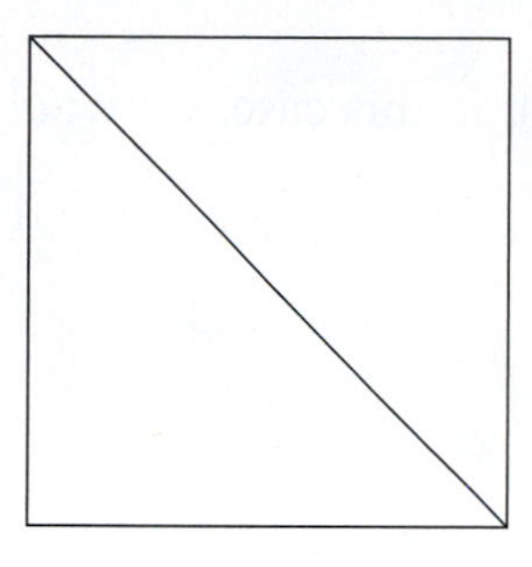

2. Use a calculator to find the decimal representation of $\sqrt{2}$. Based on your calculator's display, does it look like $\sqrt{2}$ is rational or irrational? Why? Can you tell *for sure* just by looking at your calculator's display?

3. What is the decimal representation of

$$\frac{1{,}414{,}213{,}562}{999{,}999{,}999}?$$

Is this number rational or irrational? Compare with Problem 2.

4. Suppose that it were somehow possible to write the square root of two as a fraction $\frac{A}{B}$, where A and B are counting numbers:

$$\sqrt{2} = \frac{A}{B}.$$

Show that, in this case, we would get the equation

$$A^2 = 2 \times B^2$$

5. Suppose A is a counting number, and imagine factoring it into a product of prime numbers. For example, if A is 30, then you factor it as

$$A = 2 \times 3 \times 5$$

Now think about factoring A^2 as a product of prime numbers. For example, if $A = 30$, then

$$A^2 = 2 \times 3 \times 5 \times 2 \times 3 \times 5$$

Could A^2 have an odd number of prime factors? Make a general qualitative statement about the number of prime factors that A^2 has.

6. Now suppose that B is a counting number, and imagine factoring the number $2 \times B^2$ into a product of prime numbers. For example, if $B = 15$, then

$$2 \times B^2 = 2 \times 3 \times 5 \times 3 \times 5$$

Could $2B^2$ have an even number of prime factors? Make a general qualitative statement about the number of prime factors that $2 \times B^2$ has.

7. Now use your answers in Problems 5 and 6 to explain why a number in the form A^2 can never be equal to a number in the form $2 \times B^2$, when A and B are counting numbers.

8. What does Problem 7 lead you to conclude about the assumption in Problem 4 that it is somehow possible to write the square root of two as a fraction, where the numerator and denominator are counting numbers? Now what can you conclude about whether $\sqrt{2}$ is rational or irrational?

Class Activity 12U: Pattern Tiles and the Irrationality of the Square Root of 3

You will need a set of pattern tiles like the ones shown in the next figure for this activity. If a set of pattern tiles is not available, cut out the paper pattern tiles on pages 677, 679, and 681.

One fascinating aspect of mathematics is that there are connections between simple kindergarten activities and much more advanced ideas that children might not encounter until high school or college math classes. This Class Activity will show you one such connection involving pattern tiles.

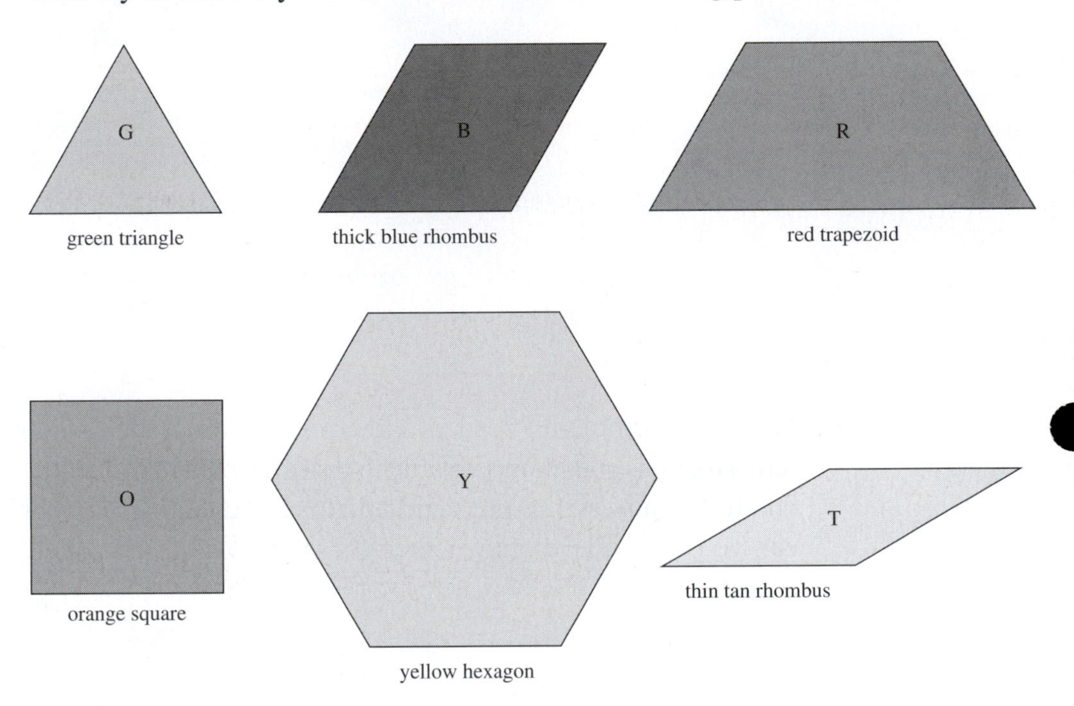

1. In early elementary school, children sometimes use pattern tiles to fill shapes. Find several different ways to fill each of the shapes in Figures 12U.1 and 12U.2 with pattern tiles.

   Even though there are many different ways to fill the shapes, can any of the shapes in Figure 12U.1 be filled by the use of one or more squares or thin rhombuses?

   Can any of the shapes in Figure 12U.2 be filled without either squares or thin rhombuses?

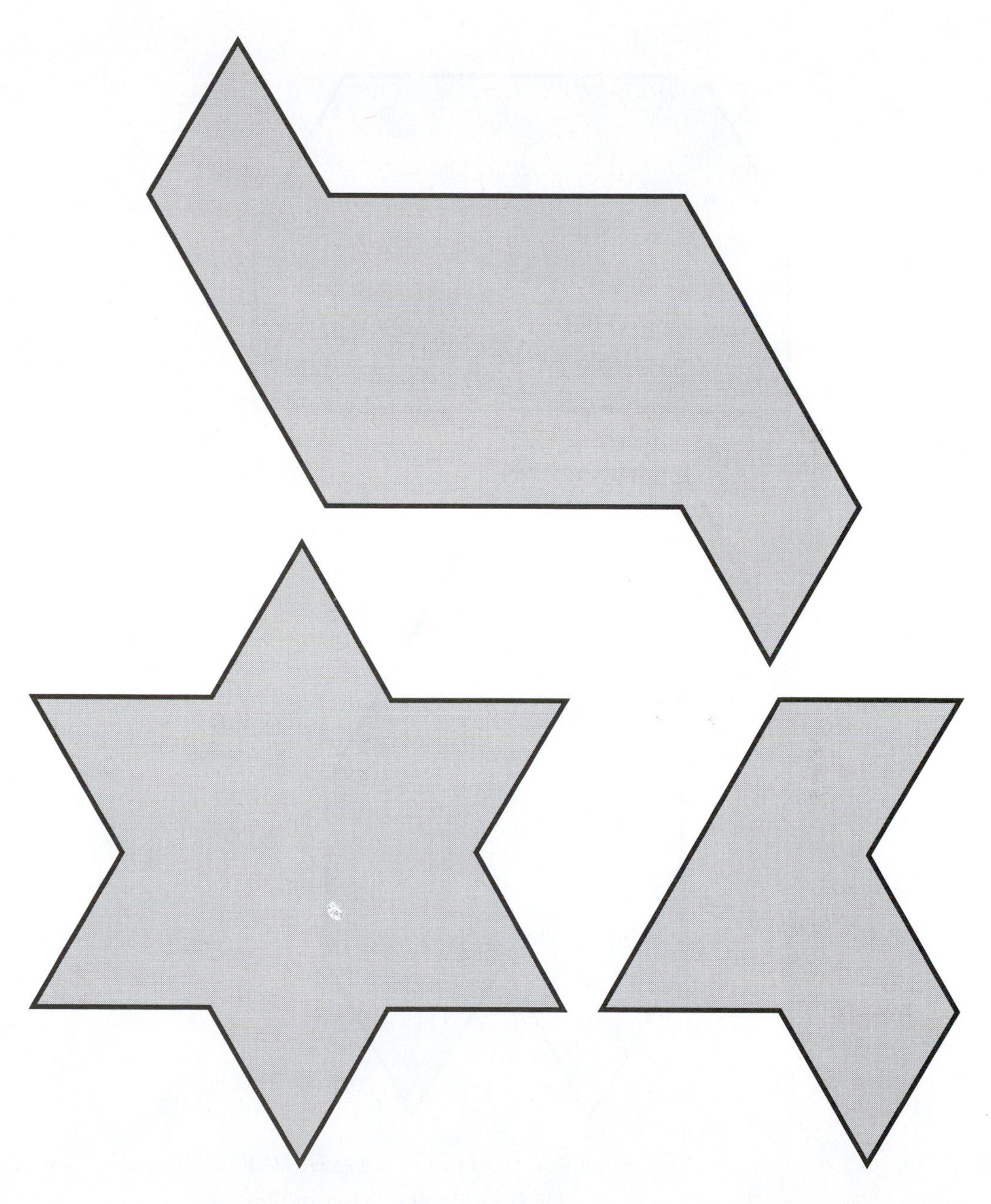

Figure 12U.1
Fill These Shapes with Pattern Tiles

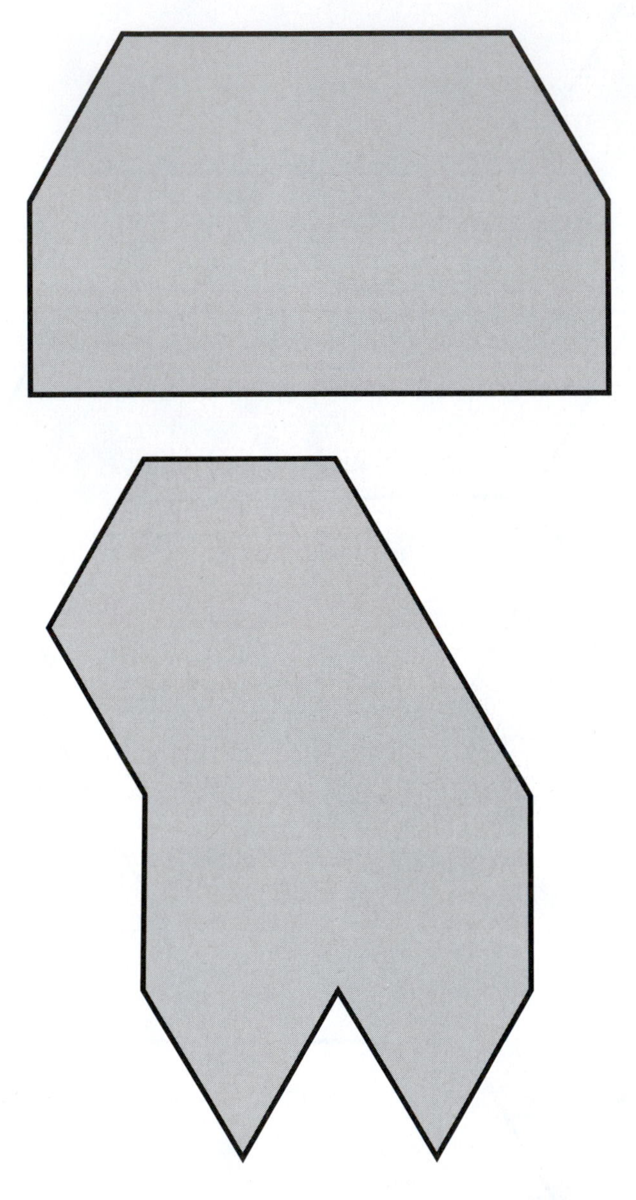

Figure 12U.2
Fill These Shapes with Pattern Tiles

2. When children try to fill shapes with pattern tiles in different ways, they may begin to notice relationships among the different tiles. Find as many relationships among the tiles as you can. Express these relationships by using the symbols G, Y, R, B, T, and O, where

- G stands for the area of the green triangle.
- Y stands for the area of the yellow hexagon.
- R stands for the area of the red trapezoid.
- B stands for the area of the thick blue rhombus.
- T stands for the area of the thin tan rhombus.
- O stands for the area of the orange square.

For example, because two green triangles fit together to make a thick blue rhombus,

$$2G = B$$

3. If you have not yet done so, find a relationship between T and O, the areas of the thin tan rhombus and the orange square. *Hint*: Use triangles in addition to the thin tan rhombus and the orange square.

4. Can you find any way to relate G, Y, R, and B with O, the area of the orange square?

Can you find any way to relate G, Y, R, and B with T, the area of the thin tan rhombus?

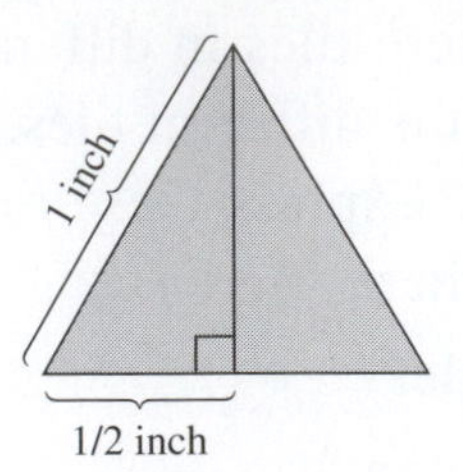

Figure 12U.3
The Green Triangle

5. Use the Pythagorean theorem and Figure 12U.3 to help you determine the area, in square inches, of the green triangle. Your answer should involve a square root of 3.

6. Using the relationships that you found in Problem 2, and the area of the green triangle that you found in Problem 5, determine Y, R, and B, the areas of the yellow hexagon, the red trapezoid, and the thick blue rhombus. All these areas will involve the square root of 3. Fill in the values for G, Y, R, and B in Figure 12U.4.

7. The orange square is 1 inch wide and 1 inch long; therefore, its area is 1 square inch. Use the relationship between O and T that you found in Problem 3 to determine T. Fill in the values for O and T in Figure 12U.4.

8. Looking at the values you wrote in Figure 12U.4, how are the areas of the green triangle, the yellow hexagon, the red trapezoid, and the thick blue rhombus qualitatively different from the areas of the orange square and the thin tan rhombus? Given this, and the fact that $\sqrt{3}$ is irrational, is it surprising that you could not find a relationship between G, Y, R, B, and O or G, Y, R, B, and T in Problem 4?

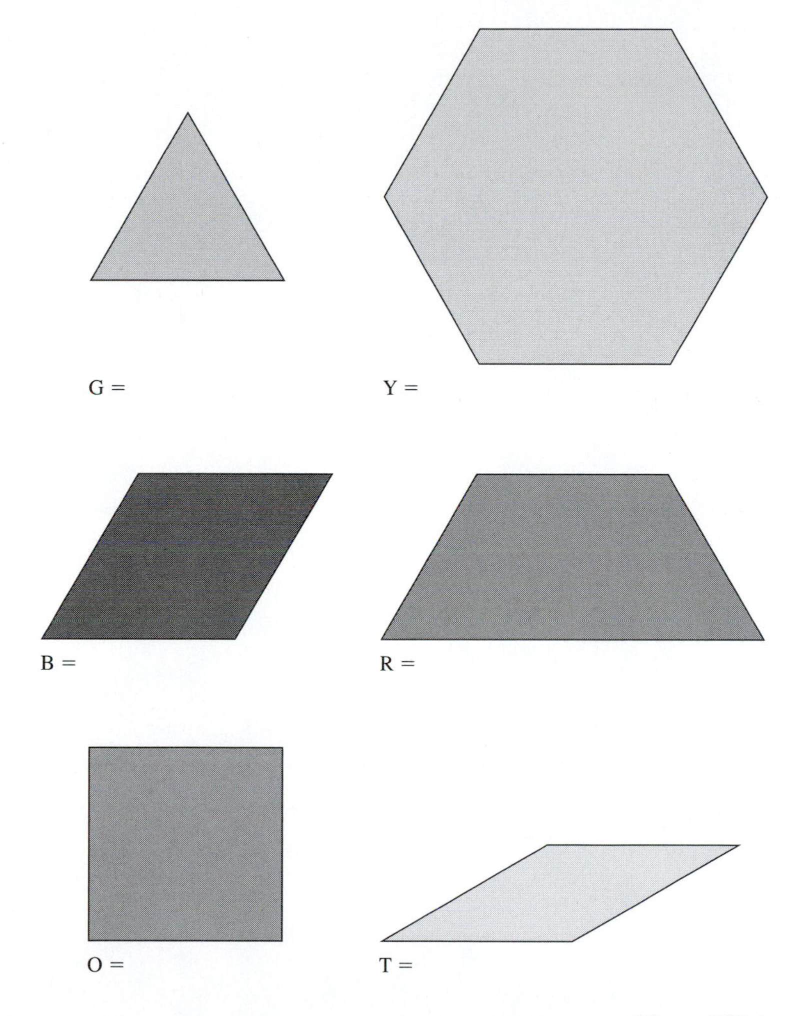

Figure 12U.4
Fill in the Areas of These Shapes

13

Functions and Algebra

13.1 Mathematical Expressions, Formulas, and Equations

Class Activity 13A: Writing Expressions and a Formula for a Flower Pattern

1. For each flower design in Figure 13A.1, write an expression for the total number of dots in the design. Each expression should involve both multiplication and addition.

2. Fill like parts in the large flower in Figure 13A.2 with the same number of dots. Then write an expression for the total number of dots in your flower design. Write equations in which you evaluate this expression, determining the total number of dots in your flower design.

3. Now consider general flower designs like the ones in Figure 13A.1 and the one you created in Problem 2? If M, N, and P are any counting numbers, we can imagine a flower design that has M dots in the center circle, N dots in each of the circles surrounding the center circle, and P dots in each of the petals of the flower. Write a formula, in terms of M, N, and P, for the number of dots in the flower design.

4. Create a design that illustrates the formula

$$M + 2N + 2P$$

by imagining different portions of the design filled with different numbers of dots. Explain why your design illustrates the formula $M + 2N + 2P$.

Figure 13A.1
Write Expressions for the Total Number of Dots in Each Flower

Figure 13A.2
Fill Like Parts with Equal Numbers of Dots and Write an Expression for the Total Number of Dots

Class Activity 13B: Expressions in Geometric Settings

1. For each of Design 1 and Design 2, write at least 3 different expressions for the total number of small shapes that the design is made of.

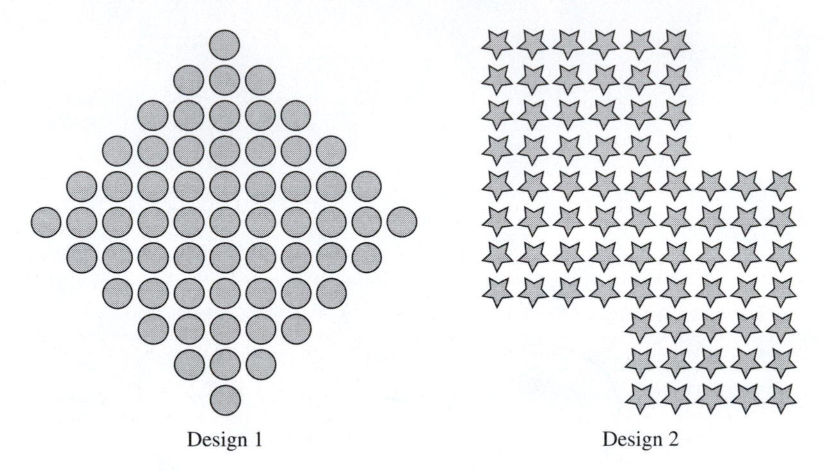

2. Write (at least) two expressions for the area and two expressions for the perimeter of the floor plan in the next figure. Make clear which is which. (Assume that all the angles are right angles.)

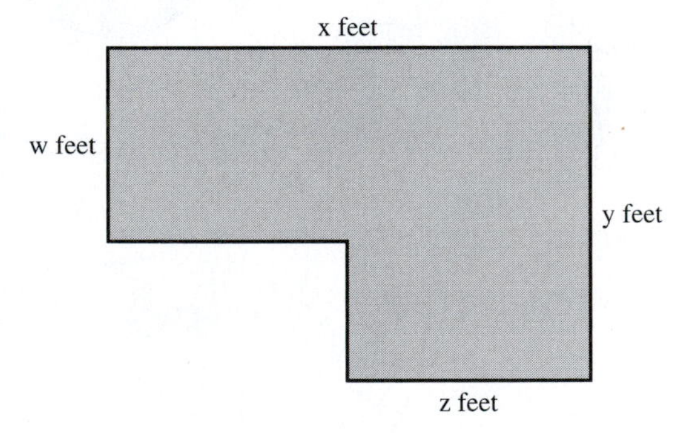

3. Explain in two different ways why your two area expressions in Problem 2 are equal. (One way should use the fact that the area is the same no matter how you calculate it, and the other way should use properties of arithmetic.)

Class Activity 13C: Expressions in 3D Geometric Settings

1. Write (at least) two expressions for the total number of 1-cm-by-1-cm-by-1-cm cubes it would take to build a 6-cm-tall prism (tower) over Base 1. Each expression should use both multiplication and addition (or subtraction). Explain briefly.

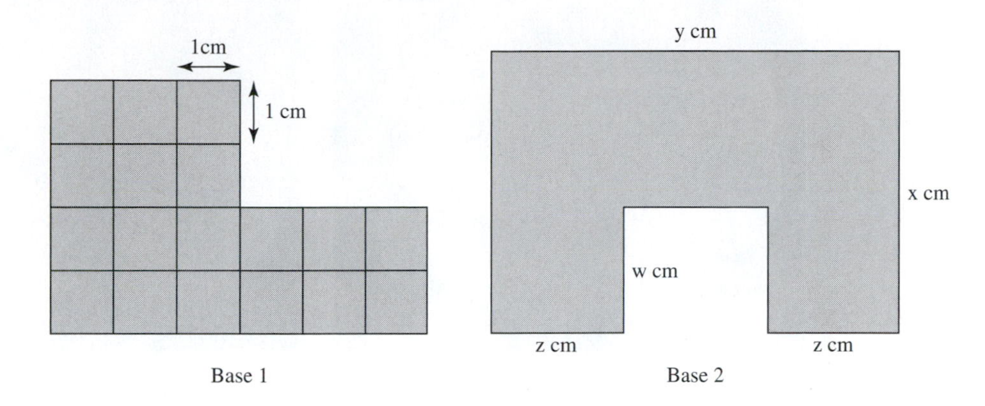

2. Write (at least) two expressions for the total number of 1-cm-by-1-cm-by-1-cm cubes it would take to build a v-cm-tall prism (tower) over Base 2. (Assume that all the angles are right angles.) Each expression should be in terms of v, w, x, y, and z. Explain briefly. Explain in two different ways why your expressions must be equal.

3. The next pictures show two views of a structure built with marshmallows. Write as many expressions as you can for the total number of marshmallows in the structure; evaluate your expressions to determine this number.

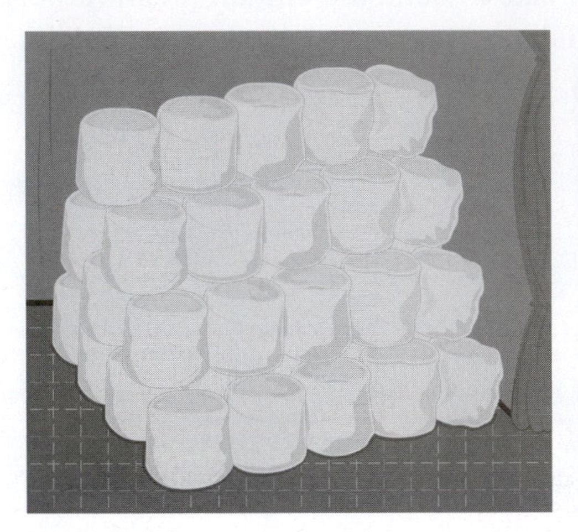
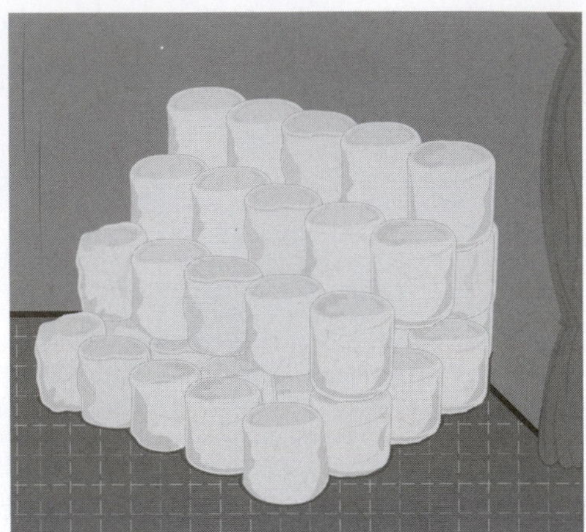

4. The next pictures show three views of a structure built with marshmallows. Write as many expressions as you can for the total number of marshmallows in the structure; evaluate your expressions to determine this number.

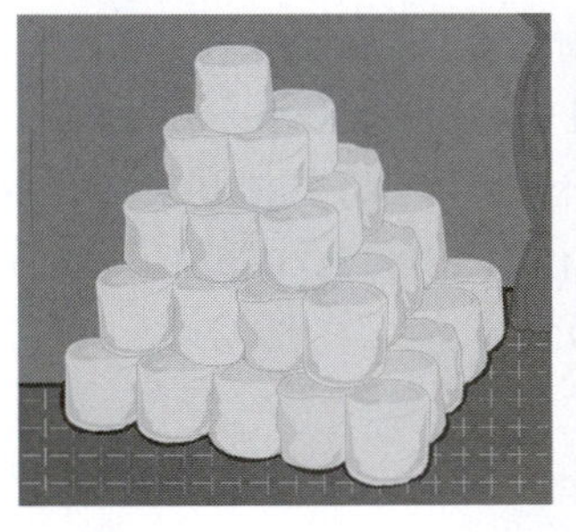
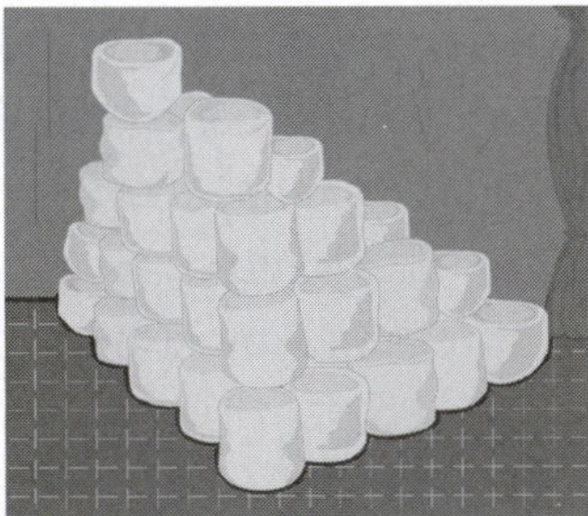

Class Activity 13D: Equations Arising from Rectangular Designs

1. Explain why design (a) gives rise to the equation

$$4 \cdot 4 + 4 \cdot 3 + 2 \cdot 4 + 2 \cdot 3 = 6 \cdot 7$$

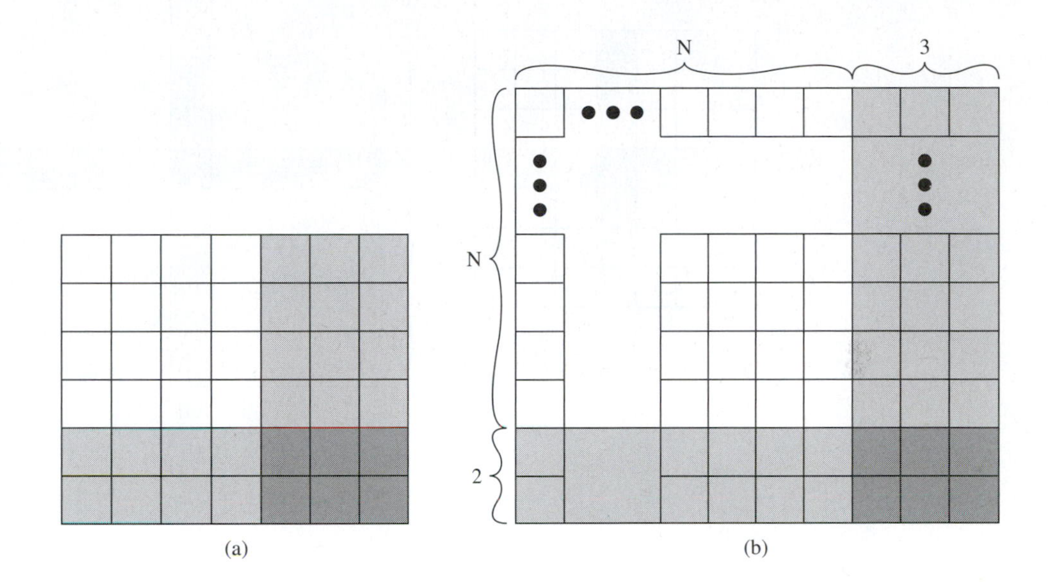

2. Design (b) represents an enlarged version of design (a). Find an equation in terms of N that design (b) gives rise to. Explain.

3. Other than using a rectangular design to explain why your equation in Problem 2 is true, how else can you see that the equation is true? (Think back to properties of arithmetic you have studied.)

4. By determining the total number of small squares in design (c) in two different ways, find an equation that design (c) gives rise to.

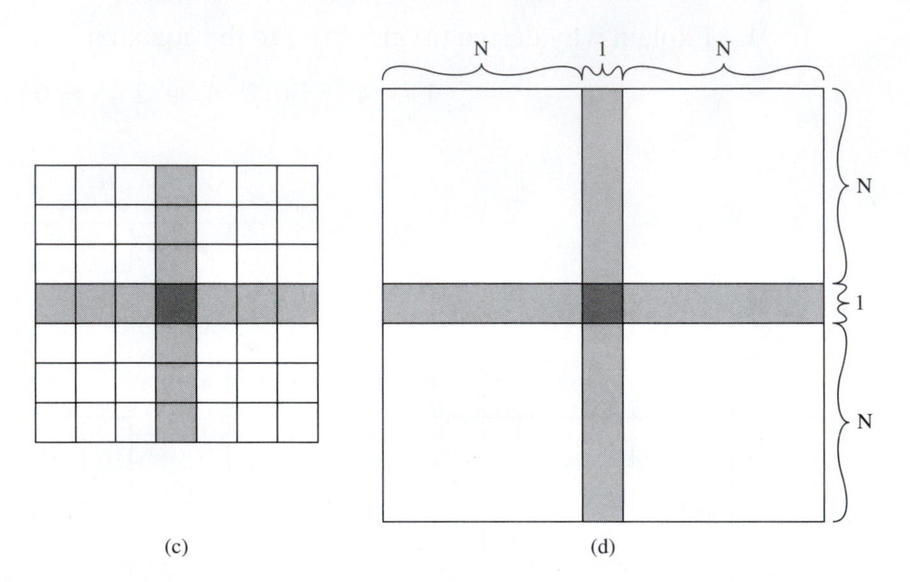

(c) (d)

5. Design (d) represents an enlarged version of design (c). Find an equation in terms of N that design (d) gives rise to. Explain.

6. Other than using a square design to explain why your equation in Problem 5 is true, how else can you see that the equation is true?

7. Draw, label, and shade a square so that it gives rise to the equation

$$(N + 1)^2 = N^2 + 2N + 1$$

Explain briefly. Other than using your square, how else can you see that the equation is true?

8. Draw, label, and shade a rectangle so that it gives rise to the equation

$$(A + 1) \cdot (B + 2) = AB + 2A + B + 2$$

Explain briefly. Other than using your rectangle, how else can you see that the equation is true?

Class Activity 13E: Expressions with Fractions

1. For each of the next two rectangles, write an expression using multiplication and addition (or subtraction) for the fraction of the area of the rectangle that is shaded. You may assume that parts that appear to be the same size really are the same size.

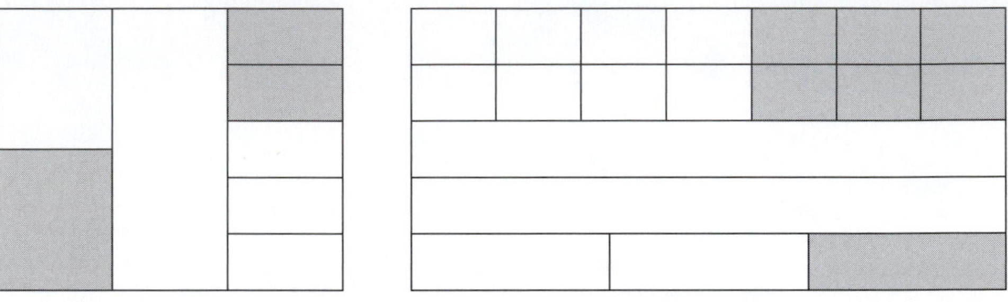

2. Shade

$$\frac{2}{3} \cdot \frac{4}{5} + \frac{1}{4} \cdot \frac{1}{10}$$

of a rectangle in such a way that you can tell the correct amount is shaded without evaluating the expression.

Class Activity 13F: Evaluating Expressions with Fractions Efficiently and Correctly

1. In order to evaluate

$$\frac{8}{35} \cdot \frac{35}{61}$$

we can cancel thus:

$$\frac{8}{\cancel{35}} \cdot \frac{\cancel{35}}{61} = \frac{8}{61}$$

Discuss the following equations and explain why they demonstrate that the canceling shown previously is legitimate:

$$\frac{8}{35} \cdot \frac{35}{61} = \frac{8 \cdot 35}{35 \cdot 61} = \frac{8 \cdot 35}{61 \cdot 35} = \frac{8}{61} \cdot \frac{35}{35} = \frac{8}{61}$$

2. Write equations to demonstrate that the canceling shown in the following equations is legitimate:

$$\frac{\overset{2}{\cancel{18}}}{5} \cdot \frac{7}{\underset{11}{\cancel{99}}} = \frac{2}{5} \cdot \frac{7}{11} = \frac{14}{55}$$

3. Which of the cancelations in Parts (a) through (d) are correct, and which are incorrect? Explain your answers.

a.

$$\frac{\overset{6}{3\cancel{6}}\cdot\overset{16}{9\cancel{6}}}{\underset{1}{\cancel{6}}} = \frac{6\cdot 16}{1} = 96$$

b.

$$\frac{\overset{6}{3\cancel{6}}\cdot 96}{\underset{1}{\cancel{6}}} = \frac{6\cdot 96}{1} = 576$$

c.

$$\frac{\overset{6}{3\cancel{6}}+\overset{16}{9\cancel{6}}}{\underset{1}{\cancel{6}}} = \frac{6+16}{1} = 22$$

d.

$$\frac{\overset{6}{3\cancel{6}}+96}{\underset{1}{\cancel{6}}} = \frac{6+96}{1} = 102$$

Class Activity 13G: Expressions for Story Problems

1. Write a story problem so that some quantity in the story situation can be expressed as $4x + 2$. Explain why $4x + 2$ is the appropriate formula for the quantity.

2. Write a story problem so that some quantity in the story situation can be expressed as $4x - 2$. Explain why $4x - 2$ is the appropriate formula for the quantity.

3. There are T tons of sand in a pile.

   a. Assume that $\frac{1}{4}$ of the sand in the pile is removed from the pile and, after that, another $\frac{2}{3}$ of a ton of sand is dumped onto the pile. Write a formula in terms of T for the number of tons of sand that are in the pile now.

   b. Evaluate your formula from Part (a) when $T = 1\frac{2}{3}$.

   c. Starting with T tons of sand, assume that $\frac{2}{3}$ of a ton of sand is dumped onto the pile and, after that, $\frac{1}{4}$ of the sand in the new, larger pile is removed. Write a formula in terms of T for the number of tons of sand that are in the pile now. Is this formula the same as the formula in Part (a)?

   d. Evaluate your formula from Part (c) when $T = 1\frac{2}{3}$.

Class Activity 13H: Writing Equations for Story Situations

For each of the following story situations, write the corresponding equation:

1. Markus had M dollars in his bank account. After removing $\frac{1}{5}$ of the money in the account and then putting in another \$200, Markus now has \$800.

2. Keisha had K dollars in her bank account. After removing \$200 and then removing $\frac{1}{5}$ of the remaining money, Keisha now has \$800.

3. Originally, there were L liters of liquid in a container. After $\frac{2}{3}$ of the liquid was poured out, another $2\frac{1}{2}$ liters of liquid were poured into the container. When $\frac{1}{4}$ of the liquid was poured out, 4 liters remained.

4. There are 2 times as many pencils in Maya's pencil box as in David's. Be sure to define your variables with care. Draw a picture to help you explain why your equation is correct.

Class Activity 13I: Writing Story Problems for Equations

1. Write a story problem for the equation

$$x - \frac{1}{4}x + 30 = 150$$

2. Write a story problem for the equation

$$x - \frac{1}{4} + 30 = 150$$

3. Write a story problem for the equation

$$(x + 30) - \frac{1}{4}(x + 30) = 150$$

4. Write a story problem for the equation

$$\frac{2}{3}(x - 60) + 20 = 80$$

5. Write a story problem for the equation

$$3x = y$$

13.2 Solving Equations Using Number Sense, Strip Diagrams, and Algebra

Class Activity 13J: Solving Equations Using Number Sense

Solve each of the equations by using your understanding of numbers and operations. Do not use any standard algebraic techniques for solving equations that you may know. Explain your reasoning in each case.

1. $382 + 49 = x + 380$

2. $12 \cdot 84 = 2 \cdot 84 + A$

3. $14Z = 7 \cdot 48$

4. $7 \cdot 36 + T = 8 \cdot 37$

Class Activity 13K: Solving Equations Algebraically and with a Pan Balance

Solve $5x + 1 = 2x + 7$ in two ways, with equations and with pictures of a pan balance. Relate the two methods.

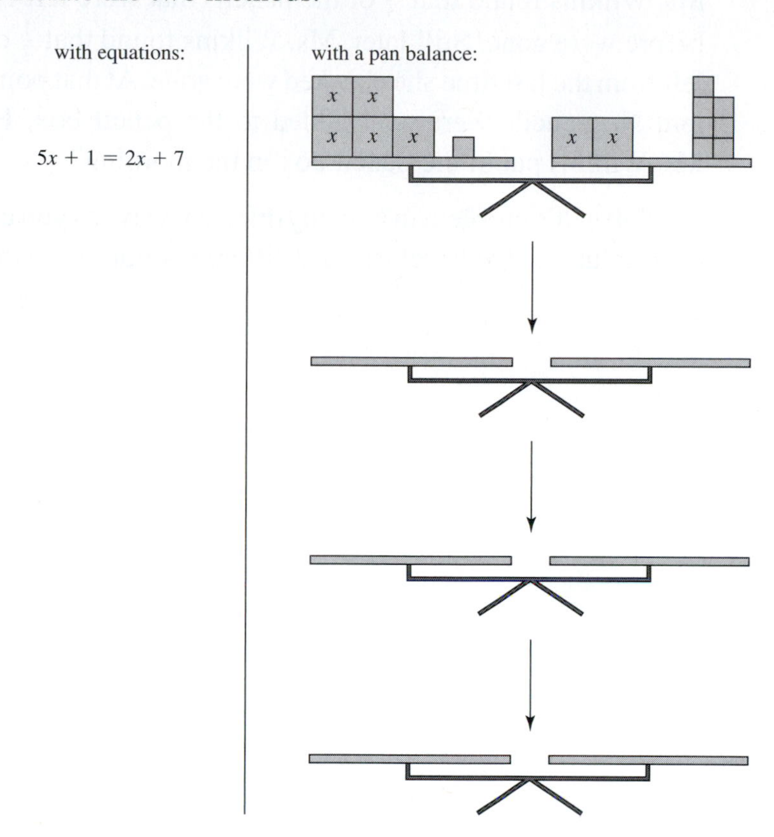

Class Activity 13L: How Many Pencils Were There?

In the morning, Ms. Wilkins put some pencils for her students in a pencil box. After a while, Ms. Wilkins found that $\frac{1}{2}$ of the pencils were gone. A little later, Ms. Wilkins found that $\frac{1}{3}$ of the pencils that were left from when she checked before were gone. Still later, Ms. Wilkins found that $\frac{1}{4}$ of the pencils that were left from the last time she checked were gone. At that point there were 15 pencils left. No pencils were ever added to the pencil box. How many pencils did Ms. Wilkins put in the pencil box in the morning?

Solve this problem in as many different ways as you can think of, and explain each solution. Try to relate your different solution methods to each other.

Class Activity 13M: Solving Story Problems with Strip Diagrams and with Equations

The problems in this activity were inspired by problems in the mathematics textbooks used in Singapore in grades 4–6 (see [12], volumes 4A–6B).

1. At a store, a hat costs 3 times as much as a t-shirt. Together, the hat and t-shirt cost $35. How much does the t-shirt cost?

   Solve this problem in two ways: by using the strip diagram shown here and with equations. Explain both solution methods, and discuss how they are related.

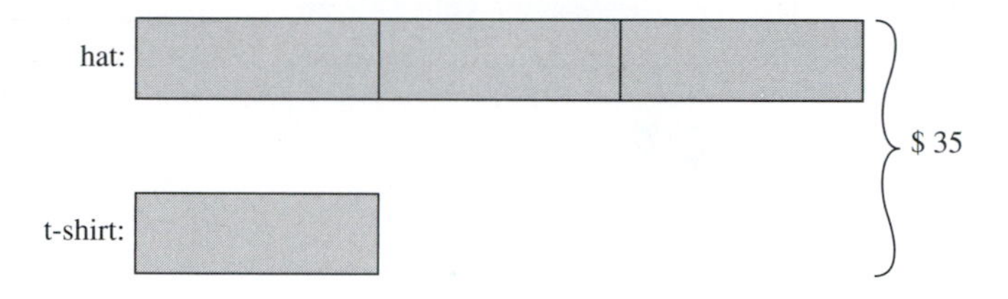

2. There are 180 blankets at a shelter. The blankets are divided into two groups. There are 30 more blankets in the first group than in the second group. How many blankets are in the second group?

   Solve this problem in two ways: by using the strip diagram shown here and with equations. Explain both solution methods, and discuss how they are related.

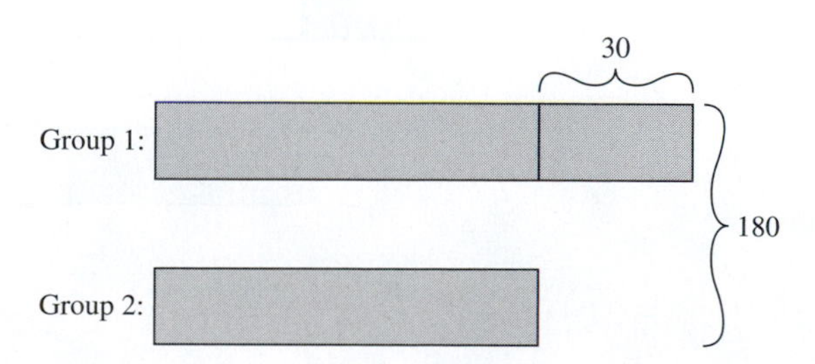

3. On a farm, $\frac{1}{7}$ of the sheep are grey, $\frac{2}{7}$ of the sheep are black, and the rest of the sheep are white. There are 36 white sheep. How many sheep in all are on the farm?

Solve this problem in two ways: by using the strip diagram shown here and with equations. Explain both solution methods, and discuss how they are related.

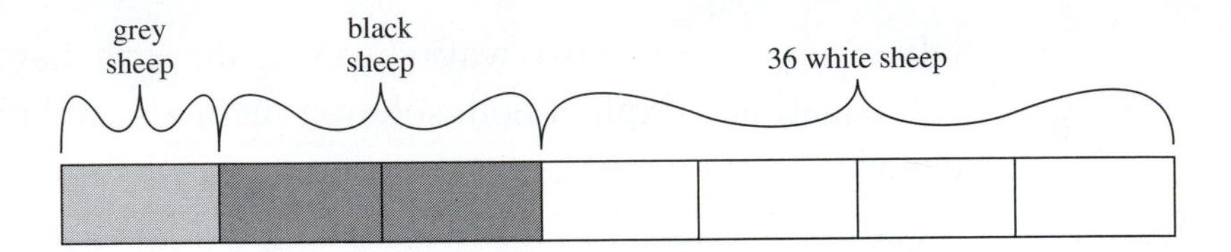

4. Ms. Jones gave $\frac{1}{4}$ of her money to charity and $\frac{1}{2}$ of the remainder to her mother. Then Ms. Jones had \$240 left. How much money did Ms. Jones have at first?

Solve this problem in two ways: by using the strip diagram shown here and with equations. Explain both solution methods, and discuss how they are related.

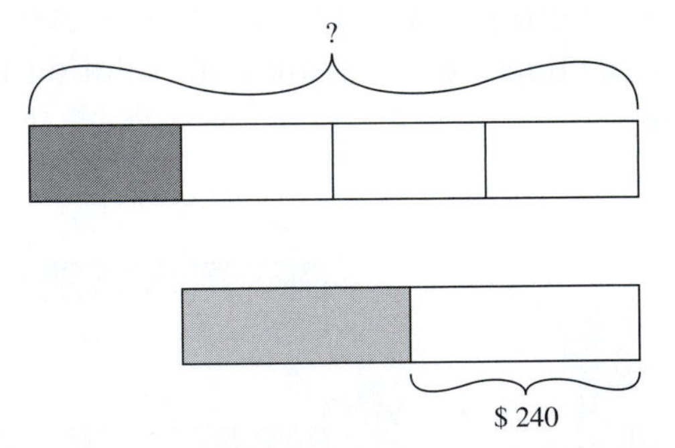

5. When a box of chocolates was full, it weighed 1.1 kilograms. After $\frac{1}{2}$ of the chocolates were eaten, the box (with the remaining chocolates) weighed 0.7 kilograms. How much did the box weigh without the chocolates?

   Solve this problem in two ways: with the aid of a diagram and with equations. Explain both solution methods, and discuss how they are related.

6. Quint had 4 times as many math problems to do as Agustin. After Quint did 20 problems and Agustin did 2 problems, they each had the same number of math problems left to do. How many math problems did Quint have to do at first?

   Solve this problem in two ways: by using the strip diagram shown here and with equations. Explain both solution methods, and discuss how they are related.

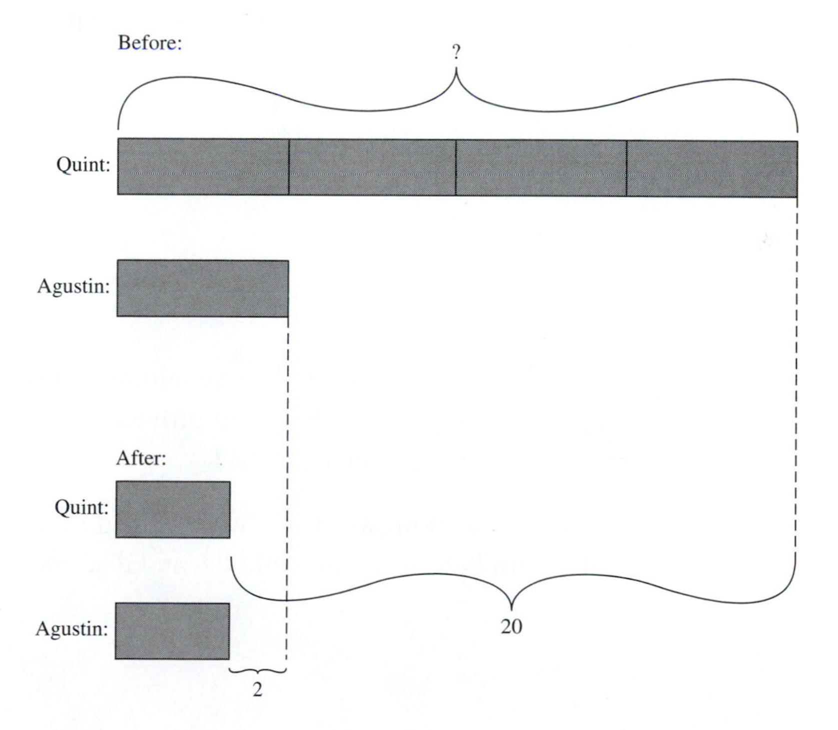

7. Carmen spent $\frac{1}{6}$ of her money on a CD. Then Carmen had $45 left. How much money did Carmen have at first?

Solve this problem in two ways: with the aid of a diagram and with equations. Explain both solution methods, and discuss how they are related.

8. There were 25 more girls than boys at a party. All together, 105 children were at the party. How many boys were at the party? How many girls were at the party?

Solve this problem in two ways: with the aid of a diagram and with equations. Explain both solution methods, and discuss how they are related.

9. A bakery sold $\frac{3}{5}$ of its muffins. The remaining muffins were divided equally among the 3 employees. Each employee got 16 muffins. How many muffins did the bakery have at first?

Solve this problem in two ways: with the aid of a diagram and with equations. Explain both solution methods, and discuss how they are related.

Class Activity 13N: Modifying Problems

1. Recall Problem 2 from Class Activity 13M:

There are 180 blankets at a shelter. The blankets are divided into two groups. There are 30 more blankets in the first group than in the second group. How many blankets are in the second group?

a. Suppose you want to modify the blanket problem for your students by changing the numbers 180 and 30 to different numbers. Can you change the numbers any way you want and still have a sensible problem? Explain.

b. Suppose you want to modify the blanket problem for your students so that the blankets will be divided into 3 unequal groups instead of 2 groups. Write such a modified problem, making sure that it can be solved. Show two different ways to solve your problem.

2. Recall the pencil problem from Class Activity 13L:

In the morning, Ms. Wilkins put some pencils for her students in a pencil box. After a while, Ms. Wilkins found that $\frac{1}{2}$ of the pencils were gone. A little later, Ms. Wilkins found that $\frac{1}{3}$ of the pencils that were left from when she checked before were gone. Still later, Ms. Wilkins found that $\frac{1}{4}$ of the pencils that were left from the last time she checked were gone. At that point there were 15 pencils left. No pencils were ever added to the pencil box. How many pencils did Ms. Wilkins put in the pencil box in the morning?

a. Suppose you want to modify the pencil problem for your students by changing the number 15 to a different number. Which numbers could you replace the 15 in the problem with and still have a sensible problem (without changing anything else in the problem)? Explain.

b. Experiment with changing some or all of the fractions $\frac{1}{2}$, $\frac{1}{3}$, and $\frac{1}{4}$ in the problem to some other "easy" fractions. When you make a change, do you also need to change the number 15? Which changes make the problem harder? Which changes make the problem easier?

Class Activity 13O: Solving Story Problems

The story problems in this activity are taken from Class Activity 13H. Solve each problem, either with the aid of a diagram or by using equations. Explain your reasoning.

1. Initially, Markus had M dollars in his bank account. After removing $\frac{1}{5}$ of the money in the account and then putting in another \$200, Markus now has \$800. How much money did Markus have initially?

2. Keisha had K dollars in her bank account. After removing \$200 and then removing $\frac{1}{5}$ of the remaining money, Keisha now has \$800. How many dollars did Keisha have in her bank account before removing any money?

3. Originally, there were L liters of liquid in a container. After $\frac{2}{3}$ of the liquid was poured out, another $2\frac{1}{2}$ liters of liquid were poured into the container. When $\frac{1}{4}$ of the liquid was poured out, 4 liters remained. What was the original amount of liquid in the container?

13.3 Sequences

Class Activity 13P: Arithmetic Sequences of Numbers Corresponding to Sequences of Figures

1. In the following sequence of figures made up of small squares, assume that the sequence continues by adding four shaded squares to the top of a figure in order to get the next figure in the sequence:

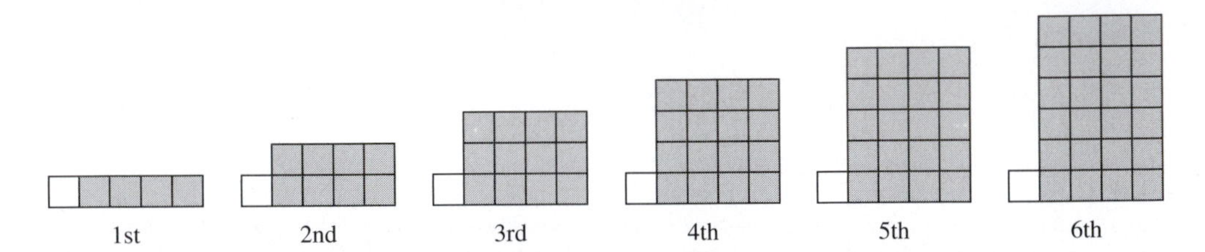

a. In the next table, write the number of small squares making up each figure.

Imagine that the sequence of figures continues forever, so that for each counting number N, there is an Nth figure. What is a formula for the number of small squares in the Nth figure? Add this information to the table.

POSITION OF FIGURE	NUMBER OF SMALL SQUARES IN FIGURE
1st	
2nd	
3rd	
4th	
5th	
6th	
7th	
⋮	⋮
Nth	

b. Relate the structure of the formula you found in Part (a) to the structure of the previous figures. Explain why your formula makes sense by relating your formula to the structure of the figures.

c. How many small squares will make up the 25th figure in the sequence? How can you tell?

d. Will there be a figure in the sequence that is made of 250 small squares? If yes, which one? If no, why not? Answer these questions in two ways: with algebra and in a way that a student in elementary school might be able to.

e. Will there be a figure in the sequence that is made of 85 small squares? Will there be a figure in the sequence that is made of 403 small squares? If yes, which one? If no, why not? Answer these questions in two ways: with algebra and in a way that a student in elementary school might be able to.

2. In the following sequence of figures made of small circles, assume that the sequence continues by adding a white circle to the end of each of the three "arms" of a figure in order to get the next figure in the sequence.

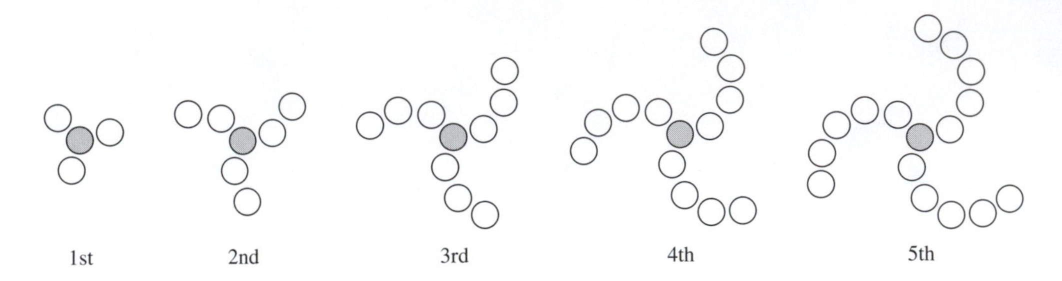

a. In the next table, write the number of small circles that the previous figures are made of.

Imagine that the sequence of figures continues forever, so that for each counting number N, there is an Nth figure. What is a formula for the number of small circles in the Nth pattern? Add this information to your table.

POSITION OF FIGURE	NUMBER OF SMALL CIRCLES IN FIGURE
1st	
2nd	
3rd	
4th	
5th	
6th	
⋮	⋮
Nth	

b. Relate the structure of the formula you found in Part (a) to the structure of the figures. Explain why your formula makes sense by relating your formula to the structure of the figures.

c. How many small circles will the 38th figure in the sequence be made of? How can you tell?

d. Will there be a figure in the sequence that is made of 100 small circles? If yes, which one? If no, why not? Answer these questions in two ways: with algebra and in a way that a student in elementary school might be able to.

e. Will there be a figure in the sequence that is made of 125 small circles? If yes, which one? If no, why not? Answer these questions in two ways: with algebra and in a way that a student in elementary school might be able to.

Class Activity 13Q: Deriving Formulas for Arithmetic Sequences

1. The next table shows some entries for an arithmetic sequence whose first entry is 5 and that increases by 3.

Entry number	Entry
1st	5
2nd	8
3rd	11
4th	14
5th	17
Nth	

a. If there were a 0th entry, what would it be? Put it in the previous table.

b. Fill in the blanks to describe how to get entries in the sequence by **starting from the 0th entry**.

- To find the 1st entry: Start at ___ and add ___ 1 time.
- To find the 2nd entry: Start at ___ and add ___ 2 times.
- To find the 3rd entry: Start at ___ and add ___ 3 times.
- To find the 4th entry: Start at ___ and add ___ 4 times.
- To find the 5th entry: Start at ___ and add ___ 5 times.
- To find the Nth entry: Start at ___ and add ___ N times.

c. For each bullet in Part (b), write an expression (using addition and multiplication) that corresponds to the description for finding the entry in the sequence.

1st entry =

2nd entry =

3rd entry =

4th entry =

5th entry =

Nth entry =

2. The next table shows some entries for an arithmetic sequence whose first entry is 1 and that increases by 4.

Entry number	Entry
1st	1
2nd	5
3rd	9
4th	13
5th	17
Nth	

a. If there were a 0th entry, what would it be? Put it in the previous table.

b. Fill in the blanks to describe how to get entries in the sequence by **starting from the 0th entry**.

- To find the 1st entry: Start at ___ and add ___ 1 time.
- To find the 2nd entry: Start at ___ and add ___ 2 times.
- To find the 3rd entry: Start at ___ and add ___ 3 times.
- To find the 4th entry: Start at ___ and add ___ 4 times.
- To find the 5th entry: Start at ___ and add ___ 5 times.
- To find the Nth entry: Start at ___ and add ___ N times.

c. For each bullet in part (b), write an expression (using addition and multiplication) that corresponds to the description for finding the entry in the sequence.

1st entry =

2nd entry =

3rd entry =

4th entry =

5th entry =

Nth entry =

Class Activity 13R: Sequences and Formulas

This activity concerns the following three sequences:

The first sequence is the one whose Nth entry is $2N + 3$.

The second sequence is the one whose Nth entry is $4N + 1$.

The third sequence is the one whose Nth entry is N^2.

1. Write the first seven numbers in each of the three sequences.

2. Informally discuss the following:

   In what way are the first two sequences similar? In what way are they different? In what way are the first two sequences different from the third sequence?

   How do the first two sequences grow? In other words, how do these sequences change in going from one entry to the next? Contrast the way the first two sequences grow with the way the third sequence grows.

3. Given that 729 is one of the entries in the first sequence, can you predict the next entry without using the formula? How?

   Given that 729 is one of the entries in the second sequence, can you predict the next entry without using the formula? How?

   Given that 729 is one of the entries in the third sequence, is there an easy way to predict the next entry without using the formula? Why is this harder than for the other two sequences?

4. How is the way the first two sequences grow related to their formulas? Why does that make sense?

5. Draw a sequence of figures made of small circles so that the Nth figure in the sequence is made of $2N + 3$ small circles.

Draw another sequence of figures made of small circles so that the Nth figure in the sequence is made of $4N + 1$ small circles.

Draw another sequence of figures made of small circles or small squares so that the Nth figure in the sequence is made of N^2 small circles or small squares.

6. How is the structure of the figures you drew in Problem 5 related to the formulas for the number of small circles they are made of?

Class Activity 13S: Geometric Sequences

1. Write a formula for a sequence whose first few entries are the number of small squares in the next sequence of figures. Explain why you wrote your formula as you did.

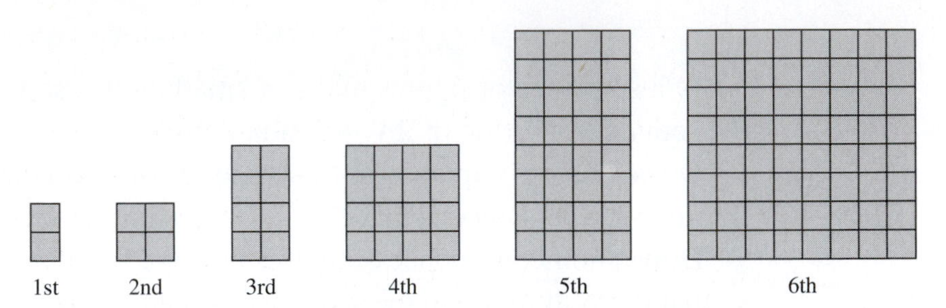

2. Write a formula for a sequence whose first few entries are the number of small squares in the next sequence of figures. Explain why you wrote your formula as you did.

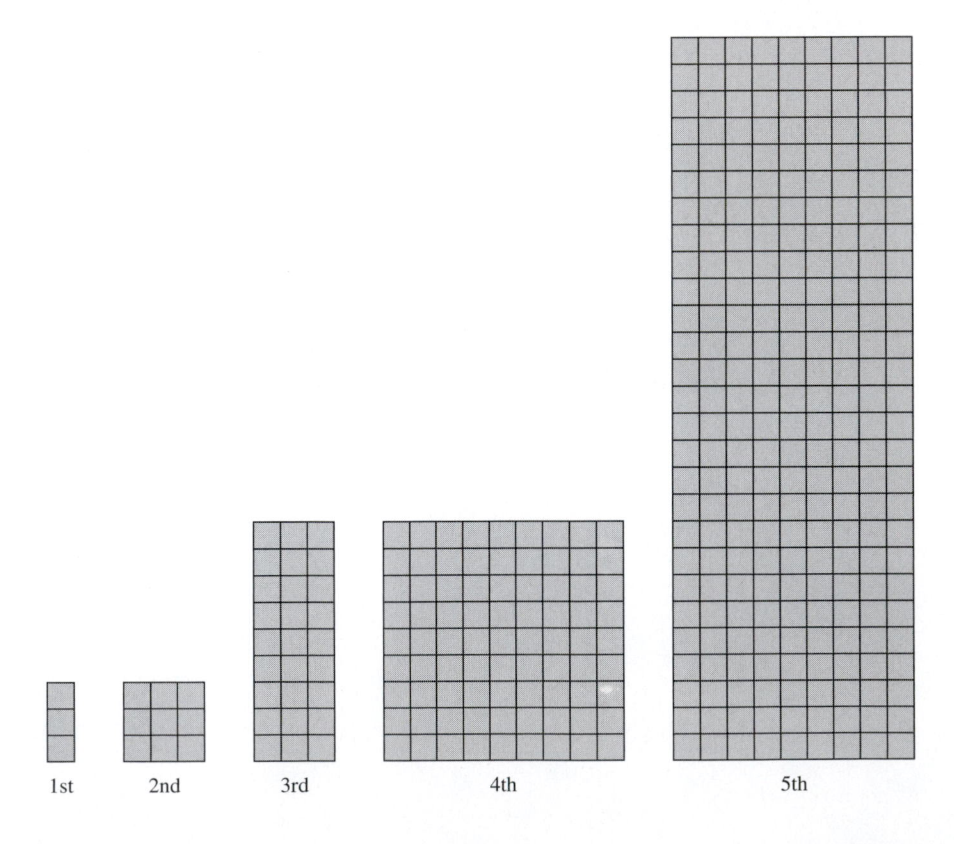

3. Suppose you owe $1000 on your credit card and that the credit card company charges you 1.4708% interest every month. That is, at the end of each month an additional 1.4708% of the amount you owe is added to the amount you owe. (This is what would happen if your credit card charged an annual percentage rate of 17.6496%.) Let's assume that you do not pay off any of this debt or the interest that is added to it. Let's also assume that you don't add any more debt other than the interest that you are charged.

a. Fill in the next table to show how much you owe at the end of 1 month, 2 months, and so on, up to 12 months. Explain why you can get each entry from the previous entry by multiplying by 1.014708.

MONTH	AMOUNT YOU OWE AT END OF MONTH
1	
2	
3	
4	
5	
6	
7	
8	
9	
10	
11	
12	

b. Determine how much you will owe at the end of 24 months without finding how much you will owe at the end of 13 months, 14 months, 15 months, and so on.

c. Write a formula for the amount of money you will owe at the end of N months. Explain why your formula is valid.

d. Use your formula from Part (c) to determine how much money you will owe after 4 years.

Class Activity 13T: Repeating Patterns

1. Even very young children can work with repeating patterns. For example, see the activities in *Navigating through Algebra in Prekindergarten–Grade 2* by the National Council of Teachers of Mathematics [8]. Some of the problems that follow are similar to some of the problems described in that publication.

   a. Assume that the following pattern of a square followed by 3 circles and 2 triangles continues to repeat:

   What will be the 100th shape in the pattern? Explain how you can tell.

   b. How many circles will there be among the first 150 entries of the given sequence? Explain your reasoning.

   c. To answer Part (b), Amanda says that since there are 6 circles among the first 10 shapes, and since 150 is 15 sets of 10, there will be $15 \times 6 = 90$ circles among the first 150 entries. Is Amanda's reasoning correct? Why or why not?

2. What day of the week will it be 100 days from today? Determine the answer with math. Explain your reasoning. How is this problem related to repeating patterns?

3. What time of day will it be 100 hours from now? Determine the answer with math. Explain your reasoning. How is this problem related to repeating patterns?

4. Five friends are sitting in a circle as shown. Antrice does "eenie meenie minee moe, catch a tiger by his toe, if he hollers, let him go, y-e-s spells yes and you are it," starting with Benton and going clockwise, pointing to a new person at each underlined word, word pair, or letter.

a. Who will be "it"? Explain how to predict the answer by using math.

b. If Fran comes and sits between Ellie and Antrice before Antrice does "eenie meenie minee moe...," then who will be "it"?

c. Some people use a different "eenie meenie minee moe" chant. If you know a different one, use math to determine who will be "it" with your chant.

5. What is the digit in the ones place of 2^{100}? Explain how you can tell.

Class Activity 13U: ✻ The Fibonacci Sequence in Nature and Art

You will need pine cones or pineapples for Problem 1. You will need a ruler for Problems 2 and 3. You will need graph paper for Problem 4.

The Fibonacci sequence is

$$1, \quad 1, \quad 2, \quad 3, \quad 5, \quad 8, \quad 13, \quad 21, \quad 34, \ldots$$

Each entry in the sequence is obtained by adding the previous two entries.

1. Look closely at a pine cone or pineapple. You should see two sets of "swirls." One set swirls around clockwise; the other set swirls around counterclockwise. Count the number of swirls in each set. These numbers are usually consecutive Fibonacci numbers.

2. Measure the lengths of the indicated bones in your body, rounded to the nearest inch, and record the results in the next table. Use the same finger for all finger measurements.

BONE CONNECTING	LENGTH OF BONE
shoulder and elbow	
elbow and wrist	
wrist and first knuckle	
first knuckle and second knuckle	
second knuckle and third knuckle	
third knuckle and fingertip	

Compare the lengths of your bones in inches with the Fibonacci sequence.

3. Draw a rectangle whose proportions are pleasing to your eye. Carefully measure the length and width of your rectangle. Then divide the length (the longer measurement) by the width (the shorter measurement). Is this ratio close to the ratios of the following consecutive Fibonacci numbers?

$$\frac{3}{2} = 1.5, \quad \frac{5}{3} = 1.67, \quad \frac{8}{5} = 1.6, \quad \frac{13}{8} = 1.625$$

4. The beginning of a sequence of rectangles is shown next. The sequence starts with a 1-unit-by-1-unit square. To create the next rectangle in the sequence, attach a square to the previous rectangle. Attach the square either below the rectangle or to the right of the rectangle, alternating between attaching it below and attaching it to the right.

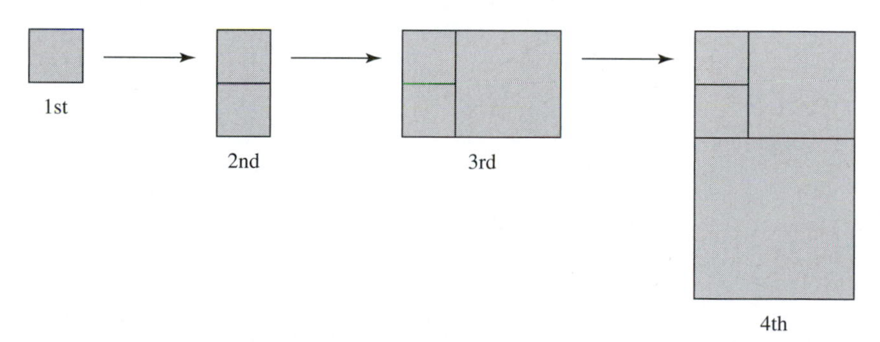

a. On graph paper, draw the next two rectangles in the sequence.

b. Do you find the proportions of these rectangles pleasing to your eye?

c. What are the lengths and widths of rectangles in this sequence?

Class Activity 13V: What's the Rule?

For each of the next sequences, find three different rules for determining the next three entries in the sequence. In each case, describe the rule you use.

1.

2, 4, 8, ___, ___, ___,

2, 4, 8, ___, ___, ___,

2, 4, 8, ___, ___, ___,

2.

2, 5, 11, ___, ___, ___,

2, 5, 11, ___, ___, ___,

2, 5, 11, ___, ___, ___,

3.

2, 3, 6, ___, ___, ___,

2, 3, 6, ___, ___, ___,

2, 3, 6, ___, ___, ___,

13.4 Series

Class Activity 13W: Sums of Counting Numbers

Is there a quick way to add a bunch of consecutive counting numbers? This activity will help you find a way by using equations that come from rectangular designs.

Imagine that the next sequence of rectangular designs is to continue indefinitely in such a way that to get the next rectangular design in the sequence, we add a row of white squares to the bottom and then a column of dark squares to the right.

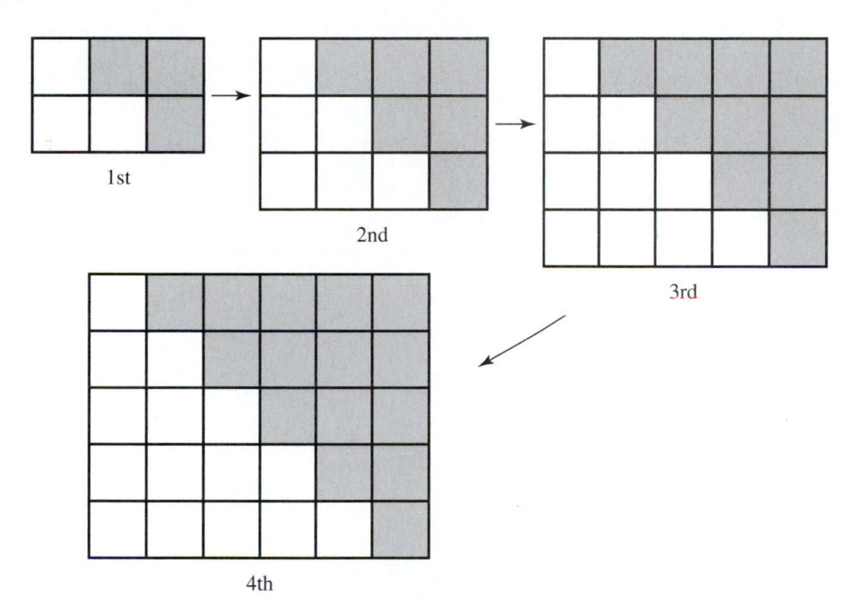

1. Discuss how the following equations are associated with the first two rectangular designs:

$$1 + 2 = \frac{2 \cdot 3}{2}$$
$$1 + 2 + 3 = \frac{3 \cdot 4}{2}$$

2. Find equations like the ones in Problem 1 that are associated with the 3rd and 4th rectangular designs in the sequence. Discuss how these equations are related to the designs.

3. Determine the sum of the following arithmetic series, and explain your reasoning:

$$1 + 2 + 3 + 4 + \ldots + 198 + 199 + 200$$

4. Write an equation that is associated with the Nth design in the previous sequence of rectangles. Explain your work.

Class Activity 13X: Sums of Odd Numbers

Is there a quick way to add a bunch of consecutive odd numbers? This activity will help you find and explain a method for adding odd numbers.

1. Calculate each of the next sums.

$$1 + 3 = \underline{\quad\quad}$$

$$1 + 3 + 5 = \underline{\quad\quad}$$

$$1 + 3 + 5 + 7 = \underline{\quad\quad}$$

$$1 + 3 + 5 + 7 + 9 = \underline{\quad\quad}$$

$$1 + 3 + 5 + 7 + 9 + 11 = \underline{\quad\quad}$$

$$1 + 3 + 5 + 7 + 9 + 11 + 13 = \underline{\quad\quad}$$

2. What is special about the solutions to the sums in Problem 1?

3. Based on your answer in Problem 2, predict the sum of the first 100 odd numbers.

4. Based on your answer in Problem 2, predict the next sum:

$$1 + 3 + 5 + 7 + 9 + \ldots + 91 + 93 + 95 + 97 + 99 = \underline{\quad\quad}$$

5. Use the next sequence of square designs to find and explain a formula for sums of consecutive odd numbers. Imagine this sequence of square designs continuing indefinitely. To get the next square design in the sequence, we always add a row and column of small squares of a new color along the bottom and right.

Write an equation that is associated with the *N*th square design. Explain.

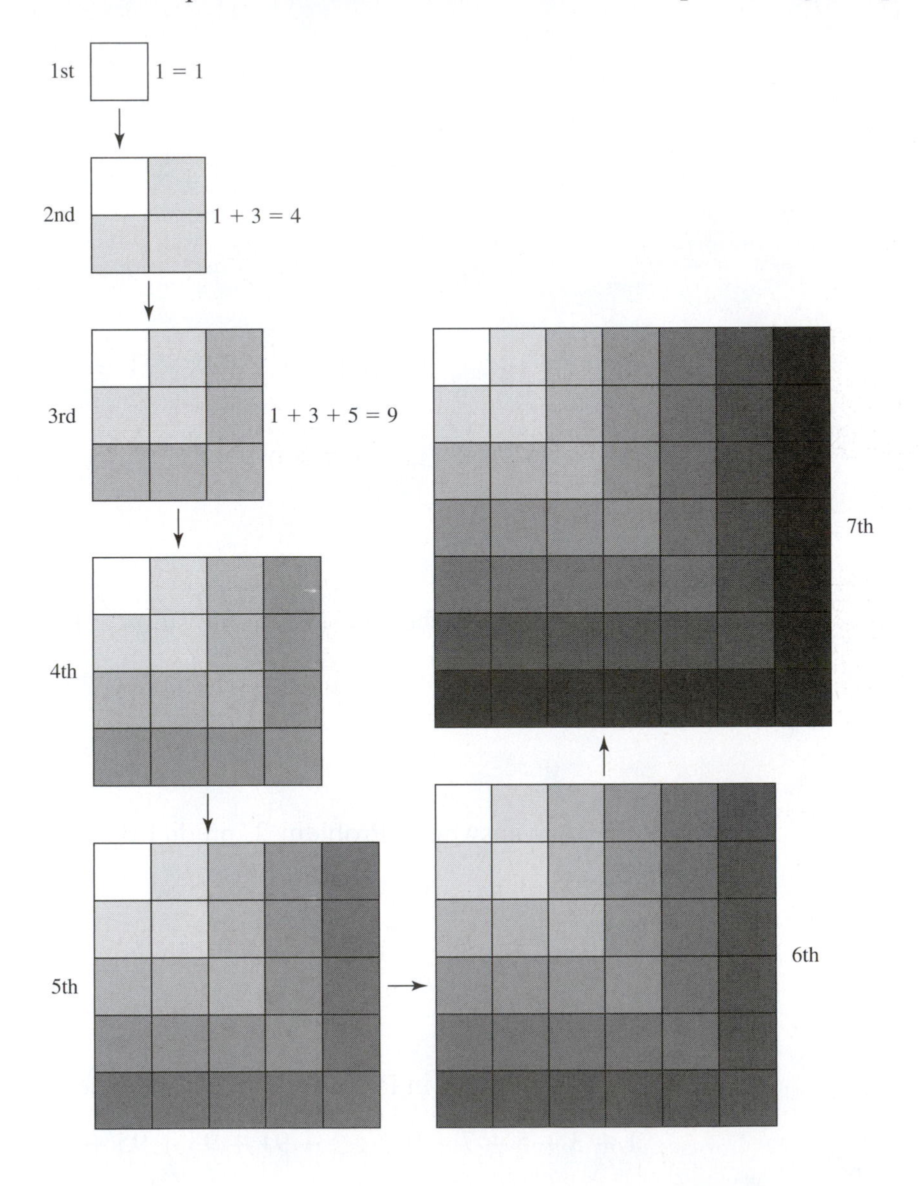

Class Activity 13Y: ∗ Sums of Squares

This activity is related to Class Activity 11AA on explaining why there is a factor of $\frac{1}{3}$ in the volume formula for pyramids and cones. In that activity, 3 oblique pyramids were put together to make a cube. In this activity, you will see that 3 oblique "step pyramids" fit together to make a rectangular prism—except that there is a gap in the prism. By describing the "step pyramids" and the "gap," you can derive a formula for a sum of squares.

Before you start this activity, you may want to review how the 3 oblique pyramids of Class Activity 11AA fit together to make a cube. Observe that the points of the 3 pyramids come together.

1. **a.** Using blocks in 3 different colors, make 3 oblique "step pyramids," as pictured in Figure 13Y.1, each of a different color. Your step pyramids should have 2 layers: the bottom layer made of $2^2 = 4$ blocks, and the top layer made of 1 block.

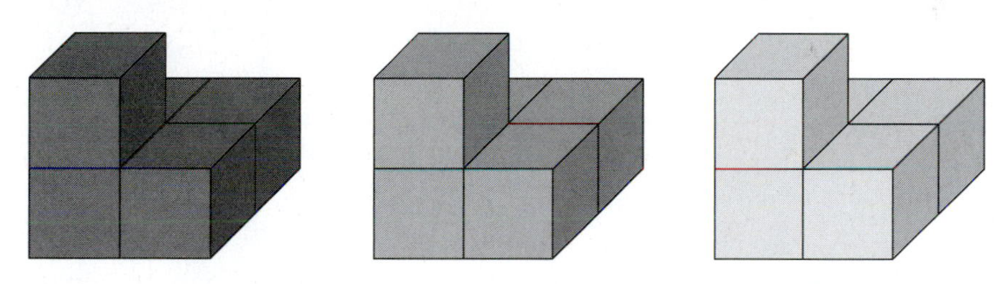

Figure 13Y.1
3 Oblique Step Pyramids

b. Put your 3 step pyramids together, so that their "points" (top layers) come together (just like the three oblique pyramids in Class Activity 11AA), and so that together they *almost* form a rectangular prism.

Observe that if you add $1+2$ blocks, you will get a rectangular prism that is 3 blocks wide, 3 blocks long, and 2 blocks tall.

c. Explain how Part (b) gives rise to the following equation:

$$3 \cdot (1 + 2^2) = 3 \cdot 3 \cdot 2 - (1 + 2)$$

2. **a.** Using blocks in 3 different colors, make 3 oblique "step pyramids," as pictured in Figure 13Y.2, each of a different color. This time your step pyramids should have 3 layers: the bottom layer made of $3^2 = 9$ blocks, the middle layer made of $2^2 = 4$ blocks, and the top layer made of 1 block.

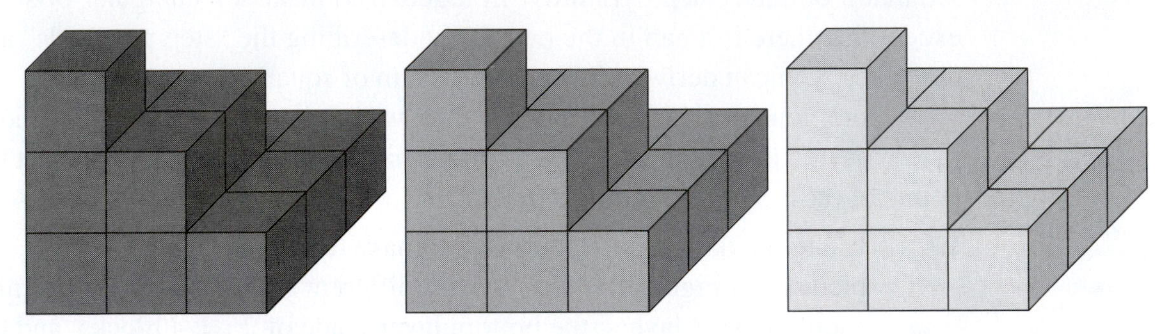

Figure 13Y.2
3 Oblique Step Pyramids

b. Put your 3 step pyramids together so that their "points" (top layers) come together (just as in Problem 1, Part (b), and just like the three oblique pyramids in Class Activity 11AA), and so that together they *almost* form a rectangular prism. It might help to secure your step pyramids with masking tape.

Observe that if you add $1 + 2 + 3$ blocks, you will get a rectangular prism that is 4 blocks wide, 4 blocks long, and 3 blocks tall.

c. Explain how Part (b) gives rise to the following equation:

$$3 \cdot (1 + 2^2 + 3^2) = 4 \cdot 4 \cdot 3 - (1 + 2 + 3)$$

3. If you can, make 3 step pyramids that each have 4 layers: the bottom layer made of 4^2 blocks, and the remaining layers made of 3^2, 2^2, and 1 block(s), respectively. Put them together as before, and explain how this gives rise to the equation

$$3 \cdot (1 + 2^2 + 3^2 + 4^2) = 5 \cdot 5 \cdot 4 - (1 + 2 + 3 + 4)$$

4. Imagine that you could continue this process of putting larger and larger step pyramids together indefinitely. Use your discoveries and the fact that

$$1 + 2 + 3 + \ldots + N = \frac{N(N+1)}{2}$$

to give a formula for

$$1 + 2^2 + 3^2 + \ldots + N^2$$

In other words, write an equation for which the left-hand side is $1 + 2^2 + 3^2 + \ldots + N^2$.

5. Based on your formula in Problem 4, calculate

$$1 + 2^2 + 3^2 + \ldots + 100^2$$

Class Activity 13Z: ✻ Sums of Powers of Two

Is there a quick way to add a bunch of consecutive powers of 2? This activity will help you find and explain a formula.

1. Calculate the next sums.

$$1 + 2 = \_\_$$
$$1 + 2 + 4 = \_\_$$
$$1 + 2 + 4 + 8 = \_\_$$
$$1 + 2 + 4 + 8 + 16 = \_\_$$

2. Based on Problem 1, predict the sum of the following geometric series without adding all the terms:

$$1 + 2 + 4 + 8 + 16 + 32 + 64 + 128 + 256 + 512$$

3. Based on Problem 1, predict a formula in terms of N for the following geometric series (fill in the blank with an appropriate expression):

$$1 + 2 + 2^2 + 2^3 + 2^4 + \ldots + 2^N = \_\_$$

4. Use Figure 13Z.1 to help you explain why your formula in Problem 3 should be true.

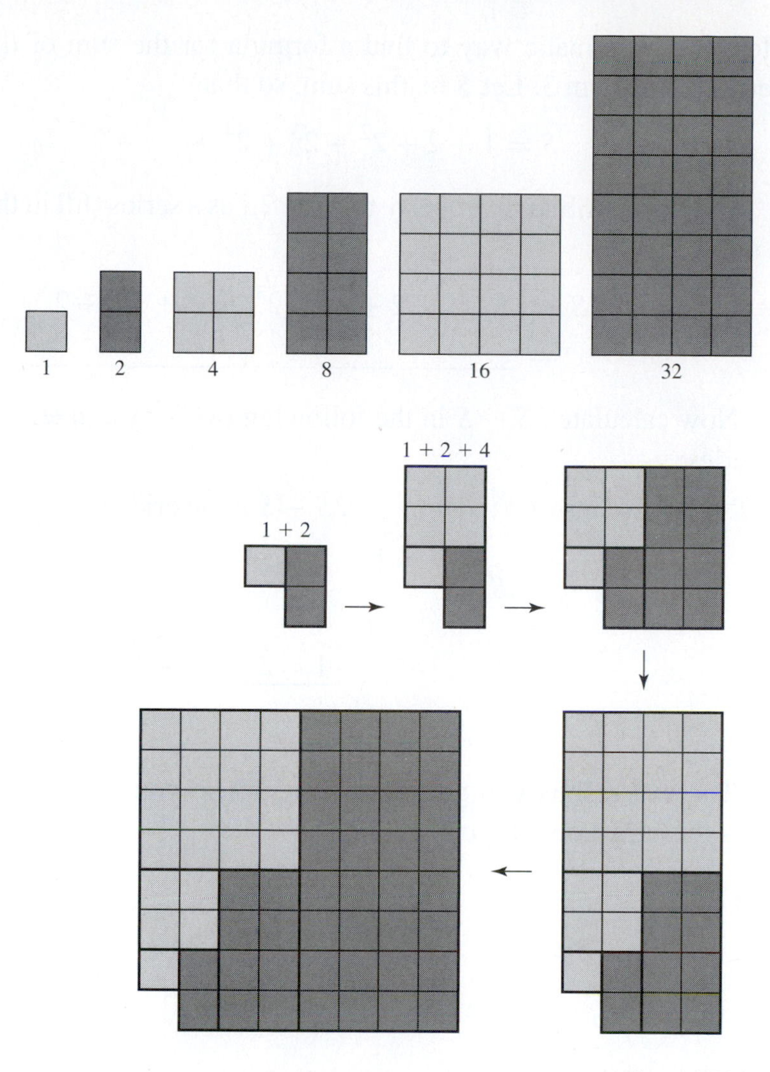

Figure 13Z.1

Sums of Powers of 2

5. Here is a systematic way to find a formula for the sum of the geometric series in Problem 3: Let S be this sum, so that

$$S = 1 + 2 + 2^2 + 2^3 + 2^4 + \ldots + 2^N$$

Use the distributive property to write $2S$ as a series (fill in the blank with a series):

$$\begin{aligned} 2S &= 2 \cdot (1 + 2 + 2^2 + 2^3 + 2^4 + \ldots + 2^N) \\ &= \underline{\hspace{6cm}} \end{aligned}$$

Now calculate $2S - S$ in the following two ways, in terms of S and as a series:

$2S - S$ in terms of S:	$2S - S$ as a series :
$2S$	
$\underline{-S}$	$\underline{-1 - 2 - 2^2 - 2^3 - 2^4 - \ldots - 2^N}$

The two results you get must be equal, so you get an equation. What does this equation tell you about S?

Class Activity 13AA: ✻ An Infinite Geometric Series

Is it possible to calculate an infinite sum of numbers? Surprisingly, the answer is "yes," as you will see in this activity.

Assume that the next sequence of partially shaded squares continues indefinitely, in such a way that we get the next square in the sequence by shading $\frac{1}{2}$ of the unshaded part of a square.

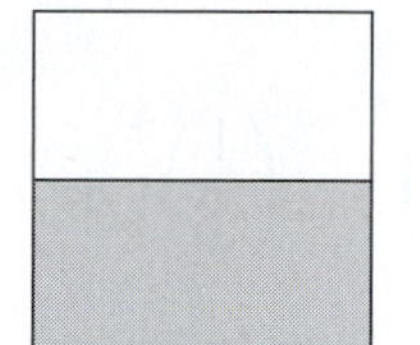 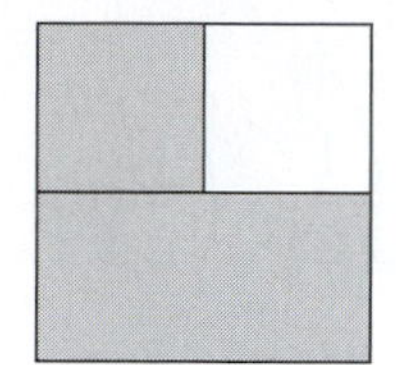 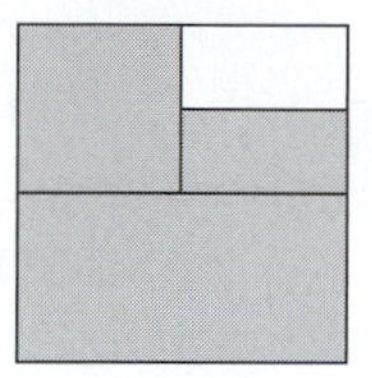 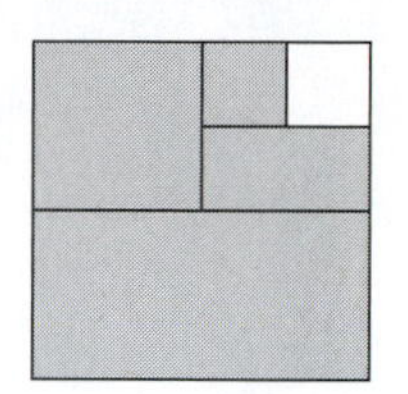

1. Explain why the shaded portions of the 2nd, 3rd, and 4th squares in the sequence are

$$\frac{1}{2} + \left(\frac{1}{2}\right)^2$$

$$\frac{1}{2} + \left(\frac{1}{2}\right)^2 + \left(\frac{1}{2}\right)^3$$

$$\frac{1}{2} + \left(\frac{1}{2}\right)^2 + \left(\frac{1}{2}\right)^3 + \left(\frac{1}{2}\right)^4$$

2. What fraction of the square is shaded in the 2nd, 3rd, and 4th figures? Give each answer in simplest form. Based on your results, predict what fraction of the Nth square is shaded.

3. Based on your work in Problem 2, when would you reach a square that is at least 99.9% shaded?

4. Based on the sequence of shaded squares, what would you expect the infinite sum

$$\frac{1}{2} + \left(\frac{1}{2}\right)^2 + \left(\frac{1}{2}\right)^3 + \left(\frac{1}{2}\right)^4 + \left(\frac{1}{2}\right)^5 + \ldots$$

to be equal to?

5. Here is a way to calculate the infinite sum in Problem 4. Let S stand for this sum. In other words,

$$S = \frac{1}{2} + \left(\frac{1}{2}\right)^2 + \left(\frac{1}{2}\right)^3 + \left(\frac{1}{2}\right)^4 + \left(\frac{1}{2}\right)^5 + \ldots$$

Assuming that there is an "infinite distributive property," write $\frac{1}{2}S$ as an infinite sum (fill in the blank):

$$\frac{1}{2}S = \frac{1}{2} \cdot \left(\frac{1}{2} + \left(\frac{1}{2}\right)^2 + \left(\frac{1}{2}\right)^3 + \left(\frac{1}{2}\right)^4 + \ldots\right)$$

$$= \underline{\hspace{6cm}}$$

Now calculate $S - \frac{1}{2}S$ in two ways, in terms of S and as a series:

in terms of S:	as a series :
S	$\frac{1}{2} + \left(\frac{1}{2}\right)^2 + \left(\frac{1}{2}\right)^3 + \left(\frac{1}{2}\right)^4 + \left(\frac{1}{2}\right)^5 + \ldots$
$-\frac{1}{2}S$	______________________________

The two results you get must be equal, so you get an equation. Solve this equation for S.

Class Activity 13BB: ✻ Making Payments into an Account

You will need a calculator for this activity.

Suppose that at the beginning of every month, you make a payment of \$200 into an account that earns 1% interest per month. (That is, the value of the account at the end of the month is 1% higher than it was at the beginning of the month.)

1. Make a guess: After making your 12th payment at the beginning of the 12th month, how much money will be in the account? (The remaining parts of this activity will help you calculate this amount exactly.)

2. Explain why the entries shown in the right-hand column in the next table are correct. Then fill in the remaining columns with series similar to those in the first three rows. (You may use "..." within your series in the last row.)

MONTH	AMOUNT AT BEGINNING OF MONTH (AFTER PAYMENT)
1	200
2	$200 + (1.01)200$
3	$200 + (1.01)200 + (1.01)^2 200$
4	
5	
6	
⋮	⋮
12	

3. Let S be the series you wrote in the last row of the right column in Problem 2. Use the distributive property to write $(1.01)S$ as a series.

4. Calculate $(1.01)S - S$ in two ways: in terms of S and as a series.

5. The two results you get in Problem 4 must be equal, so you get an equation. Solve this equation for S. Compare the result with your guess in Problem 1.

13.5 Functions

Class Activity 13CC: Interpreting Graphs of Functions

1. Items (a) through (f) are hypothetical descriptions of a population of fish. Each description corresponds to a population function, for which the input is time elapsed since the fish population was first measured, and the output is the population of fish at that time. Match the descriptions of these population functions to the graphs in Figure 13CC.1. In each case, explain why the shape of the graph fits with the description of the function.

   a. The population of fish rose slowly at first, and then rose more and more rapidly.

   b. The population of fish rose rapidly at first, and then rose more and more slowly.

   c. The population of fish rose at a steady rate.

   d. The population of fish dropped rapidly at first, and then dropped more and more slowly.

   e. The population of fish dropped slowly at first, and then dropped more and more rapidly.

   f. The population of fish dropped at a steady rate.

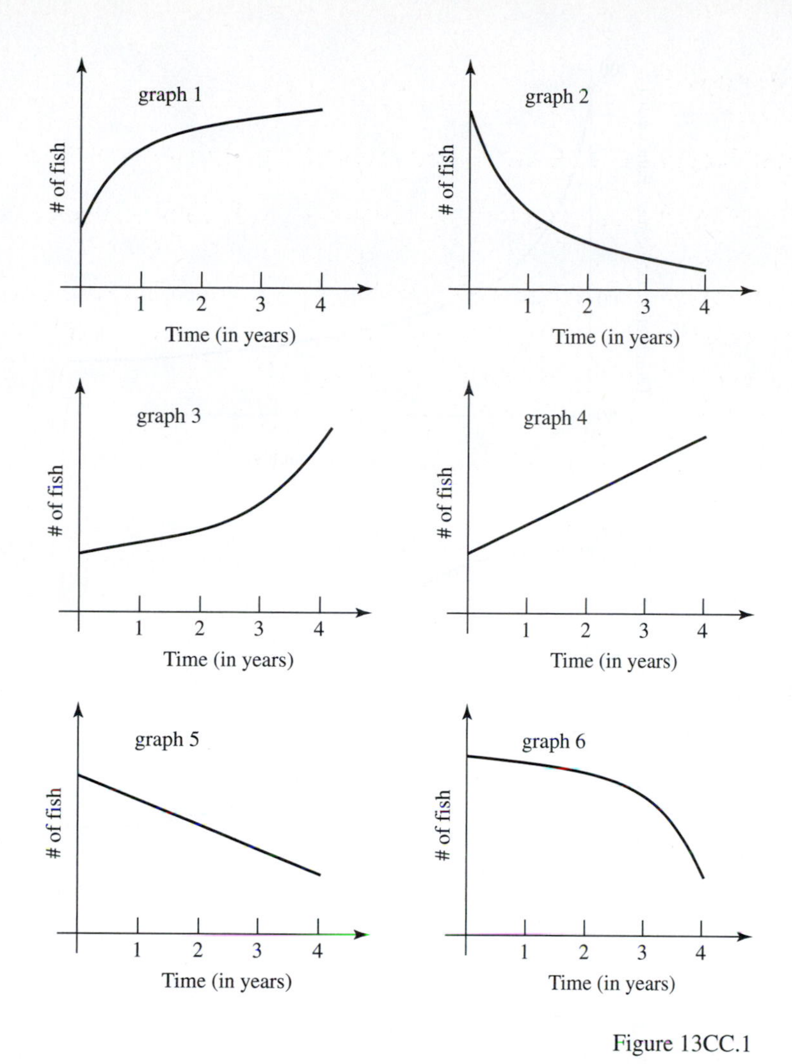

Figure 13CC.1
Fish Populations

2. Hot water is poured into a mug and left to cool. This situation gives rise to a temperature function for which the input is the time elapsed since pouring the water into the mug and the output is the temperature of the water at that time. The graph of this function is one of the three graphs shown next. Which graph do you think it is, and why? For each graph, describe how water would cool according to that graph.

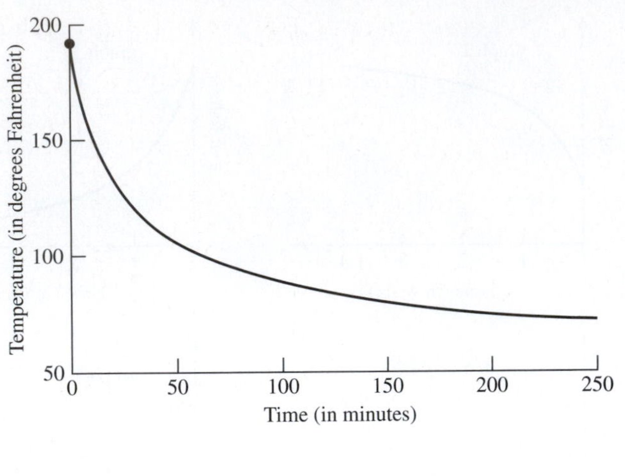
200
150
100
50
Temperature (in degrees Fahrenheit)
0
50
100
150
200
250
Time (in minutes)

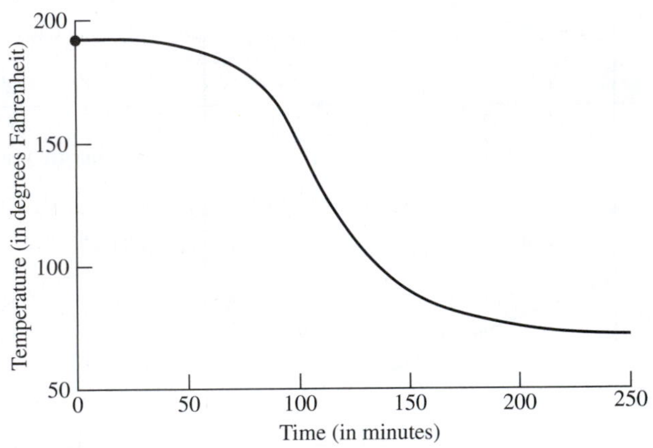
200
150
100
50
Temperature (in degrees Fahrenheit)
0
50
100
150
200
250
Time (in minutes)

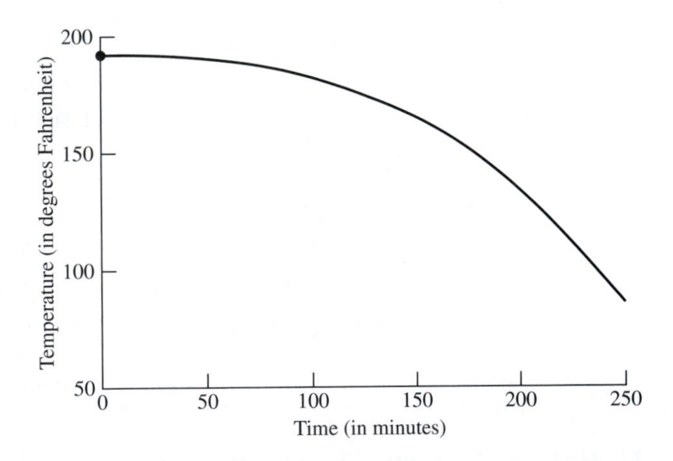
200
150
100
50
Temperature (in degrees Fahrenheit)
0
50
100
150
200
250
Time (in minutes)

Which graph is the correct one for problem 2? See page 527.

3. A tagged manatee swims up a river, away from a dock. Meanwhile, the manatee's tag transmits its distance from the dock. This situation gives rise to a distance function for which the input is the time since the manatee first swam away from the dock and the output is the manatee's distance from the dock at that time. The graph of this distance function is shown.

Write a story about the manatee that fits with this graph. Explain how features of the graph fit with your story.

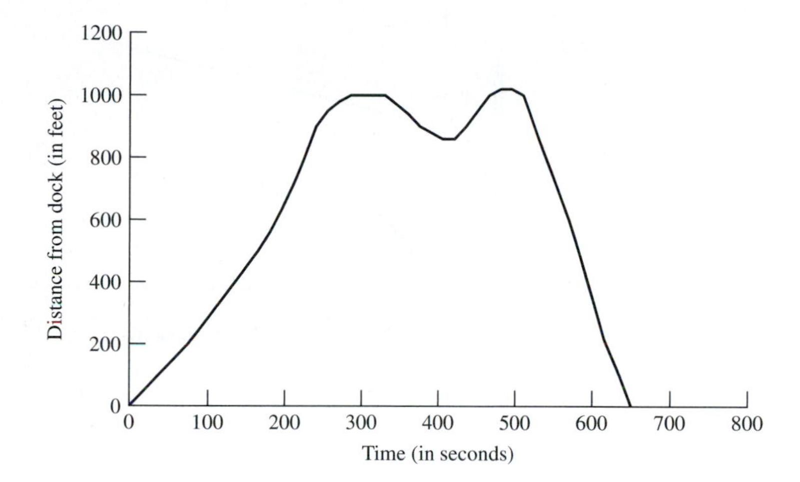

Class Activity 13DD: Are These Graphs Correct?

1. Carl started to drive from Providence to Boston, but after leaving he realized that he had forgotten something and drove back to Providence. Then Carl got back in his car and drove straight to Boston. This scenario gives rise to a distance function whose input is time elapsed since Carl first started to drive to Boston and whose output is Carl's distance from Providence. Could the next graph be the graph of the distance function described? Why or why not? If not, draw a different graph that could be the graph of the distance function. (Boston is 50 miles from Providence.)

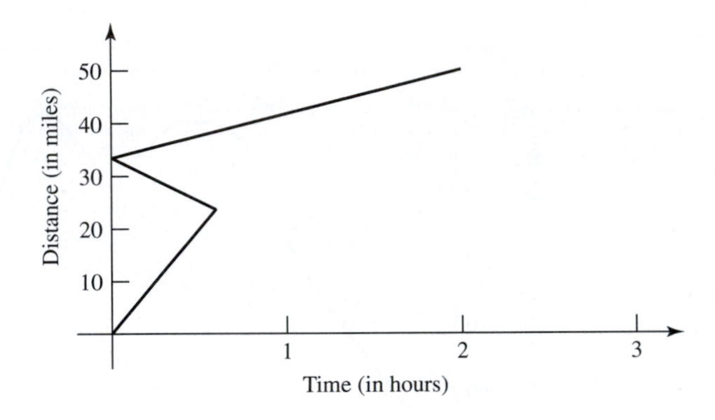

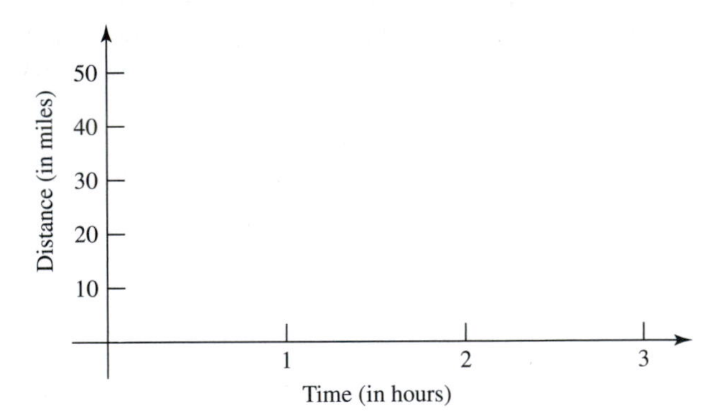

2. Here is what happened when Jenny ran a mile in 10 minutes. She got off to a good start, and ran faster and faster. Then all of a sudden, Jenny tripped. Once Jenny got back up, she started to run again, but at a slower pace. But near the end of her mile run, Jenny picked up some speed.

The next graph is supposed to fit with the story about Jenny's mile run. What is wrong with this graph?

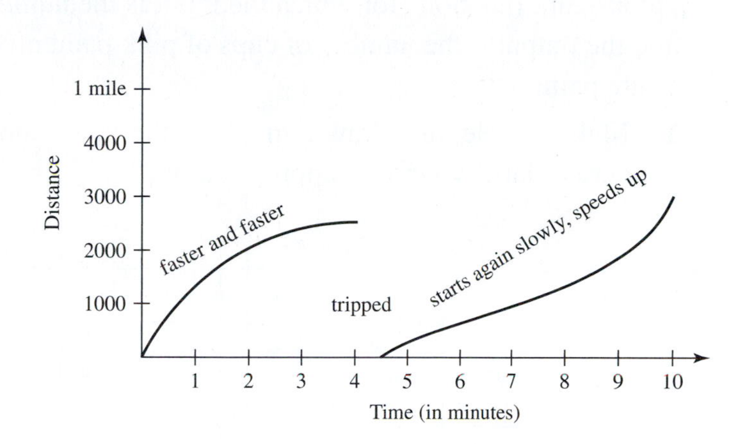

3. Describe two functions that the story about Jenny's mile run in Problem 2 could give rise to. In each case, describe the inputs and corresponding outputs. Sketch graphs that could be the graphs of these two functions.

The correct graph on p. 524 is the first graph.

13.6 Linear Functions

Class Activity 13EE: A Function Arising from Proportions

To make a certain shade of pink paint, for every 2 cups of white paint, you will produce 3 cups of pink paint by adding 1 cup of red paint to the white paint. So the ratio of white to pink paint is 2 to 3. This situation gives rise to a "pink-paint function" for which the input is the number of cups of white paint and the output is the number of cups of pink paint produced for that amount of white paint.

1. Make a table, and draw a graph of the pink-paint function described. Be sure to label your axes appropriately.

input (cups white paint)	output (cups pink paint)

2. Fill in the blank to make a true statement about the pink-paint function:
"Whenever the input increases by 2, the output ____________."
Explain how this statement is reflected in the table and the graph of the pink-paint function.

3. Fill in the blank to make a true statement about the pink-paint function:
"Whenever the input increases by 1, the output ____________."
Explain how this statement is reflected in the graph of the pink-paint function.

4. Let x stand for the number of cups of white paint. Let P be the pink-paint function, so that $P(x)$ is the number of cups of pink paint when x cups of white paint are used. Write a formula for $P(x)$, and describe how this formula is related to your answer for Problem 3.

Class Activity 13FF: Arithmetic Sequences as Functions

Every sequence can be viewed as a function that associates to the input N, the output which is the Nth entry in the sequence. Arithmetic sequences give rise to special kinds of functions.

1. Consider the arithmetic sequence

$$3,\ 5,\ 7,\ 9,\ 11,\ 13, \ldots$$

a. Make a table, find a formula, and draw a graph for the function associated with the arithmetic sequence.

input	output
x	

b. Fill in the blank to make a true statement about the function in Part (a): "Whenever the input increases by 1, the output ________________."
How is this statement reflected in the table and the graph of the function?

c. In Part (a), how are components of the formula reflected in the graph?

2. Consider the arithmetic sequence

$$2.5,\ 3,\ 3.5,\ 4,\ 4.5,\ 5, \ldots$$

a. Make a table, find a formula, and draw a graph for the function associated with the arithmetic sequence.

input	output
x	

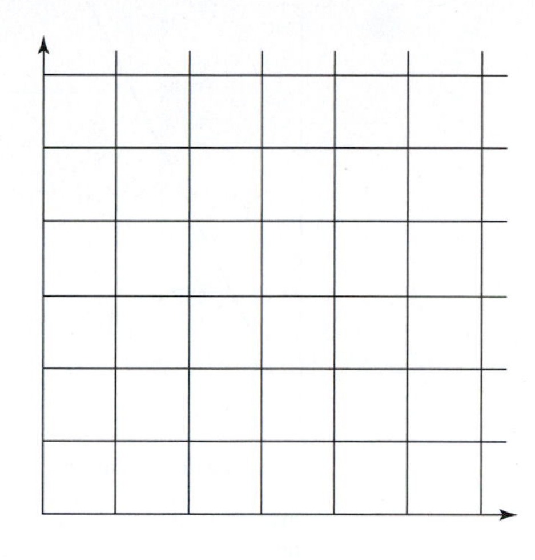

b. Fill in the blank to make a true statement about the function in Part (a):
"Whenever the input increases by 1, the output _________."
How is this statement reflected in the table and the graph of the function?

c. In Part (a), how are components of the formula reflected in the graph?

3. What do you think the graphs of arithmetic sequences all have in common?

Class Activity 13GG: Analyzing the Way Functions Change

1. You will need graph paper for parts of this activity.

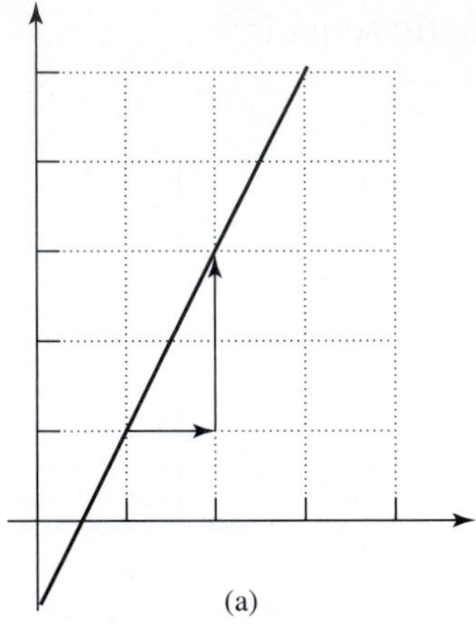

(a)

Fill in the blank to make a true statement about graph (a):
"Whenever the input increases by 1 the output ________________"
Explain why the statement is true.

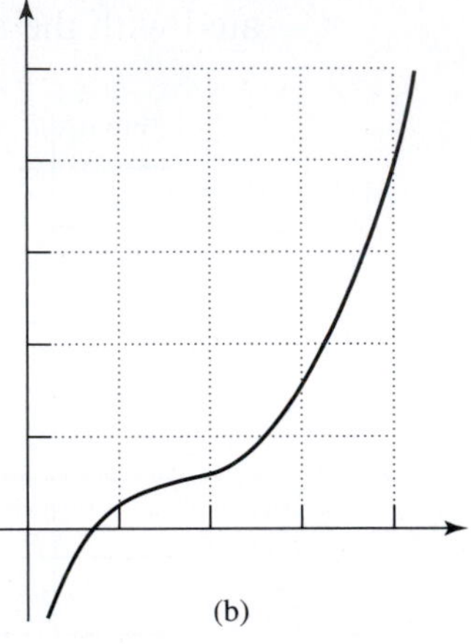

(b)

Explain why there is no number you can place in the blank to make a true statement about graph (b):
"Whenever the input increases by 1 the output increases by ________"

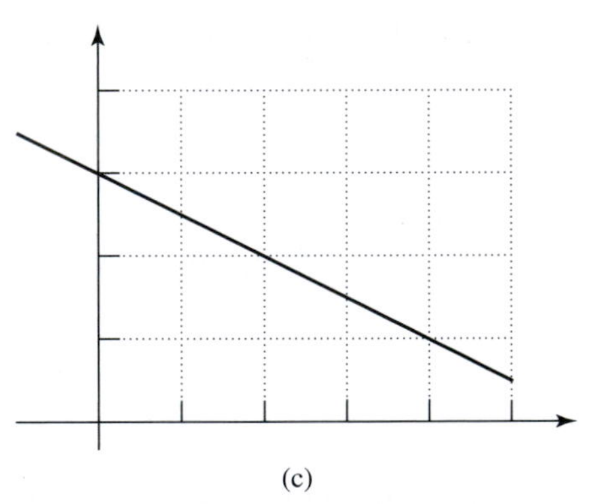

(c)

Fill in the blank to make a true statement about graph (c):
"Whenever the input increases by 1 the output ________________"
Explain why the statement is true.

2. Consider a function that has the following properties:

- When the input is 1, the output is 5.
- Whenever the input increases by 1, the output increases by 3.

a. Make a table and draw the graph of a function that has the properties described.

b. Explain how the properties in the previous two bulleted items are reflected in the graph of the function.

c. Find a formula for a function that has the properties described.

3. Consider a function that has the following properties:

- When the input is 0, the output is 0.
- Whenever the input increases by 3, the output increases by 7.

a. Make a table and draw the graph of a function that has the properties described.

b. Explain how the properties in the previous two bulleted items are reflected in the graph of the function.

c. Find a formula for a function that has the properties described.

Class Activity 13HH: Story Problems for Linear Functions

You will need graph paper for Problem 1, Part (b).

1. There will be a raffle at the fall festival. The fall festival committee spent \$45 on prizes for the raffle. Raffle tickets will be sold for \$1.50 each.

a. Describe a function that arises from this situation.

b. Make a table and draw a graph of your function in Part (a).

c. Find a formula for your function in Part (a), and explain why your formula is valid.

d. Where does the graph of your function in Part (b) cross the x-axis? What is the significance of this point in terms of the raffle?

2. Write a story problem which gives rise to a function, f, that has the formula $f(x) = 3x + 5$. Explain why your function has that formula.

3. Write a story problem which gives rise to a function, g, that has the formula $g(x) = \frac{3}{5}x$.

Class Activity 13II: Deriving the Formula for Temperature in Degrees Fahrenheit in Terms of Degrees Celsius

1. For each pair of numbers, determine the number that is halfway between the two numbers. In each case, explain why your answer is correct.

a. 0 and 14

b. 0 and 11

c. 0 and b

d. 2 and 14

e. 2 and 11

f. 2 and b

g. a and b

2. For each pair of numbers, determine the number that is $\frac{1}{3}$ of the way between the two numbers (and closer to the first number). In each case, explain why your answer is correct.

a. 0 and 12

b. 0 and 11

c. 0 and b

d. 2 and 14

e. 2 and 12

f. 2 and b

g. a and b

3. For each pair of numbers, determine the number that is 40% of the way between the two numbers (and closer to the first number). In each case, explain why your answer is correct.

a. 0 and 120

b. 0 and b

c. 20 and 170

d. 20 and b

e. a and b

4. Water freezes at 0° Celsius, which is 32° Fahrenheit, and boils at 100° Celsius, which is 212° Fahrenheit. Given that 40° Celsius is 40% of the way between 0° Celsius and 100° Celsius, what is 40° Celsius in degrees Fahrenheit? Answer this question based on your work in Problem 3.

5. Based on your work in Problem 3, write an expression for the number that is $P\%$ of the way between a and b. Explain why your expression fits with your work in Problem 3.

6. Based on your work in previous parts of this problem, write a formula for the temperature in degrees Fahrenheit of $C°$ Celsius. Your formula should be in terms of C. Explain why your formula is correct. (You may assume that C is between 0 and 100, but the correct formula will be valid for other values of C as well.)

14

Statistics

14.1 Formulating Questions, Designing Investigations, and Gathering Data

Class Activity 14A: Challenges in Formulating Survey Questions

Lincoln Elementary School would like to add soup to its school lunch menu. The school staff decides to poll the students to learn about the soups they like. Compare the questions in 1 through 5. How might the data that the school would collect be different for the different questions? What are the advantages and disadvantages of each question?

1. What is your favorite soup?
2. If you had to pick a soup to eat right now, what soup would you pick?
3. Vote for one of the following soups you like to eat:
   - tomato
   - chicken noodle
   - vegetable
   - other
4. Circle all of the following soups you like to eat:
   - tomato
   - chicken noodle
   - vegetable
5. (Write your own question.)

Class Activity 14B: Choosing a Sample

A college newspaper wants to find out how the students at the college would answer a specific question of importance to the student body. There are too many students for the newspaper staff to ask them all. So the staff decides to choose a sample of students to ask. Discuss the advantages and disadvantages of each of the following ways that the newspaper staff could gather data from a sample:

a. Ask their friends.

b. Ask as many of their classmates as they can.

c. Stand outside the buildings their classes are in and ask as many people as they can who come by.

d. Stand outside the student union or other common meeting area, and try to pick people who they think are representative of the students at their institution to ask the question.

e. Generate a list of random numbers between 1 and the number of students at the college. (Many calculators can generate random numbers; random numbers can also be generated on the Internet; see `www.aw-bc.com/beckmann`.) Pick names out of the student phone book corresponding to the random numbers (for example, for 123, pick the 123rd name), and contact that person by phone or by e-mail.

Class Activity 14C: Using Random Samples

1. At a factory that produces computer chips, a batch of 5000 computer chips has just been produced. To check the quality of the computer chips, a random sample of 100 computer chips is selected to test for defects. Out of these 100 chips, 3 were found to be defective. Based on these results, what is the best estimate you can give for the number of defective computer chips in the batch of 5000? Find several different ways to solve this problem, including ways that elementary school children might be able to develop.

2. Mr. Lawler had a bag filled with 160 plastic squares. The squares were identical, except that some were yellow and the rest were green. A student in Mr. Lawler's class randomly picked out 20 of the squares; 4 of the squares were yellow, and the rest were green. Mr. Lawler asked his students to use this information to make their best scientific estimate for the total number of yellow squares that were in the bag. Mr. Lawler's students had several different ideas. For each of the following initial ideas, discuss the idea and describe how to use it to estimate the total number of yellow squares. Which ideas are related?

a.

20	20	20	20	20	20	20	20
↓	↓	↓	↓	↓	↓	↓	↓
4	4	4	4	4	4	4	4

b.

$$\begin{aligned} 20 &\rightarrow 4 \\ 40 &\rightarrow 8 \\ 60 &\rightarrow 12 \\ 80 &\rightarrow 16 \\ 100 &\rightarrow 20 \\ 120 &\rightarrow 24 \\ 140 &\rightarrow 28 \\ 160 &\rightarrow 32 \end{aligned}$$

c. Of the 20 squares Taryn picked, $\frac{1}{5}$ were yellow.

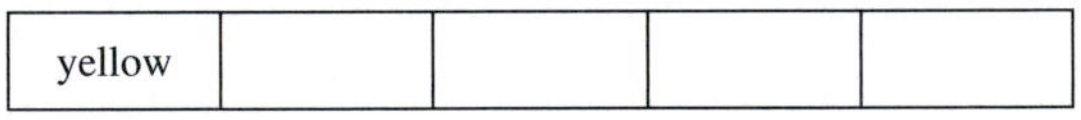

yellow				

d. Taryn picked $\frac{1}{8}$ of the squares in the bag.

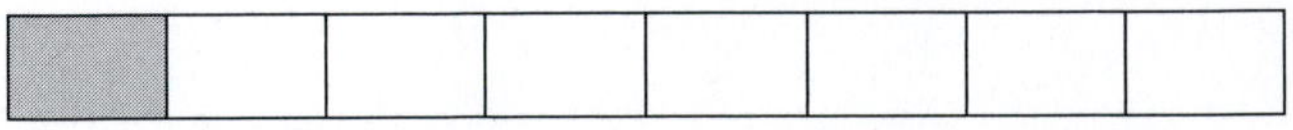

e.

$$\frac{4}{20} = \frac{8}{40} = \frac{12}{60} = \frac{16}{80} = \frac{20}{100} = \frac{24}{120} = \frac{28}{140} = \frac{32}{160}$$

f.

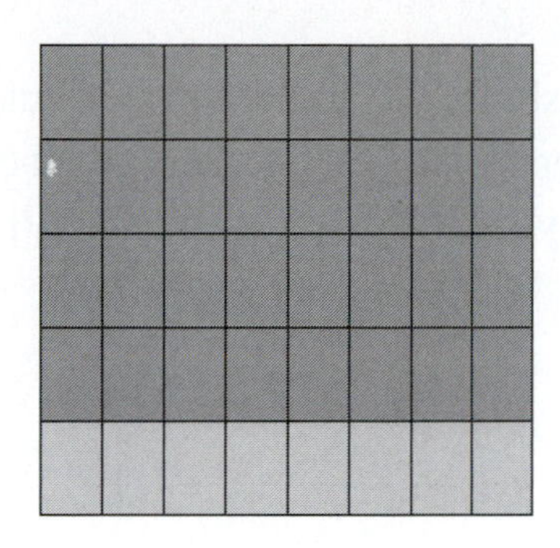

g.

$$\overset{\times 8}{\frac{4}{20} = \frac{x}{160}}$$

$\times 8$

h.

$$\times 5 \quad \frac{4}{20} = \frac{x}{160} \quad \times 5$$

Class Activity 14D: Using Random Samples to Estimate Population Size by Marking (Capture–Recapture)

You will need a bag filled with a large number (at least 100) of small, identical beans or other small objects that can be marked (such as small paper strips or beads that can be colored with a marker).

Pretend that the beans are fish in a lake. You will estimate the number of fish in the lake without counting them all by using a method called *capture–recapture*.

1. Go "fishing:" Pick between 20 and 50 "fish" out of your bag. Count the number of fish you caught, and label each fish with a distinctive mark. Then throw your fish back in the lake (the bag), and mix them thoroughly.

2. Go fishing again: Randomly pick about 50 fish out of your bag. Count the total number of fish you caught this time, and count how many of the fish are marked.

3. Use your counts from Problems 1 and 2 to estimate the number of "fish" in your bag. Explain your reasoning.

4. When Ms. Wade used the method described in Problems 1 through 3, she picked 30 "fish" at first, marked them, and put them back in the bag. Ms. Wade thoroughly mixed the fish in the bag and randomly picked out 40 fish. Out of these 40 fish, 5 were marked. The children in Ms. Wade's class had several different ideas for how to determine the total number of fish in the bag. For each of the following initial ideas, discuss the idea, and describe how to use the idea to determine approximately how many fish are in the bag. Which ideas are related?

a.

$$\begin{array}{cccccc} 40 & 40 & 40 & 40 & 40 & 40 \\ \downarrow & \downarrow & \downarrow & \downarrow & \downarrow & \downarrow \\ 5 & 5 & 5 & 5 & 5 & 5 \end{array}$$

b.

$$\begin{array}{l} 40 \rightarrow 5 \\ 80 \rightarrow 10 \\ 120 \rightarrow 15 \\ 160 \rightarrow 20 \\ 200 \rightarrow 25 \\ 240 \rightarrow 30 \end{array}$$

c. $\frac{1}{8}$ of the fish are marked.

d. $\frac{1}{6}$ of the marked fish were chosen.

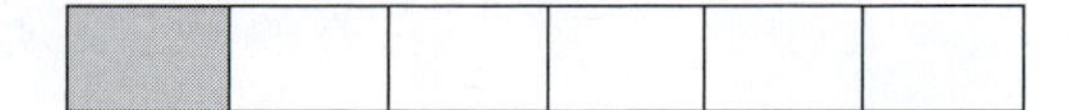

5.

$$\frac{5}{40} = \frac{10}{80} = \frac{15}{120} = \frac{20}{160} = \frac{25}{200} = \frac{30}{240}$$

6.

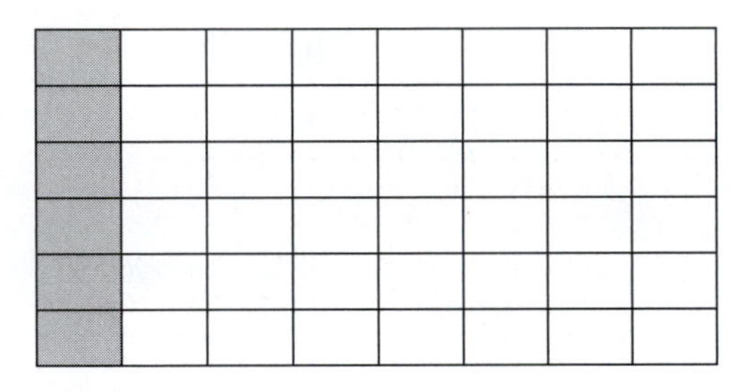

7.

$$\overset{\times 6}{\frac{5}{40} = \frac{30}{x}}$$

$\times 6$ (numerator $5 \to 30$), $\times 6$ (denominator $40 \to x$)

8.

$$\times 8 \; \frac{5}{40} = \frac{30}{x} \; \times 8$$

Class Activity 14E: Which Experiment Is Better?

Suppose a class wants to do an experiment to see if a fertilizer makes bean plants grow faster. Which of the following designs for this experiment will be better, which will be worse, and why?

1. Fill two identical trays with soil from the same bag. Mix the fertilizer with the soil in one of the trays. Plant half of the bean seeds in each tray. Give one tray to the class next door. Keep one tray in your class. Observe and water the bean plants every day. Once they sprout, measure how tall they are.

2. Same as Problem 1, except that you keep both trays in your class at windows on two different walls.

3. Same as Problem 2, except that you keep both trays side by side at the same window.

4. Fill identical paper cups with soil. Put fertilizer in half of the cups. Label each cup to show whether or not it has fertilizer. Plant seeds in the cups. Put the cups at a window, alternating between fertilized and unfertilized. Observe and water the bean plants every day, taking care to give each plant the same amount of water. Once they sprout, measure how tall they are.

5. Same as Problem 4, except that the paper cups are labeled on the bottom (so that you can't see which is which) and the cups are arranged randomly at the window.

6. Same as Problem 5, but make additional collections of cups with seeds, and put them at different windows and in different classrooms.

14.2 Displaying Data and Interpreting Data Displays

Class Activity 14F: What Do You Learn from the Display?

For each of the data displays in this activity, list several conclusions you can draw from the display. Write several questions that would require further study to answer.

1. The next bar graph was created with statistics from the U.S. Department of Education, National Center for Education Statistics, National Household Education Survey, 1993, 1999. These statistics can be found in the report *Home Literacy Activities and Signs of Children's Emerging Literacy, 1993 and 1999*. See www.aw-bc.com/beckmann.

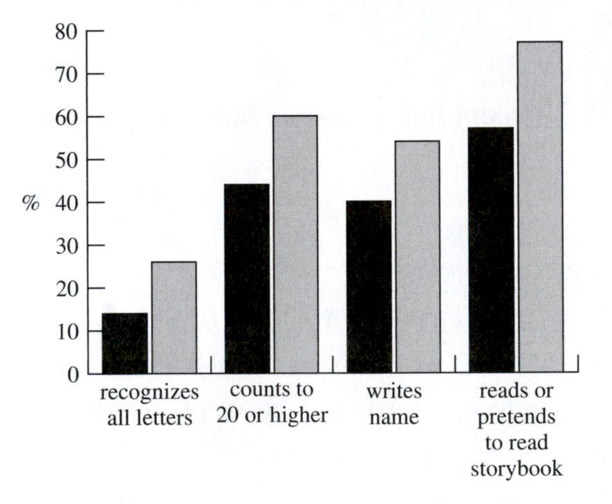

2. The next pie chart was created from a table available on the Internet based on data collected by the Current Population Survey. (This is a monthly survey conducted for the Bureau of Labor Statistics by the U.S. Census Bureau.) See www.aw-bc.com/beckmann.

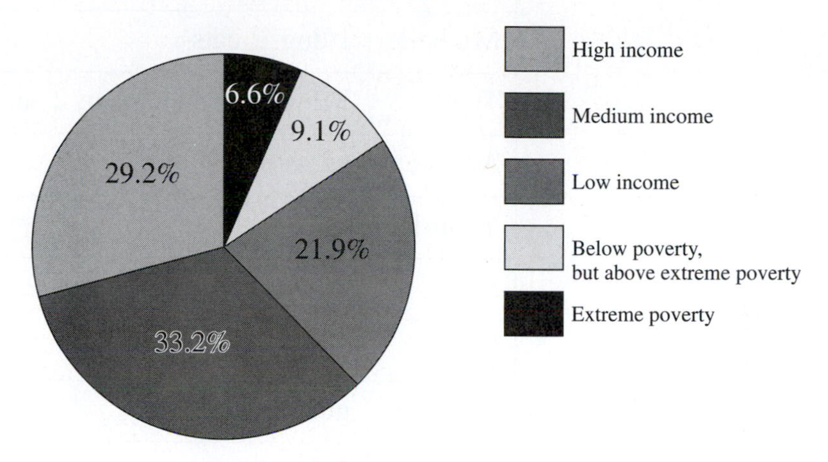

Class Activity 14G: Display These Data about Pets

1. A class collected information about the pets they have at home, as shown in the following table:

name	pets at home
Michelle	1 dog, 2 cats
Tyler	3 dogs, 1 salamander, 2 snakes
Antrice	hamster
Yoon-He	cat
Anne	none
Peter	2 dogs
Brandon	guinea pig
Brittany	1 dog, 1 cat
Orlando	none
Chelsey	2 dogs, 10 fish
Sarah	1 rabbit
Adam	none
Lauren	2 dogs
Letitia	3 cats
Jarvis	1 dog

For each of the following questions, use the data from the table to create a graph that could help the class answer the question:

a. Are dogs our most popular pet?

b. How many pets do most people have?

c. How many people have more than one pet?

d. Are most of our pets mammals?

Class Activity 14H: What Is Wrong with These Displays?

1. Ryan grabbed a handful of small plastic animals and made a pictograph like the following:

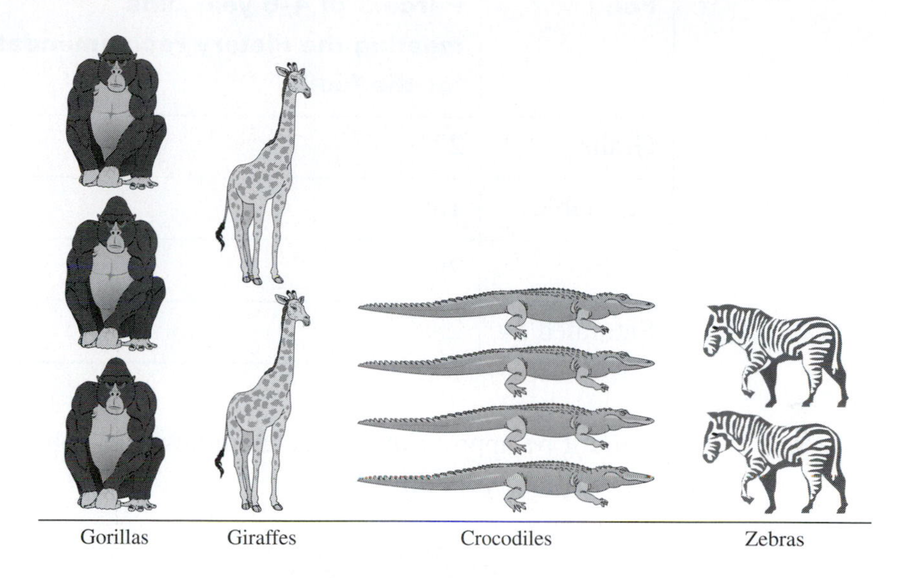

What is a problem with Ryan's pictograph?

2. The line graph for this problem was created from data about children's eating that was available on the Internet. See www.aw-bc.com/ beckmann. What is wrong with this display? Show how to fix it.

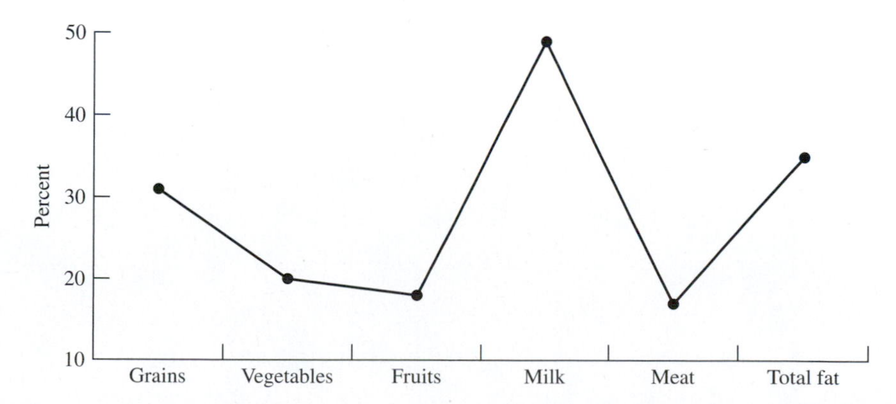

3. The next table on children's eating habits is based on a table found at `www.aw-bc.com/beckmann`.

Food	**Percent of 4–6 year olds meeting the dietary recommendation for the food**
Grains	27%
Vegetables	16%
Fruits	29%
Saturated fat	28%

Would it be appropriate to use a single pie graph to display this information? Explain your answer.

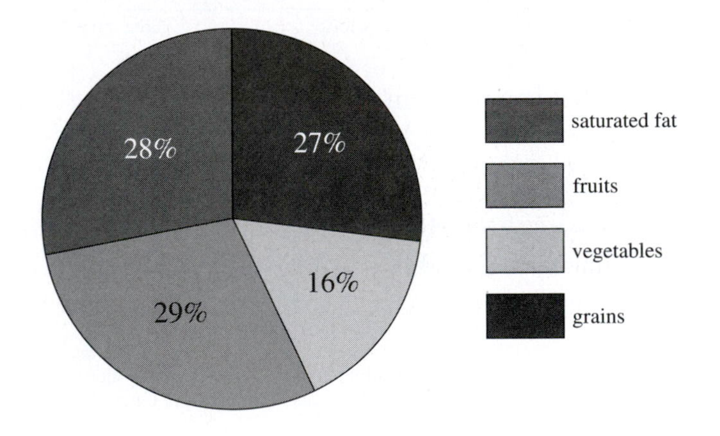

4. The line graph shown was created from data about children's smoking that were available on the Internet. See `www.aw-bc.com/ beckmann`.

Based on this display, would it be correct to say that the percentage of 8th graders who reported smoking cigarettes daily in the previous 30 days was about twice as high in 1996 as it was in 1993? Why or why not?

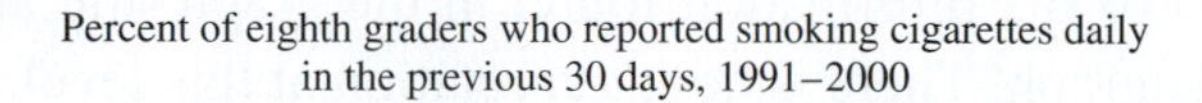

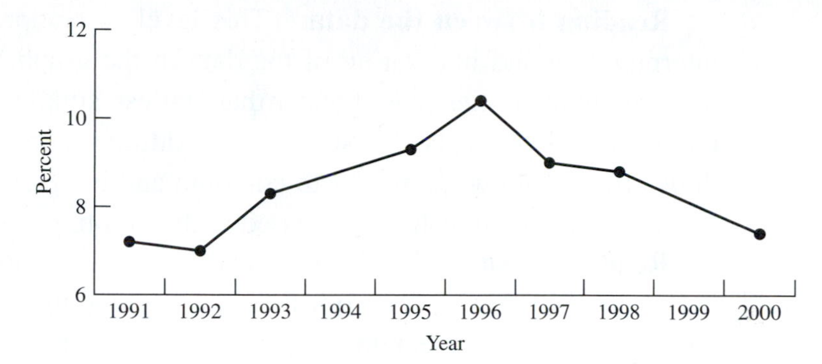

Class Activity 14I: Three Levels of Questions about Graphs

Recall that the three levels of graph comprehension described by Frances Curcio in [3, p. 7] are as follows:

Reading the data. This level of comprehension requires a literal reading of the graph. The reader simply "lifts" the facts explicitly stated in the graph, or the information found in the graph title and axes labels, directly from the graph. There is no interpretation at this level.

Reading between the data. This level of comprehension includes the interpretation and integration of the data in the graph. It requires the ability to compare quantities (e.g., greater than, tallest, smallest) and the use of other mathematical concepts and skills (e.g., addition, subtraction, multiplication, division) that allow the reader to combine and integrate data and identify the mathematical relationships expressed in the graph.

Reading beyond the data. This level of comprehension requires the reader to predict or infer from the data by tapping existing schemata (i.e., background knowledge, knowledge in memory) for information that is neither explicitly nor implicitly stated in the graph.

The following examples are questions at the different graph-reading levels. All questions are about a bar graph that shows the heights of children in a class ([3, p. 35]):

- What would be a good title for this graph? [Read between the data]
- How tall is (insert a name)? [Read the data]
- Who is the tallest of the students on the graph? [Read between the data]
- Who do you think is the oldest? Why? Can this be answered directly from the graph? [Read beyond the data]
- Who do you think has the smallest shoe size? Why? Can this be answered directly from the graph? [Read beyond the data]
- Who do you think weighs the least? Why? [Read beyond the data]

Now design your own data graphing activity that you could use with elementary school children; include a list of questions to foster the different levels of graph-reading comprehension:

1. Describe an activity in which children will gather data and create a graph.

2. Show roughly what the data display from Problem 1 might look like.

3. Write a list of questions at the three different graph-reading levels. Label each question with its level.

Class Activity 14J: The Length of a Pendulum and the Time It Takes to Swing

A 5th grader's science fair project[1] investigated the relationship between the length of a pendulum and the time it takes the pendulum to swing back and forth. The student made a pendulum by tying a heavy washer to a string and attaching the string to the top of a triangular frame, as pictured. The

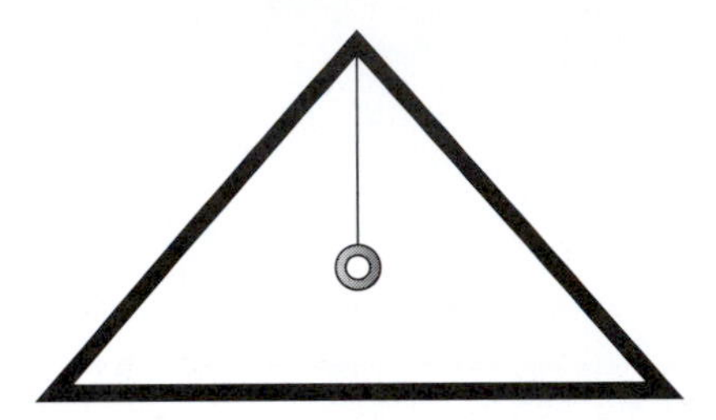

length of the string could be varied. The next table and scatterplot show how long it took the pendulum to swing back and forth 10 times for various lengths of the string. (Several measurements were taken and averaged.)

length of string in inches	time of 10 swings in seconds
1	2.61
2	2.97
3	3.04
4	3.41
5	3.96
6	4.13
7	4.22
8	4.5
9	4.64
10	5.13
11	5.32
12	5.56
13	5.62
14	5.87

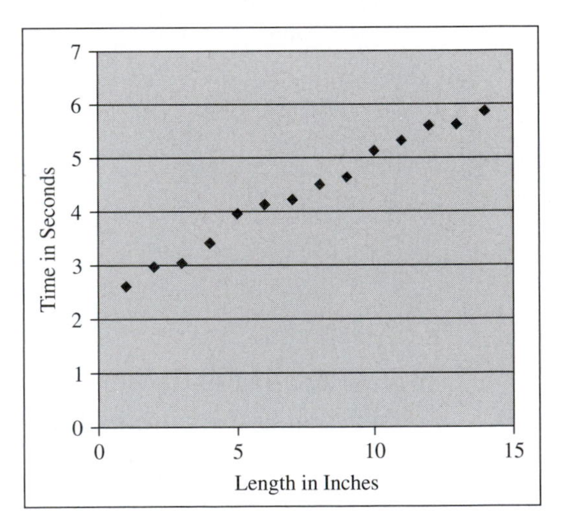

[1] Thanks to Arianna Kazez for the data and information about the project.

1. Write one or two questions about the scatterplot for each of the three graph-reading levels. Answer each question (to the extent possible).

2. Arianna used her science fair data to predict how long it would take a pendulum with 100 inches of string to swing back and forth 10 times. She started by observing that for every 4 inches of string it takes one second longer. Explain how to use this observation to determine approximately how long it might take a 100-inch pendulum to swing back and forth 10 times.*

3. Arianna used her science fair data to predict how long a string she would need so that one swing would take 1 second (like a grandfather clock). She started by observing that for every 4 inches of string it takes one second longer. She also used the fact that a 14-inch pendulum took 5.87 seconds for 10 swings. Arianna knew she needed to get to 10 seconds. Explain how to use these ideas to determine approximately how long a string is needed so that one swing will take 1 second.*

* This provides a good initial estimate, but according to physical theories, the estimate won't be fully correct.

Class Activity 14K: Investigating Small Bags of Candies

For this activity, each person, pair, or small group in the class needs a small bag of multi-colored candies. All bags should be of the same size and consist of the same type of candy. Do not open your bag of candy until after you have done the first part of this activity.

1. Do not open your bag of candy yet! Write a list of questions that the class as a whole could investigate by gathering and displaying data about the candies.

2. Open your bag of candy (but do not eat it yet!) and display data about your candies in two significantly different ways. For each display, write and answer questions at the three different graph-reading levels.

3. Together with the whole class, collect and display data about the bags of candies in order to answer some of the questions the class posed in Problem 1.

Class Activity 14L: Balancing a Mobile

For this activity, each person, pair, or small group in the class needs a drinking straw, string, tape, at least 7 paperclips of the same size, a ruler, and graph paper. You will use the straw, string, tape, and paperclips to make a simple mobile.

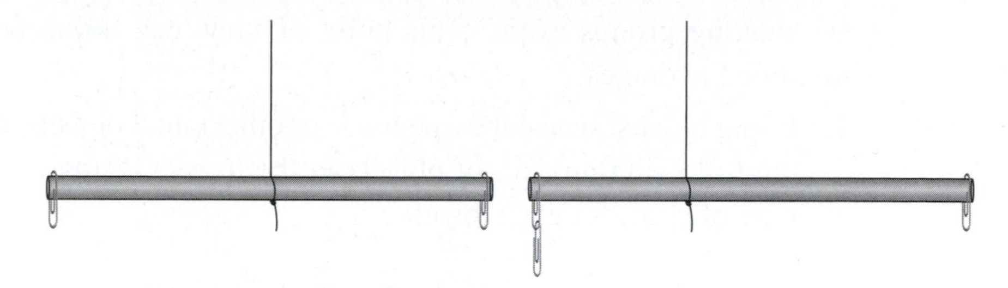

1. Tie one end of the string snugly around the straw. Tape one paperclip to each end of the straw. Hold the other end of the string so that your mobile hangs freely. Adjust the location of the string along the straw so that the straw balances horizontally. The string should now be centered on the straw, as in the picture, on the left. Measure the distance on the straw from the string to each end.
2. Repeatedly add one more paperclip to one side of the straw (but not to the other side). Every time you add a paperclip, adjust the string so that the straw balances horizontally. Each time, measure the distance on the straw from the string to the end that has multiple paperclips, and record your data.
3. Make a graphical display of your data from Problem 2. (Use graph paper.)
4. Write and answer several questions at each of the three different graph-reading levels about your graphical display in Problem 3.

14.3 The Center of Data: Mean, Median, and Mode

Class Activity 14M: The Average as "Making Even" or "Leveling Out"

In this class activity you will use physical objects to help you see the average as "making groups even." This point of view can be useful in calculations involving averages.

1. Using blocks, snap cubes, pennies, or other small objects, make towers with the following number of objects in the towers, using a different color (or type of coin) for each tower:

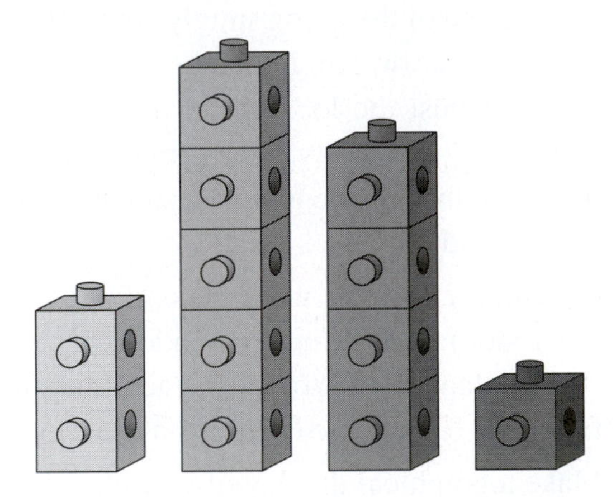

Determine the average of the list numbers 2, 5, 4, 1 by "leveling out" the block towers, or making them even. That is, redistribute the blocks among your block towers until all 4 towers have the same number of blocks in them. This common number of blocks in each of the 4 towers is the average of the list 2, 5, 4, 1.

2. Use the process of making block towers even in order to determine the averages of each of the lists of numbers shown. In some cases you may have to imagine cutting your blocks into smaller pieces.

List 1:	1,	3,	3,	2,	1
List 2:	2,	3,	4,	3,	4
List 3:	2,	3,	1,	5	
List 4:	3,	4,	4,	3,	4

3. To calculate the average of a list of numbers *numerically*, we add the numbers and divide the sum by the number of numbers in the list. So, to calculate the average of the list 2, 5, 4, 1, we calculate

$$(2+5+4+1) \div 4$$

Interpret the *numerical* process for calculating an average in terms of 4 block towers built of 2 blocks, 5 blocks, 4 blocks, and 1 block. When we add the numbers, what does that correspond to with the blocks? When we divide by 4, what does that correspond to with the blocks?

Explain why the process of determining an average physically by making block towers even must give us the same answer as the numerical procedure for calculating the average.

4. Suppose you have made 3 block towers: one 3 blocks tall, one 6 blocks tall, and one 2 blocks tall. Describe some ways to make two more towers so that there is an average of 4 blocks in all 5 towers.

Solve this problem in several ways, and explain your solutions.

5. If you run 3 miles every day for 5 days, how many miles will you need to run on the sixth day in order to have run an average of 4 miles per day over the 6 days?

Solve this problem in several ways, and explain your solutions.

6. The average of 7 numbers is 12.1. An eighth number, 12.9, is included in the list. What is the average of the 8 numbers?

Solve this problem in several ways, and explain your solutions.

7. Explain how you can quickly calculate the average of the following list of test scores without adding the numbers:

81, 78, 79, 83

8. If you run an average of 3 miles a day over one week and an average of 4 miles a day over the next two weeks, what is your average daily run distance over that 3-week period?

Before you solve this problem, explain why it makes sense that your average daily run distance over the 3-week period is *not* just the average of 3 and 4, namely, 3.5. Should your average daily run distance over the 3 weeks be greater than 3.5 or less than 3.5? Explain how to answer this without a precise calculation.

Now determine the exact average daily run distance over the 3-week period. Explain your solution.

Class Activity 14N: The Average as "Balance Point"

In this Class Activity you will use physical objects to help you see the average of a list of numbers as the "balance point" among the numbers. You will need the following:

- a 12 inch ruler
- a block or tile on which to balance the ruler—a thin rhombus from a set of pattern tiles is ideal
- small objects, such as pattern tiles, that are stable when placed on the ruler

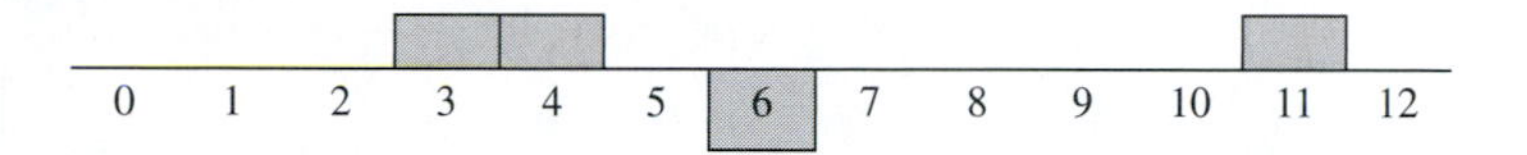

1. Make a see-saw by balancing the ruler on a tile or a block, checking that it balances at the number 6 (more or less).
2. Holding the ruler steady, place one object on the see-saw for each of the following numbers:

   3, 4, 11

   In each case, place the object centered on that number, as shown in the previous figure. Check that the see-saw still balances at the number 6 (more or less). Check that the average of the numbers is also 6.

3. Repeat 1 with other lists of numbers *that have an average of 6*, such as

   2, 10

   and

   4, 4, 6, 10

   In each case, verify that the ruler balances at 6.

   For lists of numbers whose average is not 6, the ruler will not balance at the average due to the different weights of the parts of the ruler on either side of the fulcrum (the point where the ruler balances). If it were possible to make a weightless ruler, then the ruler would always balance at the average of the numbers.

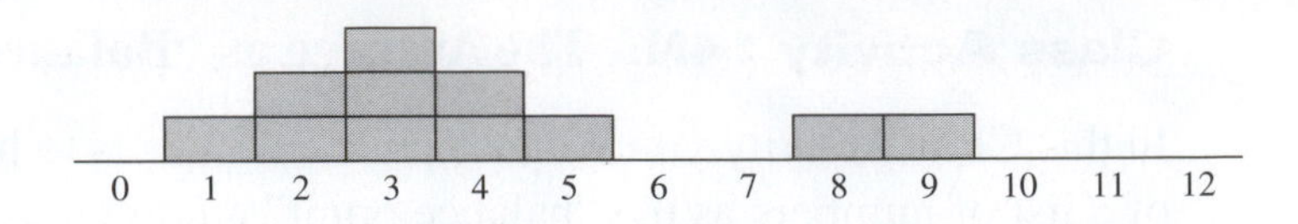

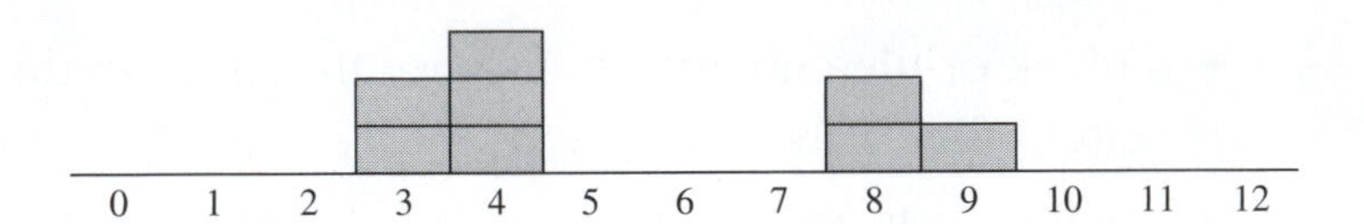

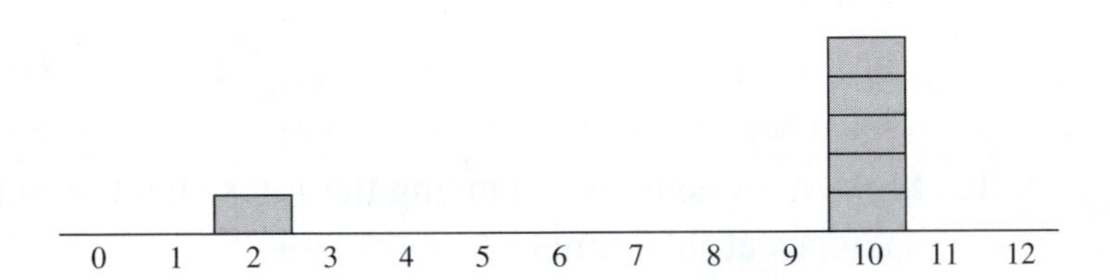

4. The three see-saws in the previous figure represent the following three lists of numbers:

List 1: 1, 2, 2, 3, 3, 3, 4, 4, 5, 8, 9

List 2: 3, 3, 4, 4, 4, 8, 8, 9

List 3: 2, 10, 10, 10, 10, 10

For each list, predict the average by guessing where the corresponding see-saw (made out of a weightless ruler) would balance. Check your predictions by calculating the averages of the lists.

Class Activity 14O: Same Median, Different Average

In most cases, the median of a list of numbers is not the same as its average. In this activity, you will alter a data set to keep the same median, but vary the average.

The pennies along the number line at the top of Figure 14O.1 are arranged to represent the following data set:

4, 5, 5, 6, 6, 6, 7, 7, 8

Arrange real pennies (or other small objects) along the number line at the bottom of Figure 14O.1 to represent the same data set.

1. Rearrange your pennies so that they represent new lists of numbers that still have median 6, but have averages *less than* 6. To help you do this, think about the average as the "balance point." Draw pictures of your penny arrangements.

2. Rearrange your pennies so that they represent new lists of numbers that still have median 6, but have averages *greater than* 6. To help you do this, think about the average as the "balance point." Draw pictures of your penny arrangements.

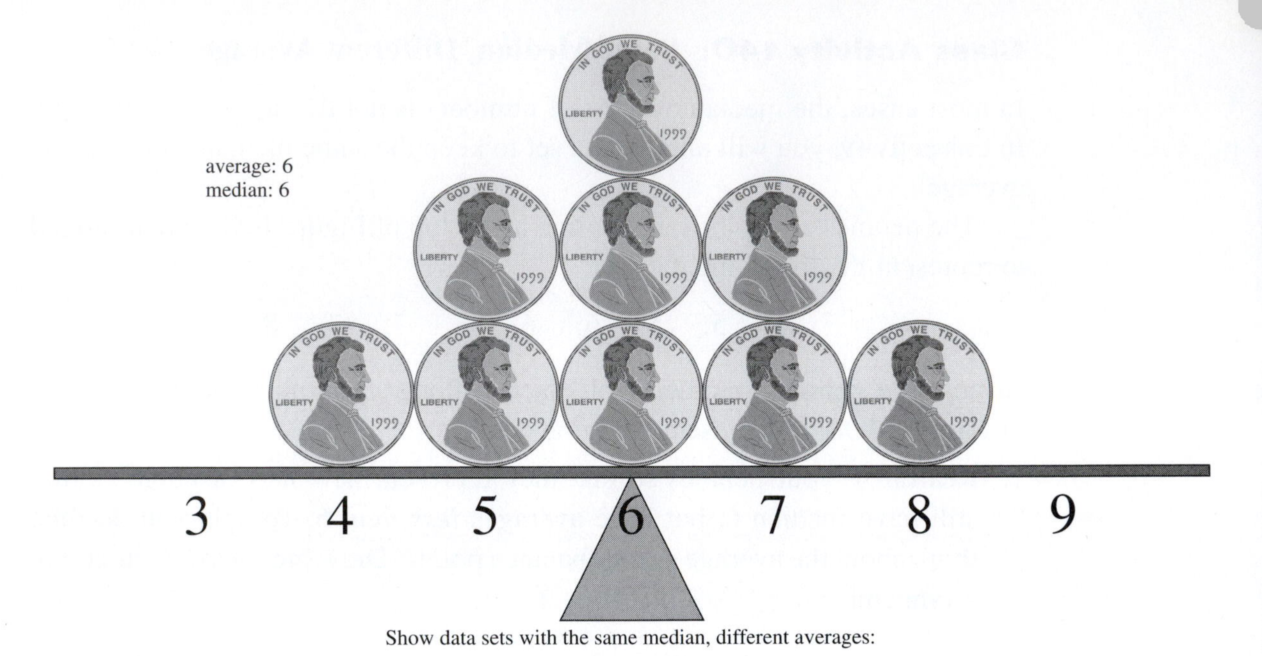

Show data sets with the same median, different averages:

3 4 5 6 7 8 9

Figure 14O.1

Same Medians, Different Averages

Class Activity 14P: Can More Than Half Be above Average?

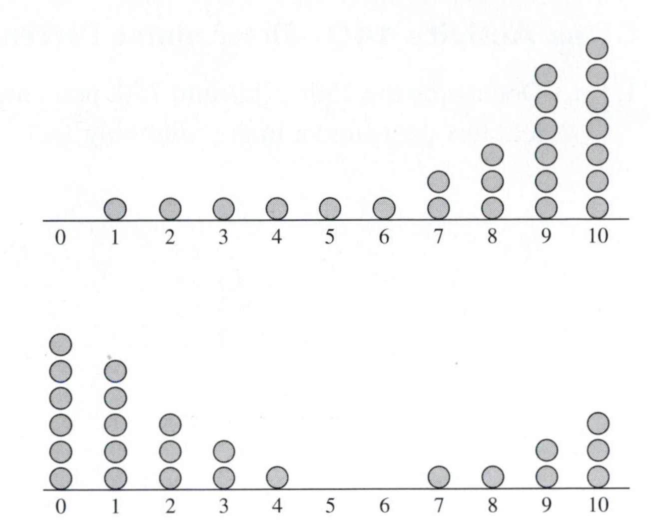

1. **a.** For each of the dot plots shown, decide which is greater without calculating: the median or the average of the data.

   b. Calculate the medians and averages of the data in the two dot plots. Verify that your answers in Part (a) were correct.

2. A teacher gives a test to a class of 20 children.

   Is it possible that 90% of the class scores above average? If so, give an example of test scores for which this is the case. If not, explain why not.

   Is it possible that 90% of the class scores below average? If so, give an example of test scores for which this is the case. If not, explain why not.

3. A radio program describes a fictional town in which "all the children are above average." In what sense is it possible that all the children are above average? In what sense is it not possible that all the children are above average?

14.4 Percentiles and the Distribution of Data

Class Activity 14Q: Determining Percentiles

1. **a.** Determine the 25th, 50th, and 75th percentiles for each of the hypothetical test data shown in the following:

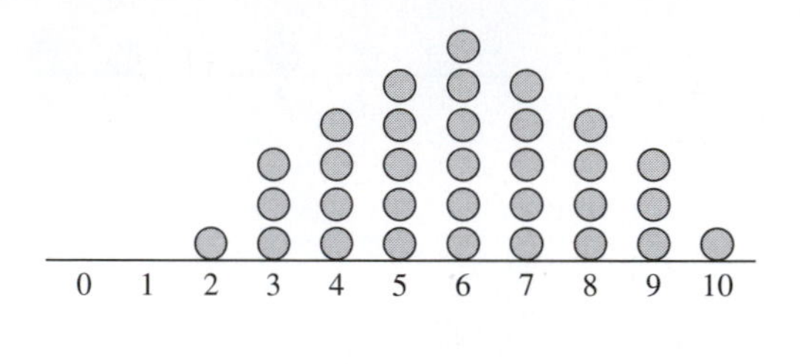

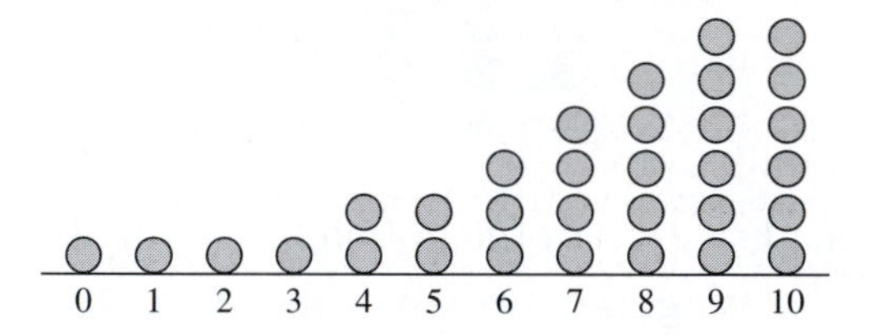

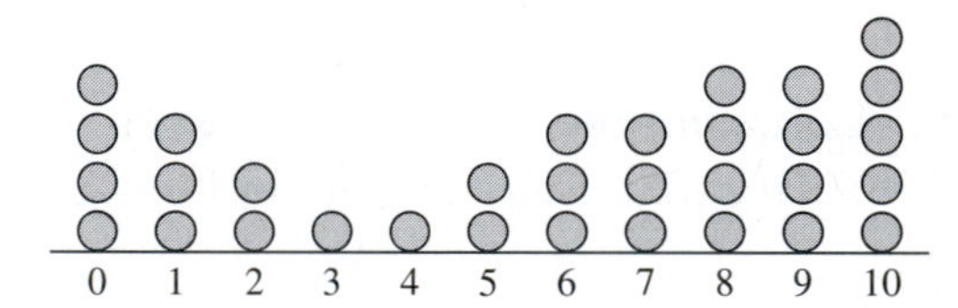

b. Suppose you only had the percentiles for each of the three data sets from Part (a) and you didn't have the dot plots. Discuss what you could tell about how the three data sets are distributed.

2. Suppose that a 400-point test is given to third graders at three different schools. The bar graphs in Figure 14Q.1 show hypothetical test results.

a. Using the bar graphs, determine approximately the 20th, 40th, 60th, and 80th percentiles at each of the schools. Explain your reasoning.

b. Describe qualitative differences in test performance at the three schools, and explain how these qualitative differences are reflected in the percentile scores.

c. At each of the three schools, consider a third grader who scored at the 60th percentile. In each case, approximately what score did such a third grader probably get? Did such third graders get 60% correct, more than 60% correct, or less than 60% correct on the test?

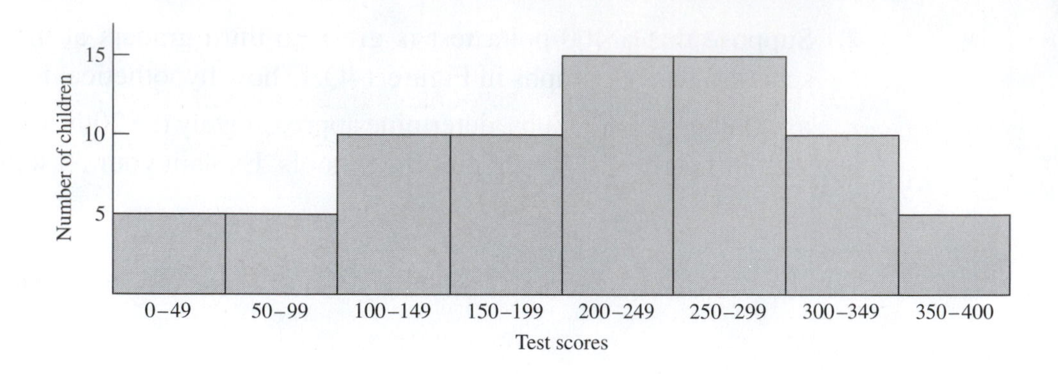

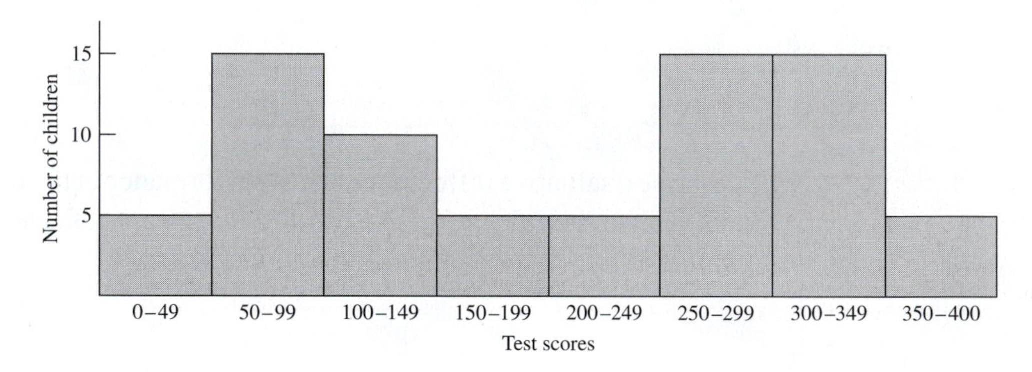

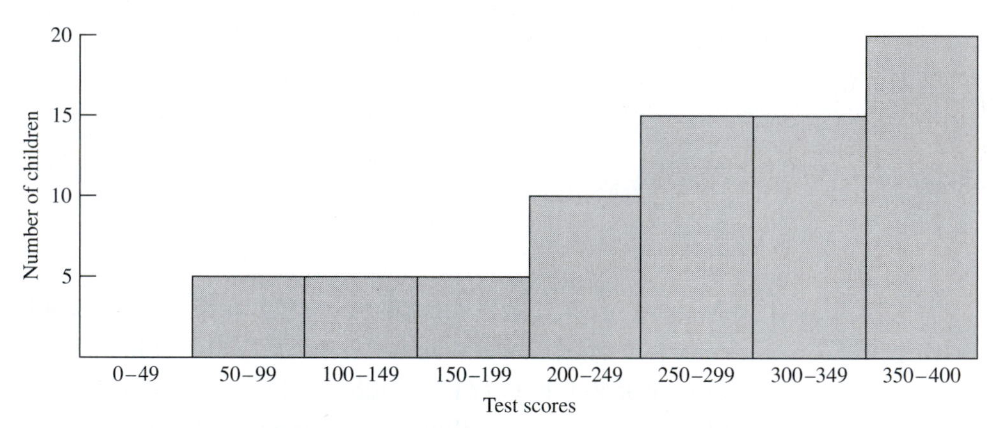

Figure 14Q.1
Bar Graphs of Hypothetical Test Scores

Class Activity 14R: Percentiles versus Percent Correct

1. What is the difference between getting 90% correct on a test and being in the 90th percentile on a test? Give some specific examples to illustrate.

2. Mrs. Smith makes an appointment to talk to Johnny's teacher. Johnny has been getting As in math, but on the standardized test he took, he was in the 80th percentile. Mrs. Smith is concerned that this means Johnny is really doing B work in math, not A work. If you were Johnny's teacher, what could you tell Mrs. Smith?

Class Activity 14S: Box-and-Whisker Plots

1. Make box-and-whisker plots for the three dot plots that follow.

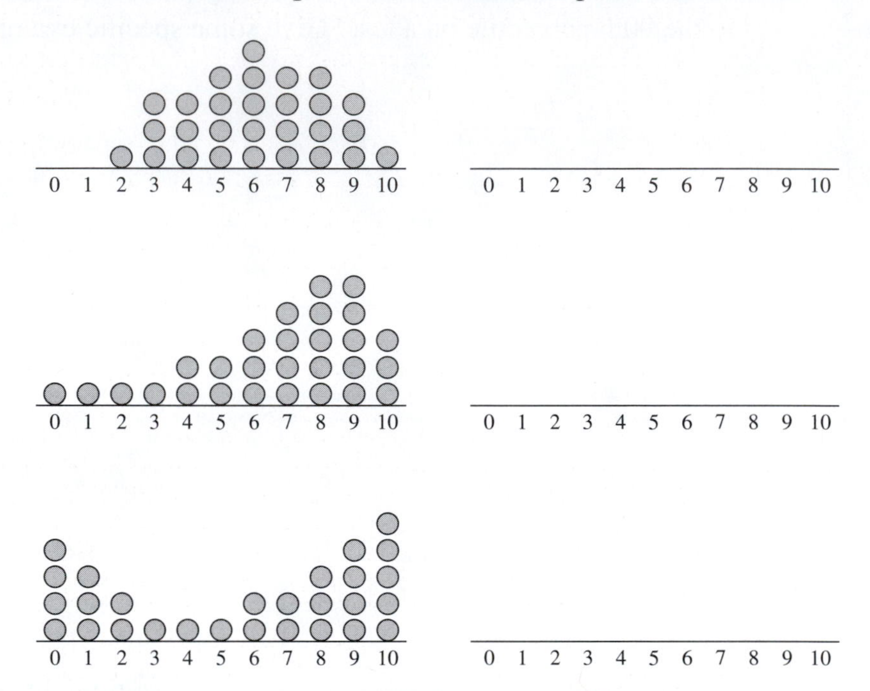

2. Suppose you had only the box-and-whisker plots from Problem 1, but you didn't have the dot plots. Discuss what you could tell about how the three data sets are distributed.

Class Activity 14T: How Percentiles Inform You about the Distribution of Data: The Case of Household Income

If you know several different percentiles for a set of data, then you have information about how the data are distributed. In particular, you can determine the approximate shape of a bar graph for the data. In this activity, you will see a way to construct an approximate bar graph based on percentile data.

The percentile data in the next tables on household incomes in the United States in 2000 and 1970 were taken from a table provided by the U.S. Census Bureau. See `www.aw-bc.com/beckmann`. All dollar values are in inflation-adjusted year 2000 dollars, so that dollar values can be compared across years.

	Household Income Limits by Percentile					
Year	10th	20th	50th	80th	90th	95th
2000	$10,600	$17,955	$42,000	$81,960	$111,602	$145,526
1970	$ 7,944	$14,245	$33,746	$56,646	$ 72,251	$ 89,553

On the next page, make approximate bar graphs for household incomes in the United States in 2000 and 1970 by making bars out of squares of the indicated size. What do your bar graphs tell you about how household income was distributed in 2000 versus in 1970?

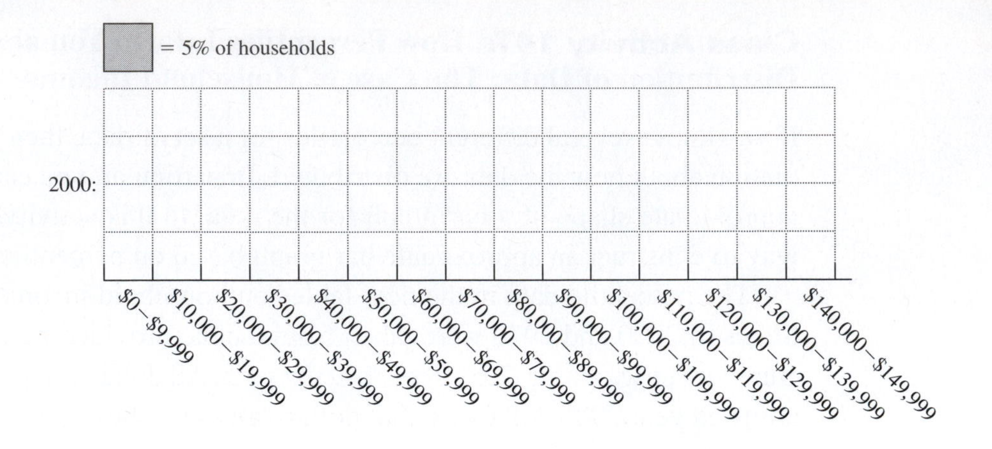
= 5% of households
2000:
$0–$9,999
$10,000–$19,999
$20,000–$29,999
$30,000–$39,999
$40,000–$49,999
$50,000–$59,999
$60,000–$69,999
$70,000–$79,999
$80,000–$89,999
$90,000–$99,999
$100,000–$109,999
$110,000–$119,999
$120,000–$129,999
$130,000–$139,999
$140,000–$149,999

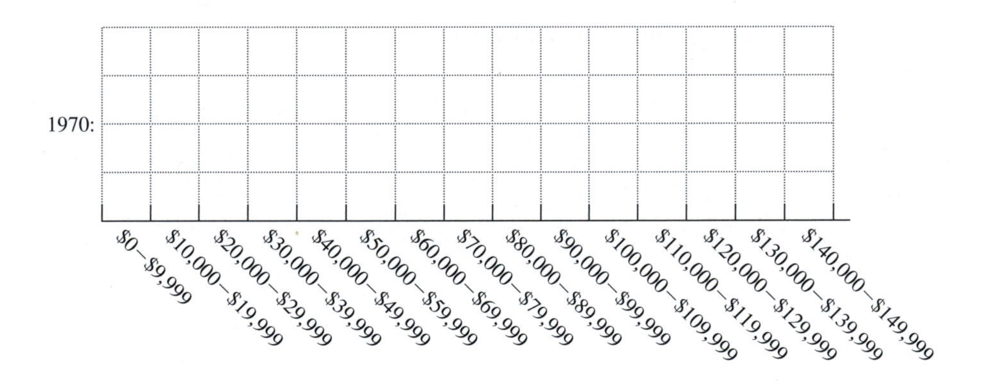
1970:
$0–$9,999
$10,000–$19,999
$20,000–$29,999
$30,000–$39,999
$40,000–$49,999
$50,000–$59,999
$60,000–$69,999
$70,000–$79,999
$80,000–$89,999
$90,000–$99,999
$100,000–$109,999
$110,000–$119,999
$120,000–$129,999
$130,000–$139,999
$140,000–$149,999

Class Activity 14U: Distributions of Random Samples

For this activity, you will need a large collection of small objects (200 or so) in a bag. The objects should be identical, except that they come in two different colors, 40% in one color and the remaining 60% in another color. The objects could be poker chips, small squares, small cubes, or even small slips of paper. Think of the objects as representing a group of voters. The 40% in one color represent "yes" votes, and the 60% in the other color represent "no" votes.

1. You will pick random samples of 10 from the bag. Each time you pick a random sample of 10, you will determine the percentage of "yes" votes, and you will plot this percent on a dot plot. But before you start picking these random samples, use the dot plot on the next page to make a prediction about what your actual dot plot will look like. Assume that you will plot about 20 dots.

   Why do you think your dot plot might turn out that way?

   How do you think the fact that 40% of the votes in the bag are "yes" votes might be reflected in the dot plot?

2. Now pick about 20 random samples of 10 objects from the bag. Each time you pick a random sample of 10, determine the percentage of "yes" votes and plot this percentage on the next dot plot. Compare your results with your prediction.

3. If possible, join your data with other people's data in order to form a dot plot with more dots. Do you see the fact that 40% of the votes in the bag are "yes" votes reflected in the dot plot? If so, how?

Predicted:

0 10 20 30 40 50 60 70 80 90 100

Percent "yes" votes

Actual:

0 10 20 30 40 50 60 70 80 90 100

Percent "yes" votes

Actual (larger number of samples):

0 10 20 30 40 50 60 70 80 90 100

Percent "yes" votes

4. Compare the two bar graphs on the next page. The first bar graph shows the percent of "yes" votes in 200 samples of 100 taken from a population of 1,000,000 in which 40% of the population votes "yes." The second graph shows the percent of "yes" votes in 200 samples of 1000 taken from the same population.

Compare the way the data are distributed in each of these bar graphs, and compare these bar graphs with your dot plots in Problems 2 and 3.

How is the fact that 40% of the population votes "yes" reflected in these bar graphs?

What do the bar graphs indicate about using samples of 100 versus samples of 1000 to predict the outcome of an election?

5. What if we made a bar graph like the ones on the next page by using the same population, but by picking 200 samples of 500 (instead of 200 samples of 100 or 1000)? How do you think this bar graph would compare with the ones on the next page? What if samples of 2000 were used?

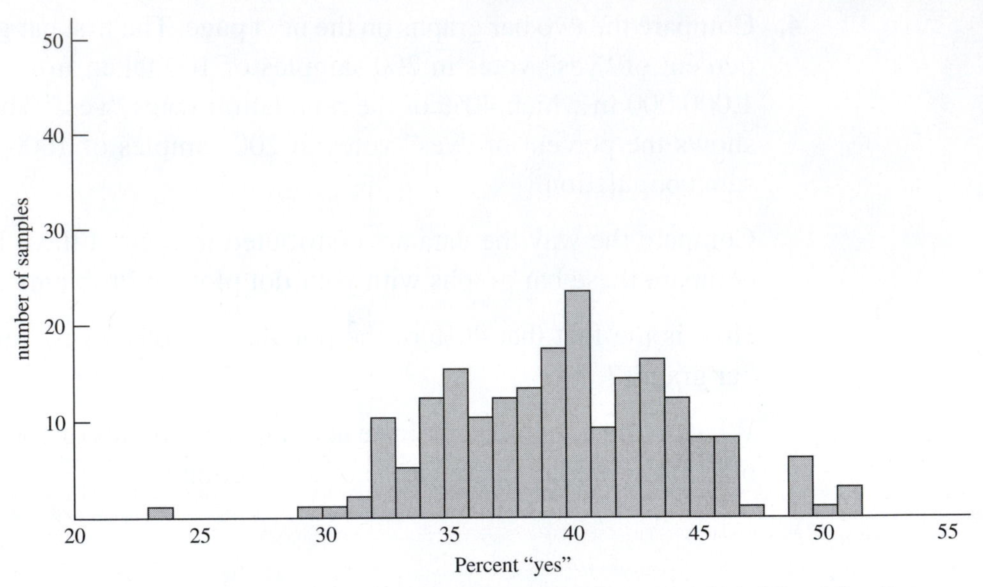

Percent of "yes" votes in **samples of 100** taken from 1,000,000 voters in which 40% vote "yes."

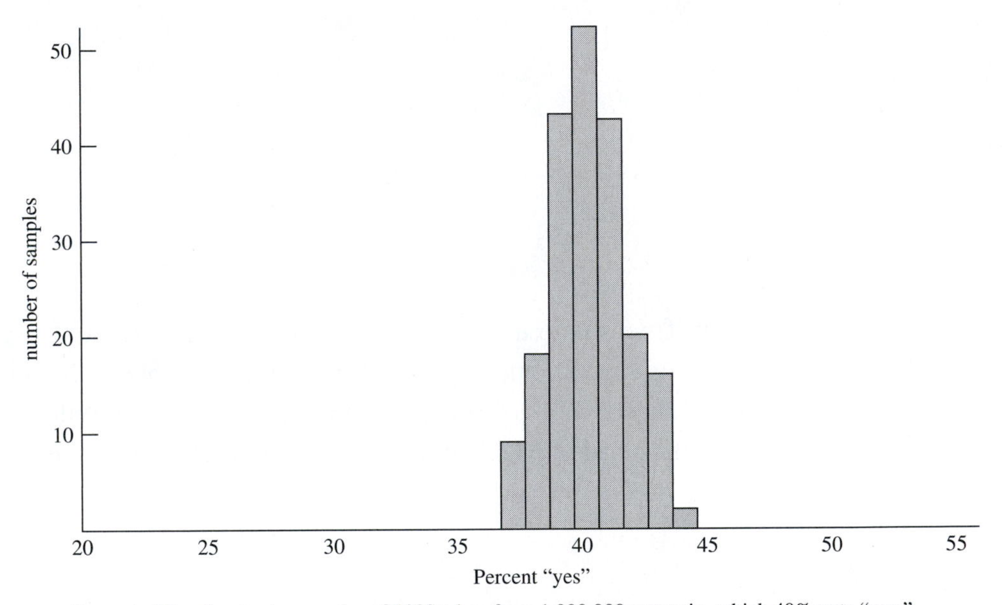

Percent of "yes" votes in **samples of 1000** taken from 1,000,000 voters in which 40% vote "yes."

15

Probability

15.1 Basic Principles and Calculation Methods of Probability

Class Activity 15A: Comparing Probabilities

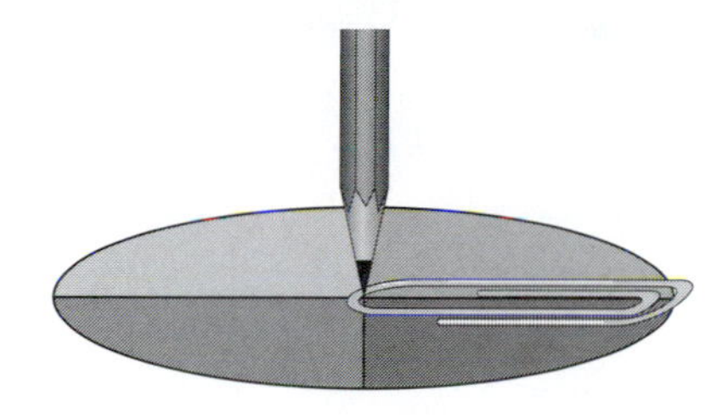

1. Many children's games use "spinners." You can make a simple spinner by placing the tip of a pencil through a paperclip and holding the pencil so that its tip is at the center of the circle as shown. The paperclip should spin freely around the pencil tip.

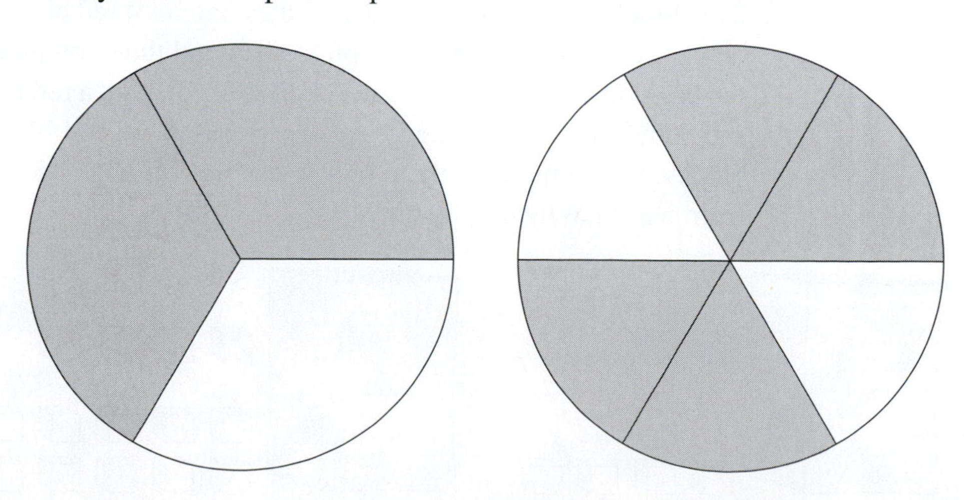

This problem is similar to a problem in an activity book for grades 1–3; see [6]:

Compare the two spinners shown on the previous page. For which spinner is a paperclip most likely to point into a shaded region? Explain your answer.

2. Compare the next two spinners. For which spinner is the paperclip that spins around a pencil point held at the indicated center point most likely to land in a shaded region? Explain your answer.

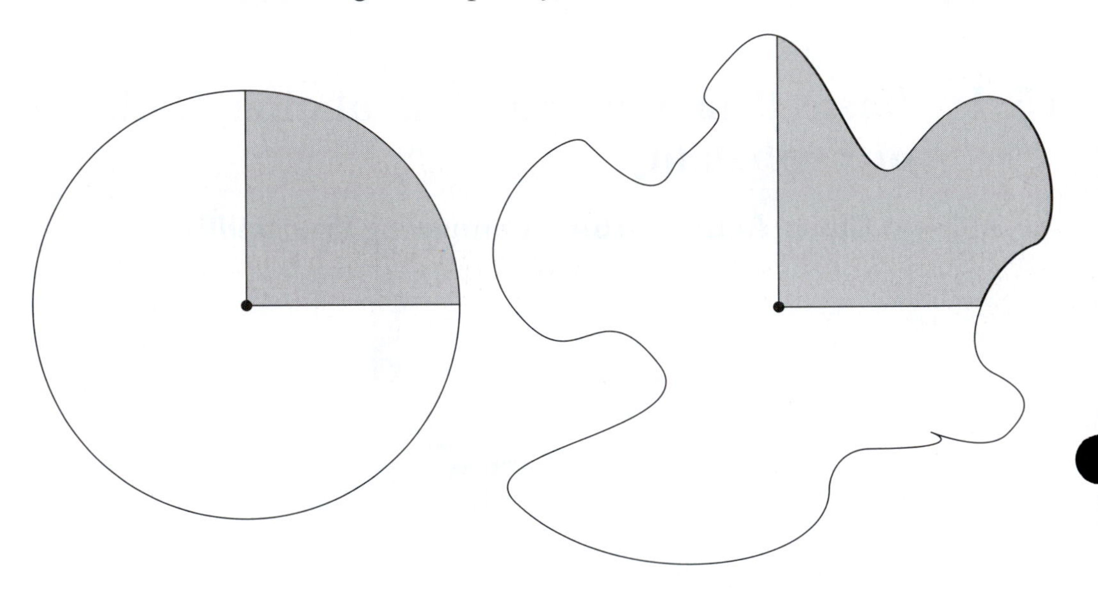

3. A family math night at school features the following game: There are two opaque bags, each containing red blocks and yellow blocks. Bag 1 contains 2 red blocks and 4 yellow blocks. Bag 2 contains 4 red blocks and 16 yellow blocks. To play the game, you pick a bag and then you pick a block out of the bag without looking. You win a prize if you pick a red block. Eva thinks she should pick from bag 2 because it has more red blocks than yellow blocks. Is Eva more likely to pick a red block if she picks from bag 2 than from bag 1? Why or why not?

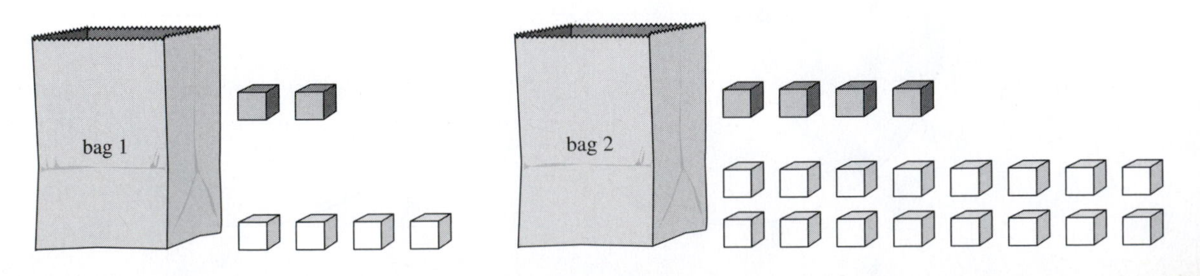

Class Activity 15B: Experimental versus Theoretical Probability: Picking Cubes from a Bag

Each person (or small group) will need an opaque bag, 3 red cubes, 7 blue cubes, and a sticky note. In this activity you will compare experimental and theoretical probabilities of picking a red cube from a bag containing 3 red and 7 blue cubes.

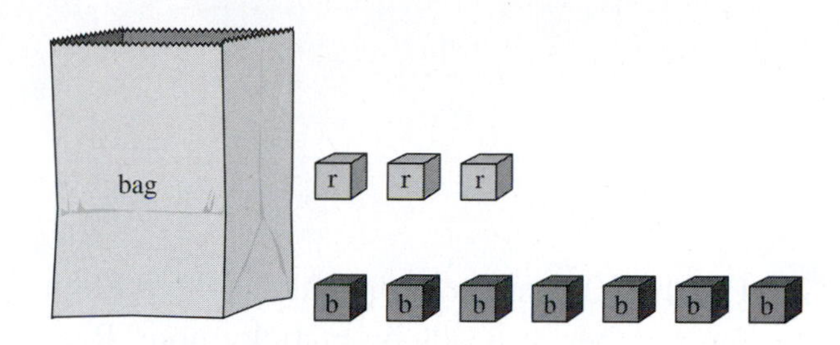

1. Put the 10 cubes in the bag, mix them up, and randomly pick a cube from the bag without looking. Record the color of the cube, and put the cube back in the bag. Repeat until you have picked 10 cubes. Record the number of red cubes you picked on your sticky note. Calculate the experimental probability of picking a red cube based on your 10 picks. Is it the same as the theoretical probability of picking red?

2. Now work with a large group (e.g., the whole class). Collect the sticky notes of Problem 1 from the full group. Determine the total number of reds picked and the total number of picks among the large group. Use these results to determine the experimental probability of picking a red cube obtained by the large group. Compare this experimental probability with the theoretical probability of picking red.

3. Use the sticky notes of Problem 1 to create a dot plot. How is the fact that there are 3 red cubes and 7 blue cubes in the bag reflected in the dot plot?

Class Activity 15C: If You Flip 10 Pennies, Should Half Come Up Heads?

You will need a bag, 10 pennies or 2-color counters, and some sticky notes for this activity.

1. Make a guess: What do you think the probability is of getting exactly 5 heads on 10 pennies when you dump the 10 pennies out of a bag?

2. Put the 10 pennies in the bag, shake them up, and dump them out. Record the number of heads on a sticky note. Repeat this for a total of 10 times, using a new sticky note each time. Out of these 10 tries, how many times did you get 5 heads? Therefore, what is the experimental probability of getting 5 heads based on your 10 trials?

3. Now work with a large group (e.g., the whole class). Collect the whole group's data on the sticky notes from Problem 2. Find a way to display these data so that you can see how often the whole group got 5 heads and other numbers of heads.

4. Is the probability of getting exactly 5 heads from 10 coins 50%? What does your data display from Problem 3 suggest?

Class Activity 15D: Number Cube Rolling Game

Maya, James, Kaitlyn, and Juan are playing a game in which they take turns rolling a pair of number cubes. Each child has chosen a "special number" between 2 and 12, and receives 8 points whenever the total number of dots on the two number cubes is their special number. (They receive their points regardless of who rolled the number cubes. Their teacher picked 8 points so that the children would practice counting by 8s.)

Maya's special number is 7.

James's special number is 10.

Kaitlyn's special number is 12.

Juan's special number is 4.

The first person to get to 100 points or more wins. The children have played several times, each time using the same special numbers. They notice that Maya wins most of the time. They are wondering why.

1. Roll a pair of number cubes many times, and record the total number of dots each time. Display your data so that you can compare how many times each possible number between 2 and 12 has occurred. What do you notice?

2. Draw an array showing all possible outcomes on each number cube when a pair of number cubes are rolled. (Think of the pair as *number cube 1* and *number cube 2*.)

   For which outcomes is the total number of dots 7? 10? 12? 4?

   What is the probability of getting 7 total dots on a roll of two number cubes? What is the probability of getting 10 total dots on a roll of two number cubes? What about for 12 and 4?

   Is it surprising that Maya kept winning?

Class Activity 15E: Picking Two Marbles from a Bag of 1 Black and 3 Red Marbles

You will need an opaque bag, 3 red marbles, and 1 black marble for this activity. Put the marbles in the bag.

If you reach in without looking and randomly pick out two marbles, what is the probability that one of the two marbles you pick is black? You will study this question in this activity.

1. Before you continue, make a guess: What do you think the probability of picking the black marble is when you randomly pick two marbles out of the four marbles (3 red, 1 black) in the bag?

2. Pick 2 marbles out of the bag. Repeat this many times, recording what you pick each time. What fraction of the times did you pick the black marble?

3. Now calculate the probability theoretically, using a tree diagram. For the purpose of computing the probability, think of first picking one marble, then (without putting this marble back in the bag) picking a second marble. From this point of view, draw a tree diagram that will show all possible outcomes for picking the two marbles. But draw this tree diagram in a special way, *so that all outcomes shown by your tree diagram are equally likely.*

Hints: The first stage of the tree should show all possible outcomes for your first pick. Remember that all branches you show should be equally likely. In the second stage, the branches you draw should depend on what happened in the first stage. For instance, if the first pick was the black marble, then the second pick must be one of the three red marbles.

a. How many total outcomes for picking two marbles, one at a time, out of the bag of four (3 red, 1 black) does your tree diagram show?

b. In how many outcomes is the black marble picked (on one of the two picks)?

c. Use your answers to Problems 3(a) and 3(b) and the basic principles of probability to calculate the probability of picking the black marble when you pick two marbles out of a bag filled with one black and three red marbles.

4. Why was it important to draw the tree diagram so that all outcomes were equally likely?

5. Here's another method for calculating the probability of picking the black marble when you pick two marbles out of a bag filled with one black and three red marbles:

a. How many pairs of marbles can be made from the four marbles in the bag?

b. How many of those pairs of marbles in Problem 5(a) contain the black marble? (Use your common sense.)

c. Use Problems 5(a) and 5(b) and basic principles of probability to determine the probability of picking the black marble when you pick two marbles out of a bag containing one black and three red marbles.

6. Compare your answers with Problems 3(a) and 5(a), and compare your answers with Problems 3(b) and 5(b). How and why are they different?

Class Activity 15F: Applying Probability

Ms. Wilkins is planning a game for her school's fall festival. She will put 2 red, 3 yellow, and 10 green plastic bears in an opaque bag. (The bears are identical except for their color.) To play the game, a contestant will pick two bears from the bag, one at a time, without putting the first bear back before picking the second bear. Contestants will not be able to see into the bag, so their choices are random. To win a prize, the contestant must pick a green bear first and then a red bear. The school is expecting about 300 people to play the game. Each person will pay 50 cents to play the game. Winners receive a prize that costs the school $2.

1. How many prizes should Ms. Wilkins expect to give out? Explain.

2. How much money (net) should the school expect to make from Ms. Wilkins's game? Explain.

Class Activity 15G: Some Probability Misconceptions

1. Simone has been flipping a coin and has just flipped 5 heads in a row. Simone says that because she has just gotten so many heads, she is more likely to get tails than heads the next time she flips. Is Simone correct? What is the probability that Simone's next flip will be a tail? Does the answer depend on what the previous flips were?

2. The probability of winning a game is $\frac{3}{1000}$. Does this mean that if you play the game 1000 times, you will win 3 times? If not, what does the probability of $\frac{3}{1000}$ stand for?

3. There are two opaque bags, each containing red blocks and yellow blocks. Bag 1 contains 1 red block and 3 yellow blocks. Bag 2 contains 3 red blocks and 12 yellow blocks. Tom says he is more likely to pick a red block out of bag 2 than out of bag 1 because bag 2 has more red blocks than bag 1. Is this correct?

4. Let's say you flip two coins simultaneously. There are three possible outcomes: Both are heads, both are tails, or one is heads and the other is tails. Does this mean that the probability of getting one head and one tail is $\frac{1}{3}$?

5. Kevin has a bag that is filled with 2 red balls and 1 white ball. Kevin says that because there are two different colors he could pick from the bag, the probability of picking the red ball is $\frac{1}{2}$. Is this correct?

15.2 Using Fraction Arithmetic to Calculate Probabilities

Class Activity 15H: Using the Meaning of Fraction Multiplication to Calculate a Probability

Use the circle in Figure 15H.1, a pencil, and a paperclip to make a spinner as follows: Put the pencil through the paperclip, and put the point of the pencil on the center of the circle. The paperclip will now be able to spin freely around the circle.

To win a game, Jill needs to spin a blue followed by a red in her next two spins.

1. What do you think Jill's probability of winning is? (Make a guess.)

2. Carry out the experiment of spinning the spinner twice in a row 20 times. (In other words, spin the spinner 40 times, but each experiment consists of 2 spins.) Out of those 20 times, how often does Jill win? What fraction of 20 does this represent? Is this close to your guess in Problem 1?

3. Calculate Jill's probability of winning theoretically as follows: Imagine that Jill carries out the experiment of spinning the spinner twice in a row many times. In the ideal, what fraction of those times should the first spin be blue? ______________ Show this by shading the rectangle on the next page.

   In the ideal, what fraction of those times when the first spin is blue should the second spin be red? ______________ Show this by further shading the rectangle on the next page.

   In the ideal, what fraction of pairs of spins should Jill spin first a blue and then a red? Therefore, what is Jill's probability of winning? ______________ Explain how you can determine this fraction from the shading of the rectangle and from the meaning of fraction multiplication. Compare your answer with Problems 1 and 2.

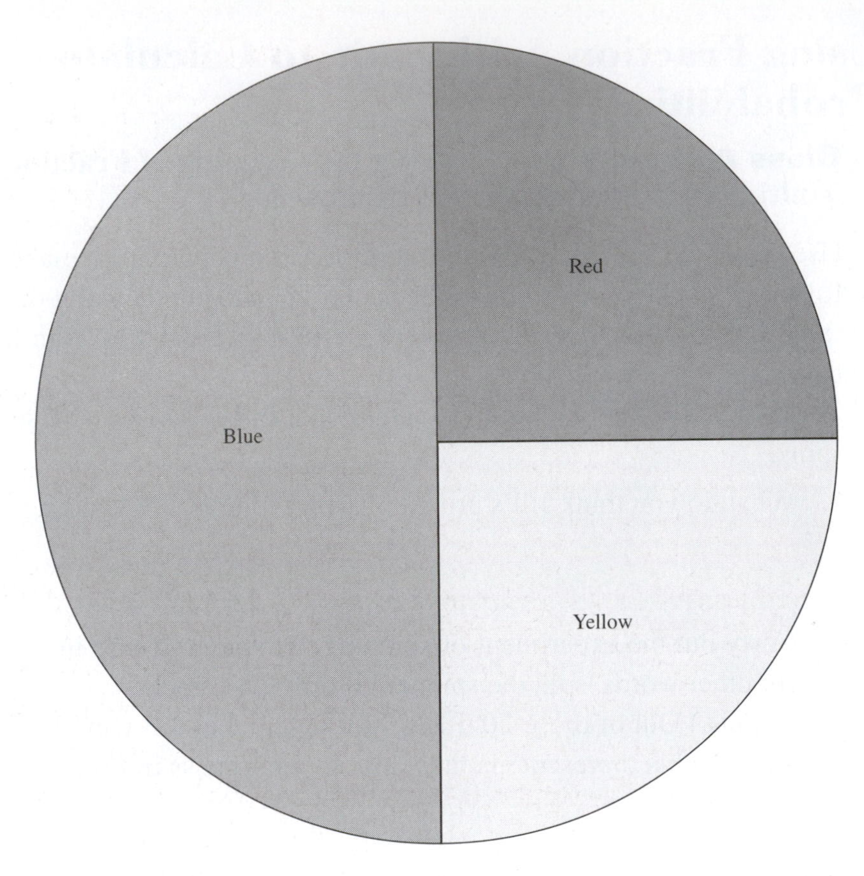

Figure 15H.1
A Spinner

This rectangle represents many pairs of spins.

Class Activity 15I: Using Fraction Multiplication and Addition to Calculate a Probability

A paperclip, an opaque bag, and blue, red, and green tiles would be helpful if available.

A game consists of spinning the spinner in Figure 15H.1 and then picking a small tile from a bag containing 1 blue tile, 3 red tiles, and 1 green tile. (All tiles are identical except for color, and the person picking a tile cannot see into the bag, so the choice of a tile is random.) To win the game, a contestant must pick the same color tile that the spinner landed on. So a contestant wins from either a blue spin followed by a blue tile or a red spin followed by a red tile.

1. Make a guess: What do you think the probability of winning the game is?
2. If the materials are available, play the game a number of times. Record the number of times you play the game (each game consists of both a spin *and* a pick from the bag), and record the number of times you win. What fraction of the time did you win? How does this compare with your guess in Problem 1?

3. To calculate the (theoretical) probability of winning the game, imagine playing the game many times. Answer the next questions in order to determine the probability of winning the game.
   a. In the ideal, what fraction of the time should the spin be blue? _______ Show this by shading the rectangle on the next page.

      In the ideal, what fraction of those times when the spin is blue should the tile that is chosen be blue? _______ Show this by further shading the rectangle on the next page.

      Therefore, in the ideal, what fraction of the time is the spin blue and the tile blue? _______ Explain how you can determine this fraction from the shading of the rectangle and from the meaning of fraction multiplication.

b. In the ideal, what fraction of the time should the spin be red? _______ Show this by shading the rectangle on this page.

In the ideal, what fraction of those times when the spin is red should the tile that is chosen be red? _______ Show this by further shading the rectangle on this page.

Therefore, in the ideal, what fraction of the time is the spin red and the tile red? _______ Explain how you can determine this fraction from the shading of the rectangle and from the meaning of fraction multiplication.

c. In the ideal, what fraction of the time should you win the game, and therefore, what is the probability of winning the game? Explain why you can calculate this answer by multiplying and adding fractions. Compare your answer with Problems 1 and 2.

This rectangle represents playing the game many times.

Bibliography

[1] George W. Bright. Helping elementary-and middle-grades preservice teachers understand and develop mathematical reasoning. In *Developing Mathematical Reasoning in Grades K–12*, pages 256–269. National Council of Teachers of Mathematics, 1999.

[2] T. P. Carpenter, E. Fennema, M. L. Franke, S. B. Empson, and L. W. Levi. *Children's Mathematics: Cognitively Guided Instruction*. Heinemann, Portsmouth, NH, 1999.

[3] Frances Curcio. *Developing Data-Graph Comprehension in Grades K–8*. National Council of Teachers of Mathematics, 2d ed., 2001.

[4] K. C. Fuson. Developing mathematical power in whole number operations. In J. Kilpatrick, W. G. Martin, and D. Schifter, eds. *A Research Companion to Principles and Standards for School Mathematics*. Reston, VA, NCTM, 2003.

[5] K. C. Fuson, S. T. Smith, and A. M. Lo Cicero. Supporting First Graders' Ten-Structured Thinking in Urban Classrooms. *Journal for Research in Mathematics Education*, vol. 28, Issue 6, pages 738–766, 1997.

[6] Graham Jones and Carol Thornton. *Data, Chance, and Probability, Grades 1–3 Activity Book*. Learning Resources, 1992.

[7] Edward Manfre, James Moser, Joanne Lobato, and Lorna Morrow. *Heath Mathematics Connections*. D. C. Heath and Company, 1994.

[8] National Council of Teachers of Mathematics. *Navigating through Algebra in Pre-kindergarten–Grade 2*. National Council of Teachers of Mathematics, 2001.

[9] National Research Council. *Adding It Up: Helping children learn mathematics*. J. Kilpatrick, J. Swafford, and B. Findell, eds. Mathematics Learning Study Committee, Center for Education, Division of Behavioral and Social Sciences and Education. National Academy Press, Washington, DC, 2001.

[10] A. M. O'Reilley. Understanding teaching/teaching for understanding. In Deborah Schifter, ed., *What's Happening in Math Class?*, volume 2: Reconstructing Professional Identities, pages 65–73. Teachers College Press, 1996.

[11] Dav Pilkey. *Captain Underpants and the Attack of the Talking Toilets*. Scholastic, 1999.

[12] Singapore Curriculum Planning and Development Division, Ministry of Education. *Primary Mathematics Workbook*, volume 1A–6B. Times Media Private Limited, Singapore, 3d ed., 2000. Available at http://www. singapore-math.com

[13] K. Stacey. Traveling the Road to Expertise: A Longitudinal Study of Learning. In Chick, H. L. and Vincent, J. L. eds. *Proceedings of the 29th Conference of the International Group for the Psychology of Mathematics Education*, vol. 1, pp. 19–36. PME, Melbourne, 2005.

[14] V. Steinle, K. Stacey, and D. Chambers. *Teaching and Learning about Decimals* [CD-ROM]: Department of Science and Mathematics Education, The University of Melbourne, 2002. Online sample at http://extranet.edfac.unimelb.edu.au/DSME/decimals/

Index

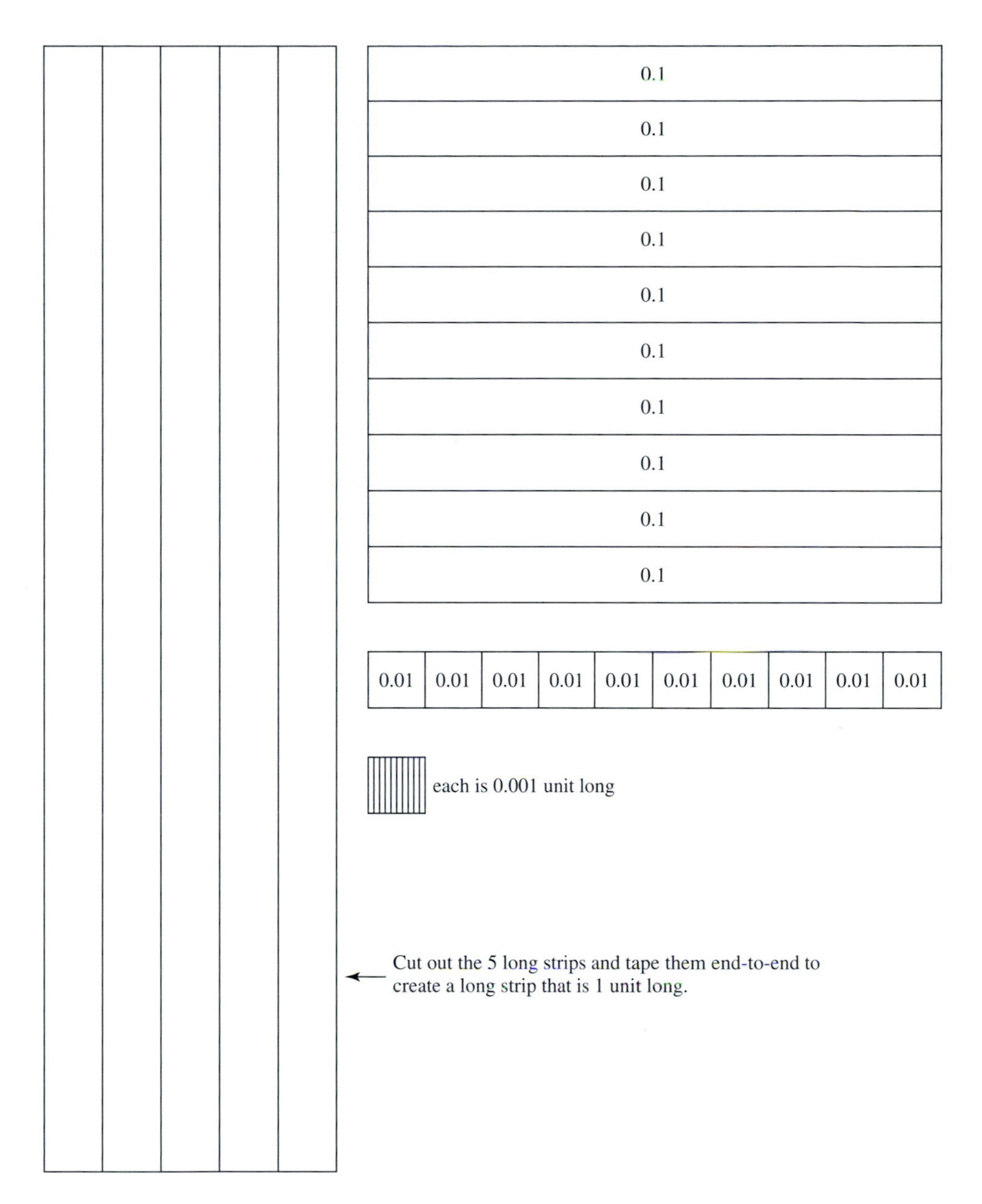

Figure A.1
Use these Strips to Represent Decimals as Lengths in Class Activity 2E on Page 16

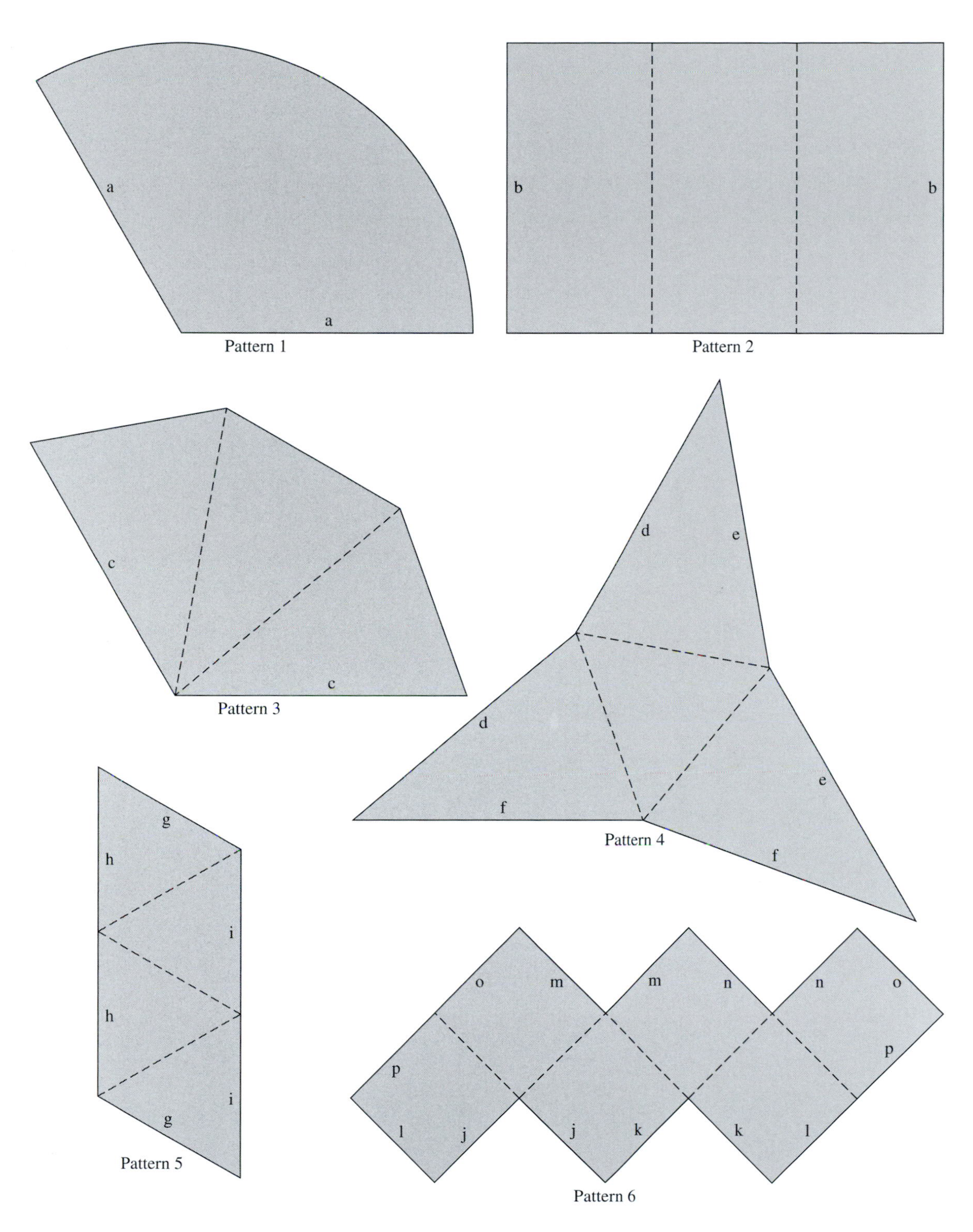

Figure A.2
Patterns for Shapes for Class Activity 8A on Page 245

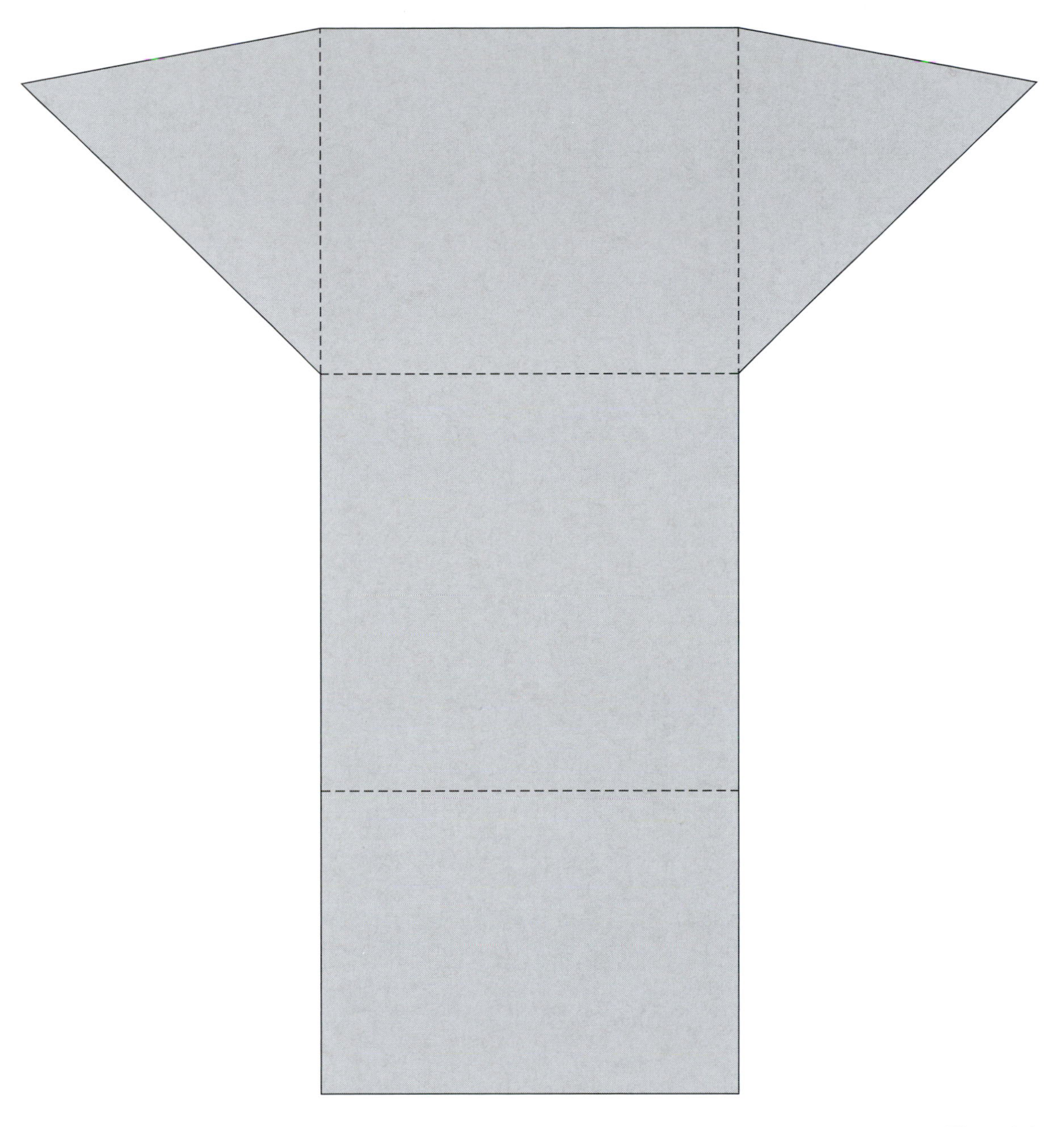

Figure A.3
For Class Activity 8FF on Page 302

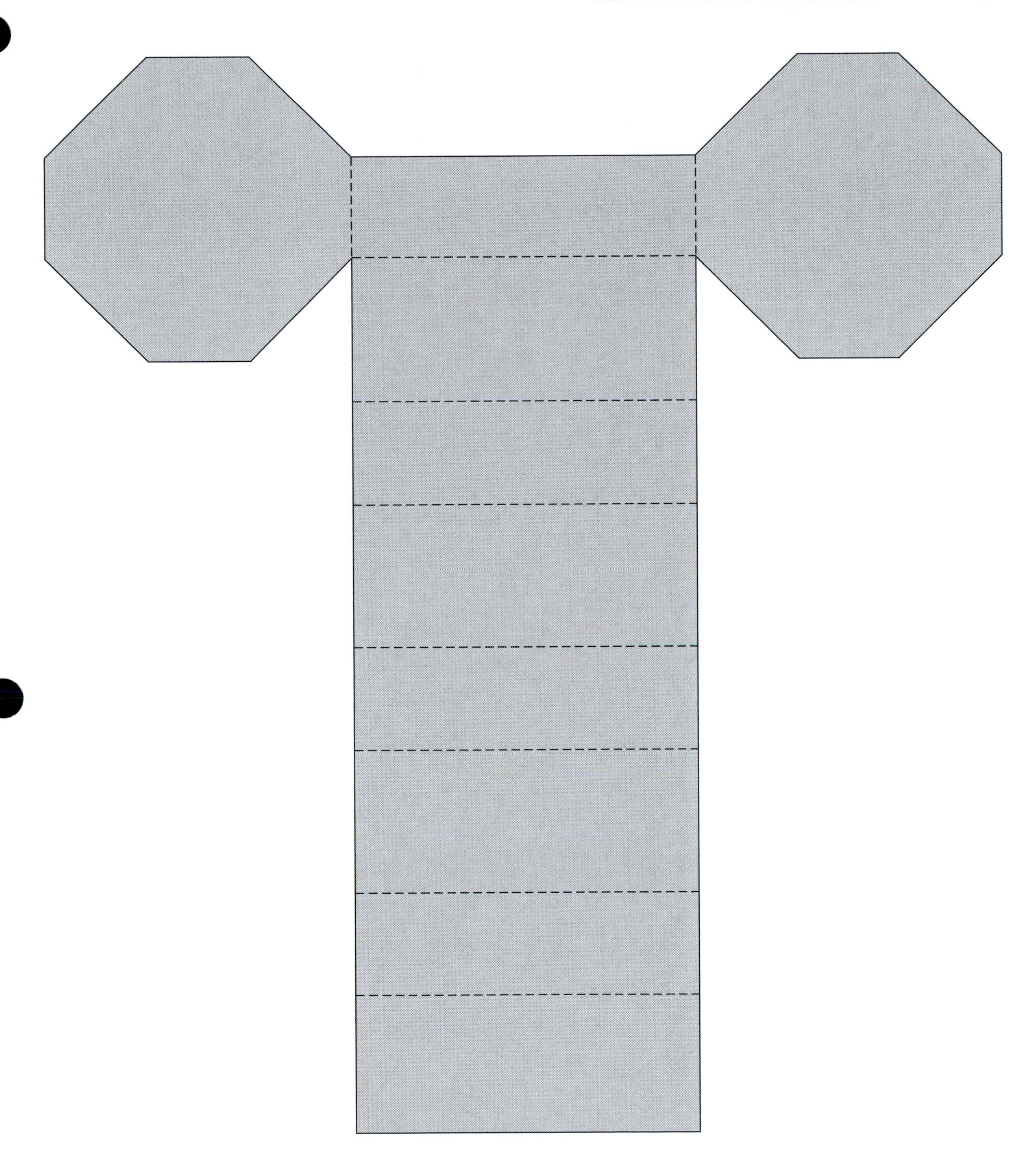

Figure A.4

For Class Activity 8FF on Page 302

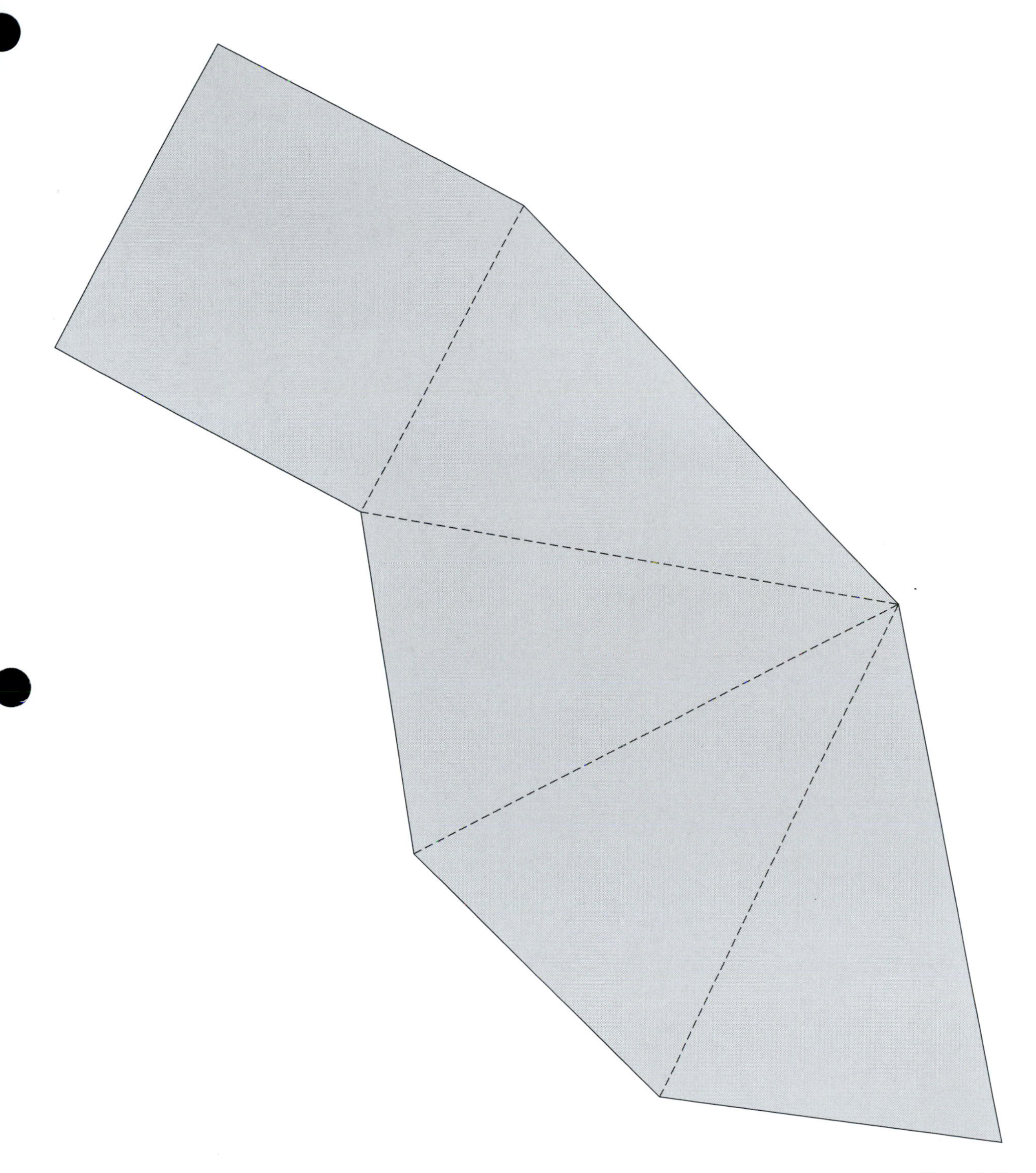

Figure A.5
For Class Activity 8FF on Page 302

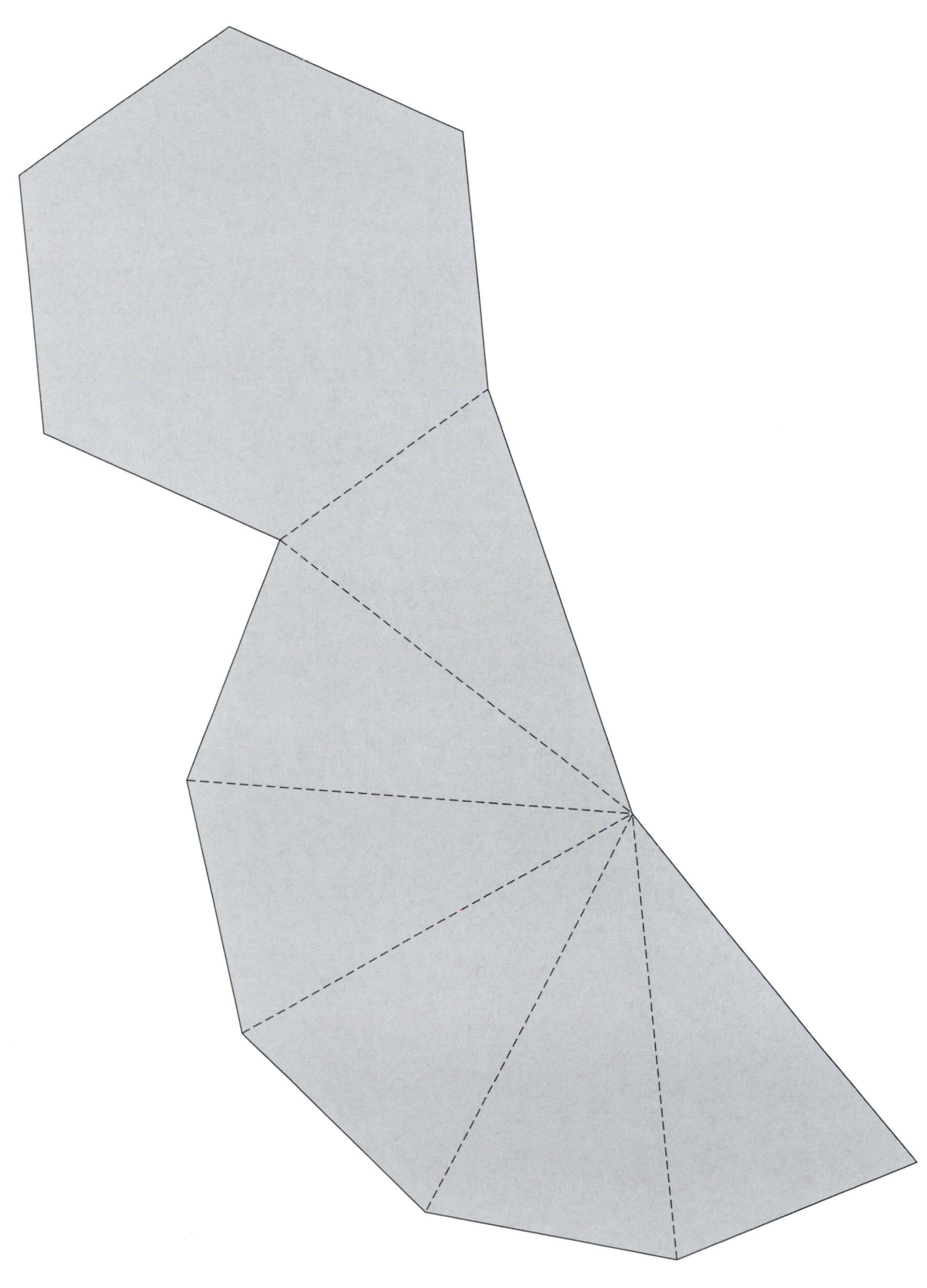

Figure A.6
For Class Activity 8FF on Page 302

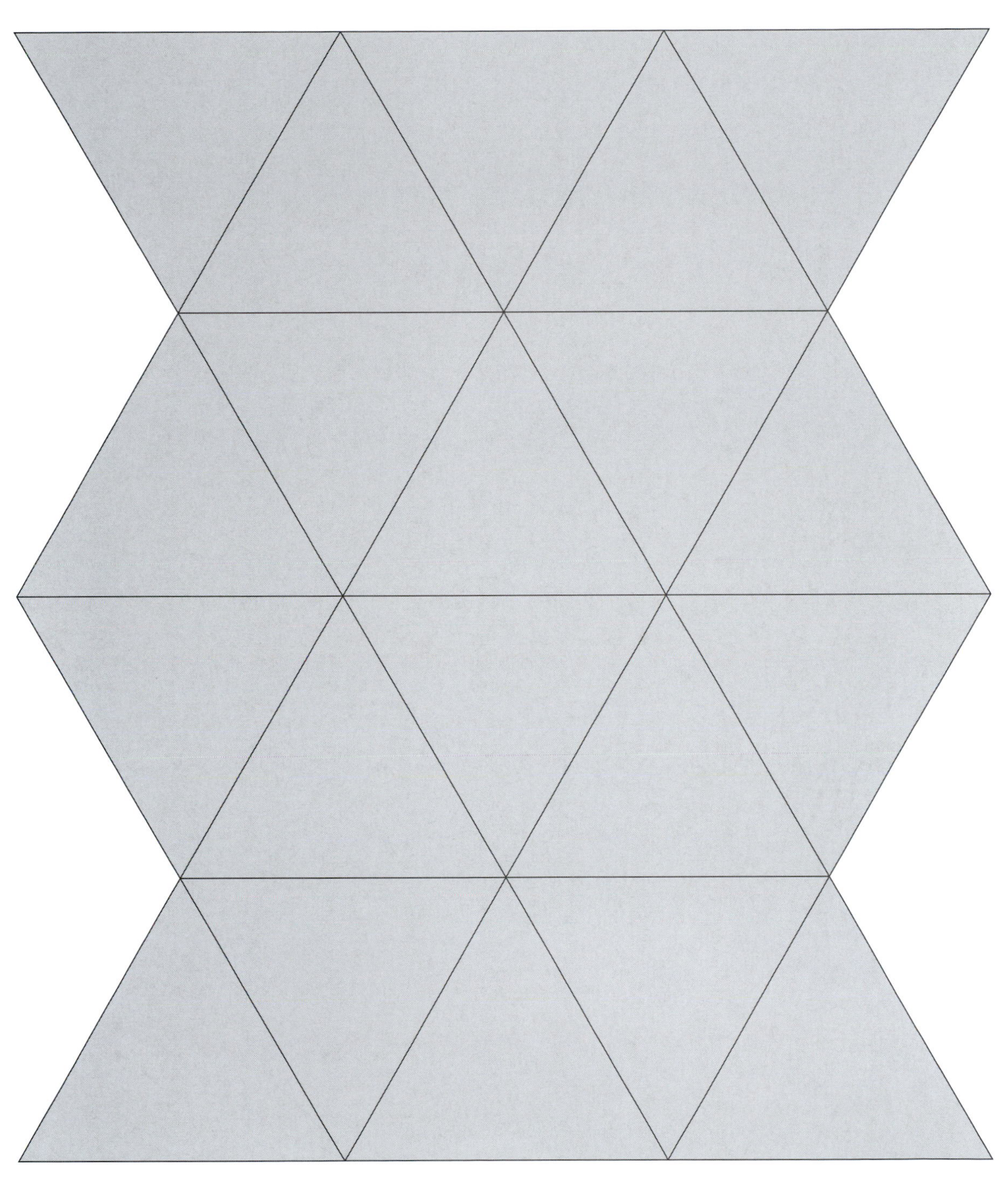

Figure A.7

For Class Activities 8KK and 8II on Pages 308 and 306

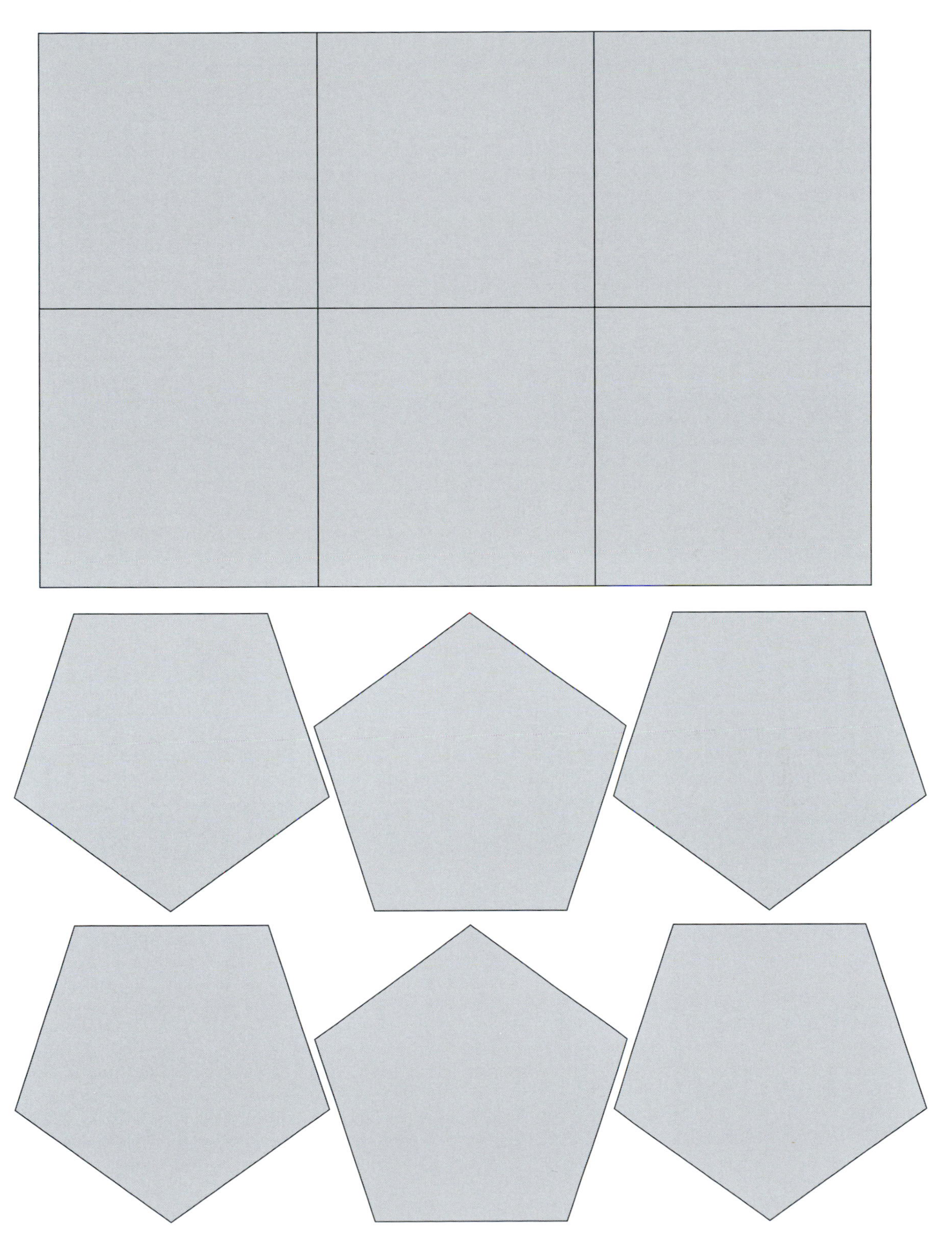

Figure A.8

For Class Activity 8KK on Page 308

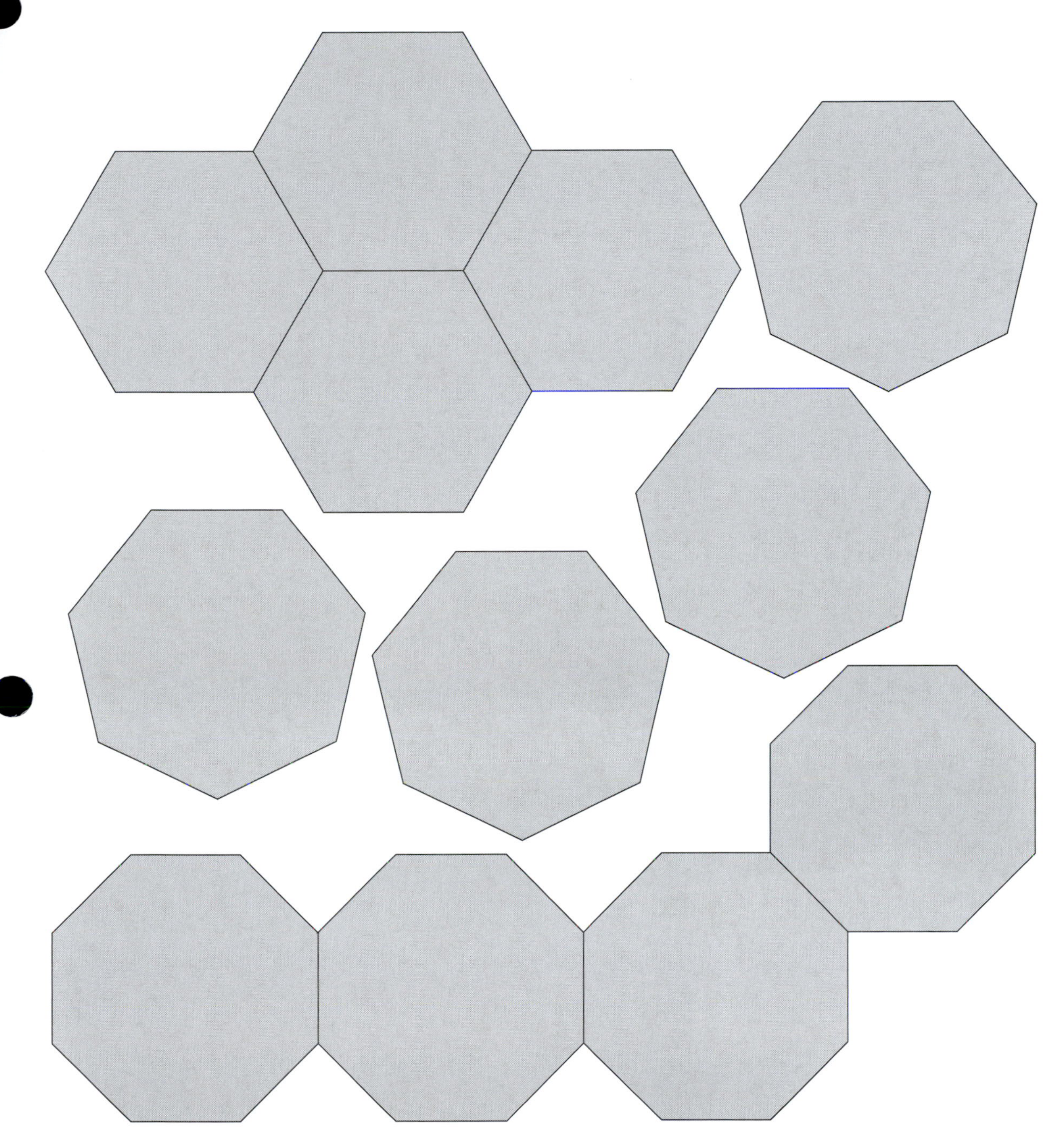

Figure A.9

For Class Activity 8KK on Page 308

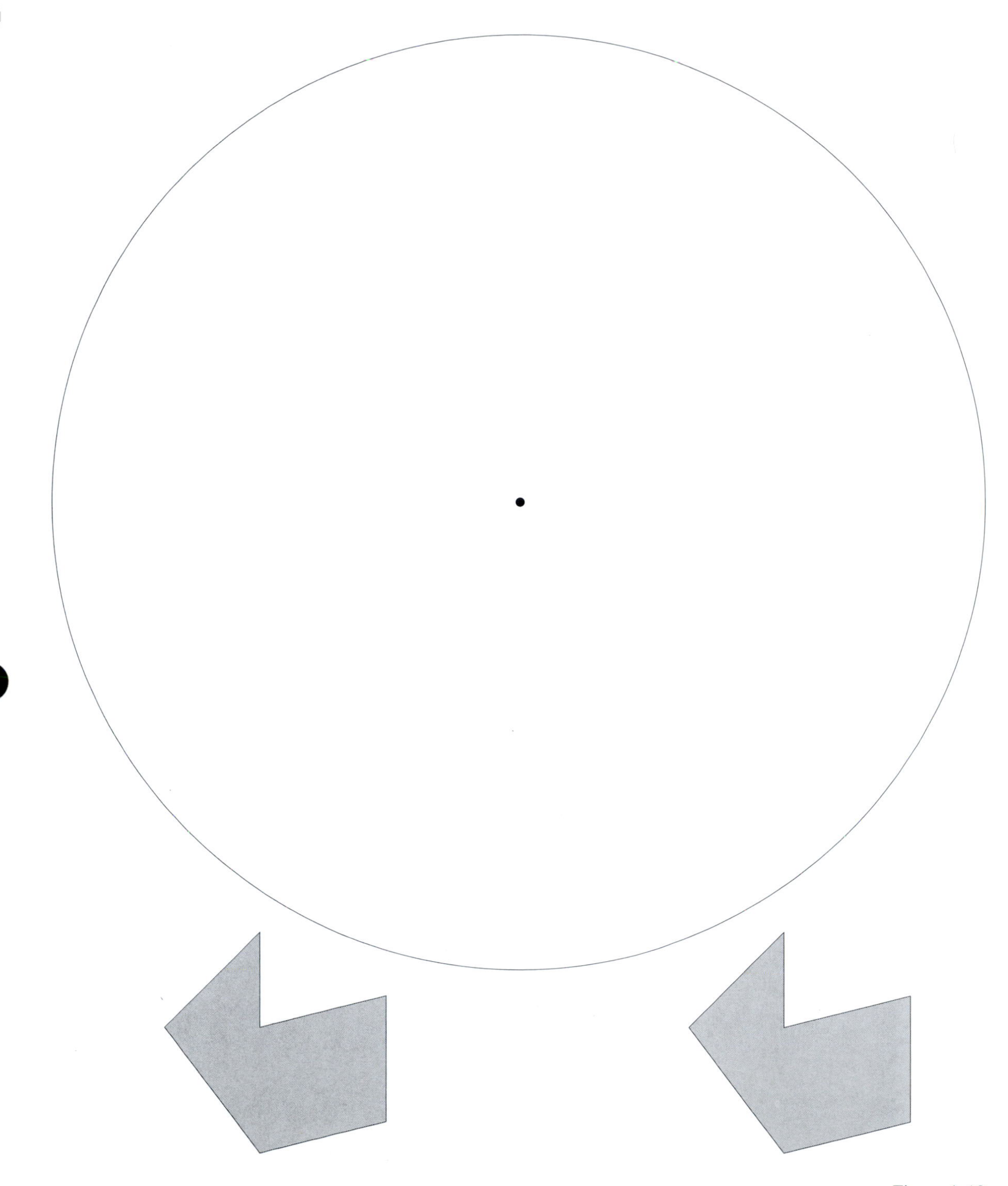

Figure A.10
A Circle and Two Shapes for Class Activities 9A and 9B on Pages 313 and 315

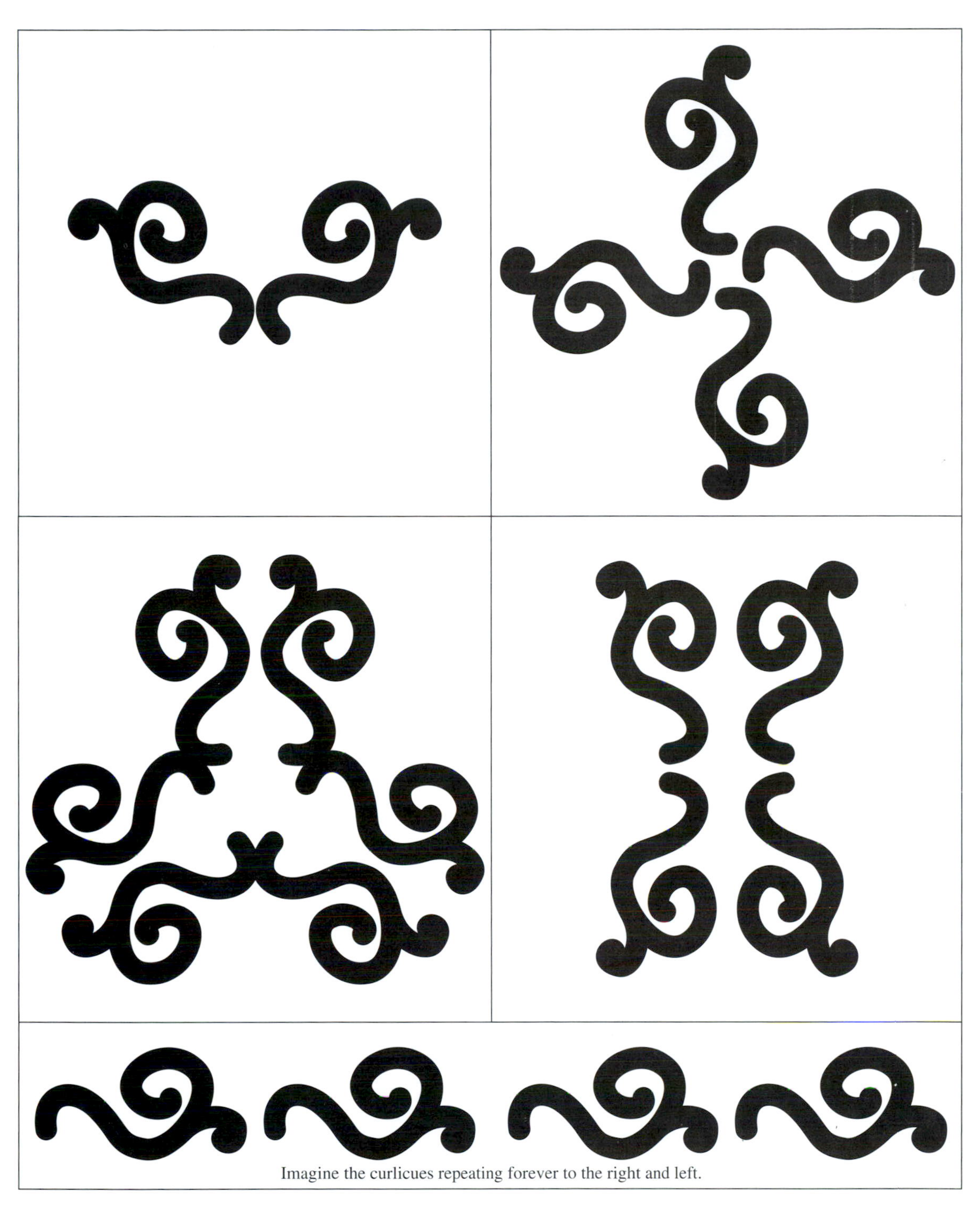

Imagine the curlicues repeating forever to the right and left.

Figure A.11
Five Designs for Class Activity 9G on Page 322

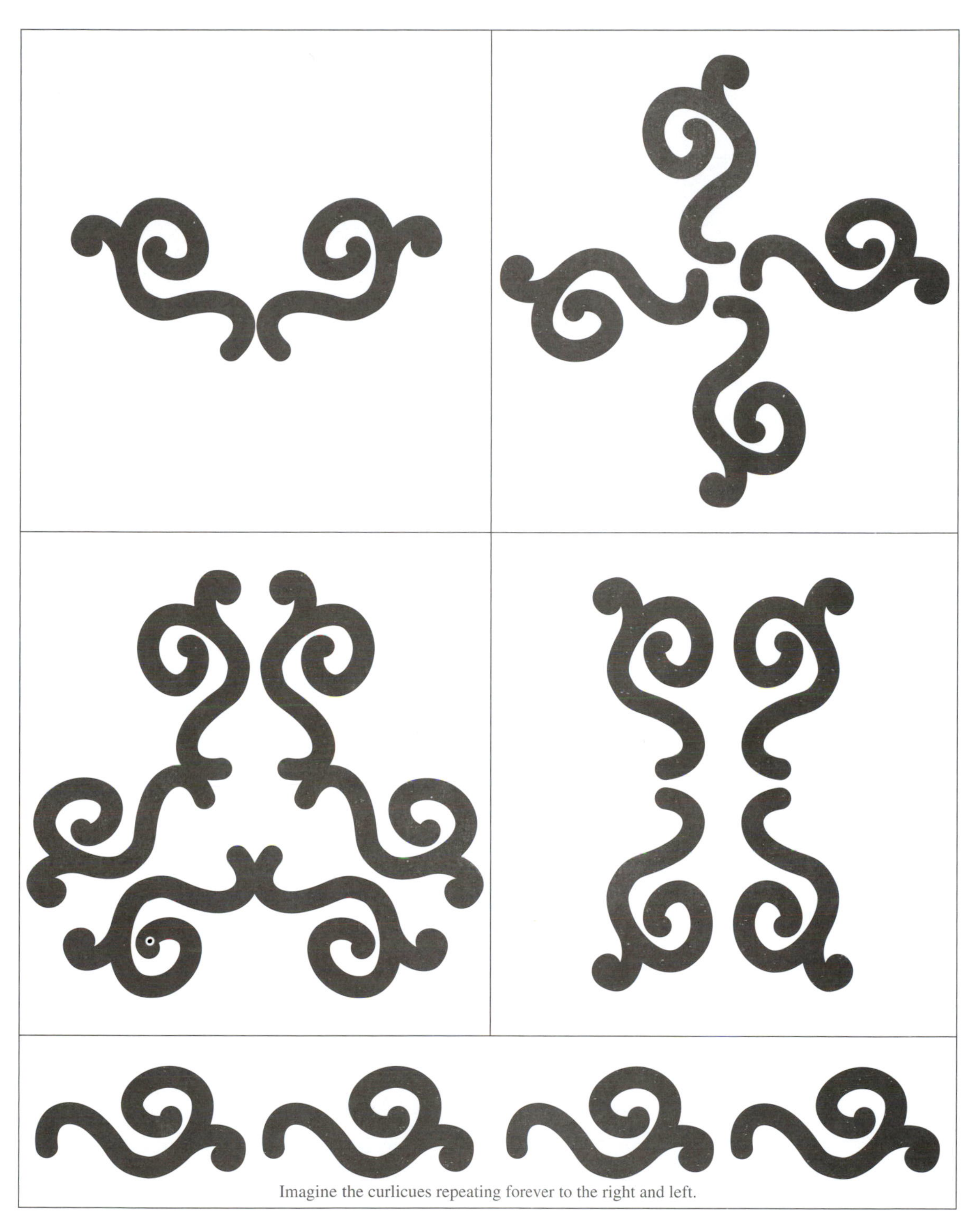

Imagine the curlicues repeating forever to the right and left.

Figure A.12
Second Copy of Five Designs for Class Activity 9G on Page 322

Figure A.13
Four Frieze Patterns for Class Activity 9H on Page 323

Figure A.14
Copy of Four Frieze Patterns for Class Activity 9H on Page 323

Figure A.15

Three Frieze Patterns for Class Activity 9H on Page 323

Figure A.16

Copy of Three Frieze Patterns for Class Activity 9H on Page 323

Figure A.17

Square Designs for Class Activity 9M on Page 329 on Analyzing Designs

Figure A.18
To Cut Out for Class Activity 10G on Page 358

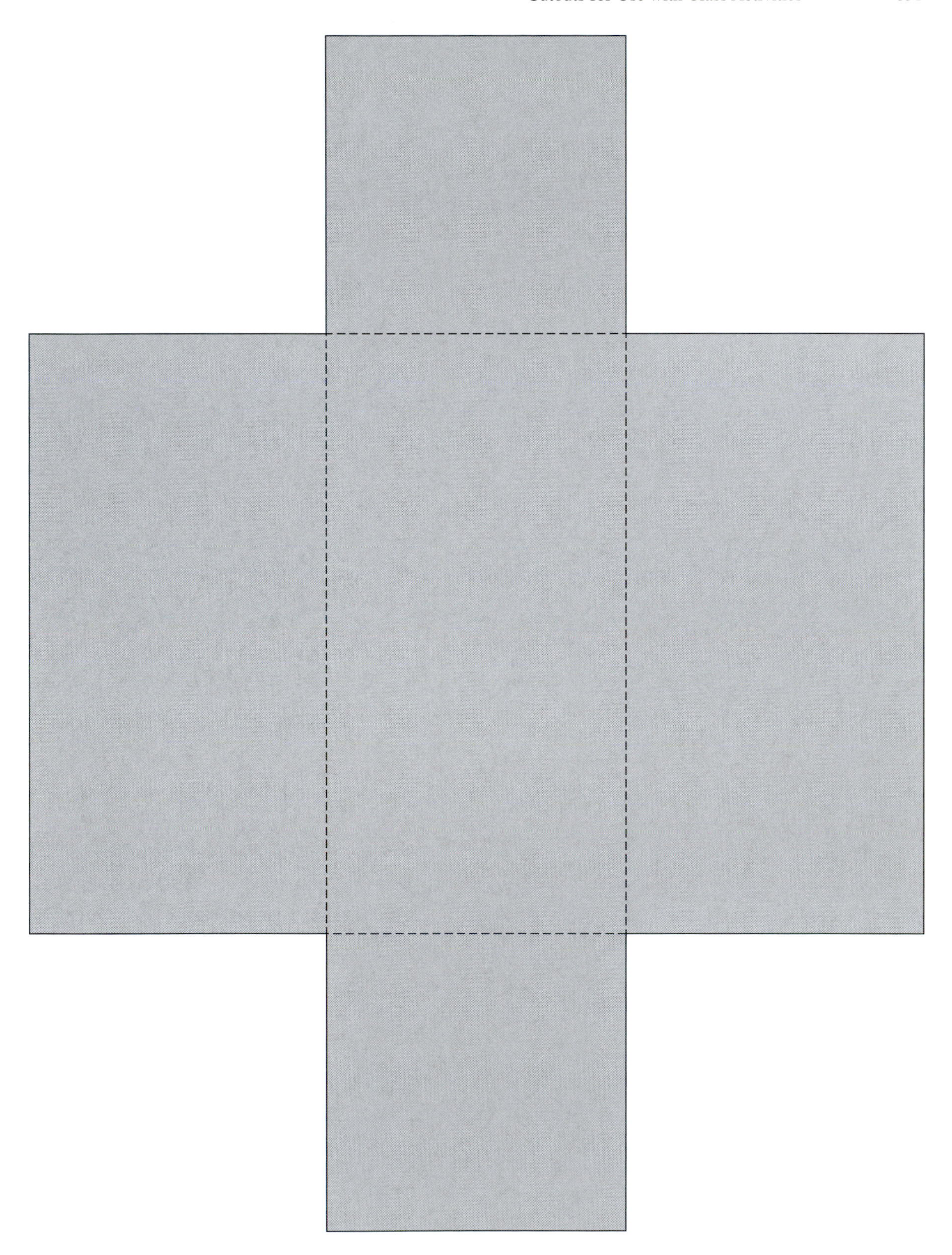

Figure A.19

Box Pattern for Class Activity 10H on Page 359

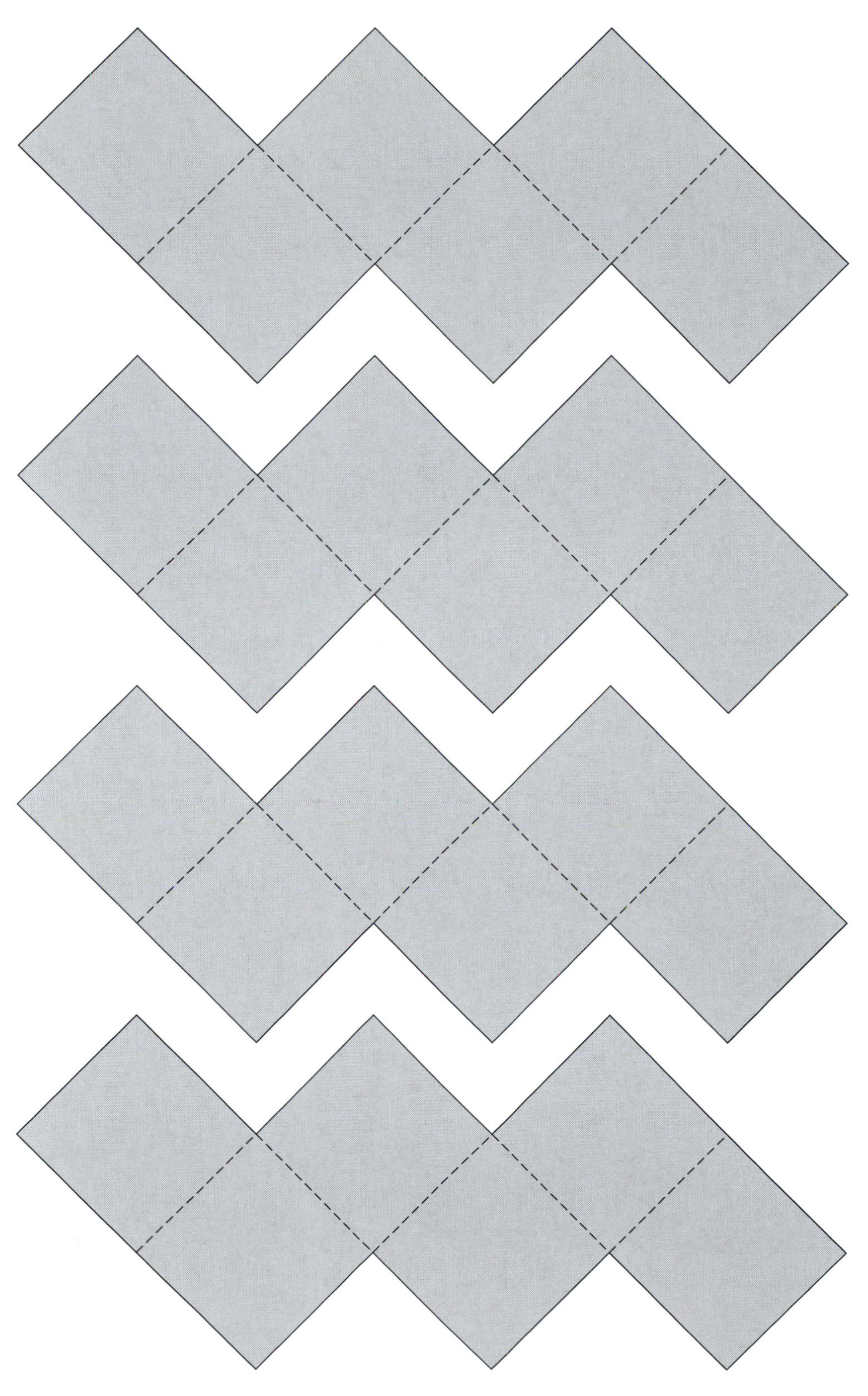

Figure A.20
Cube Patterns for Class Activity 10H on Page 359

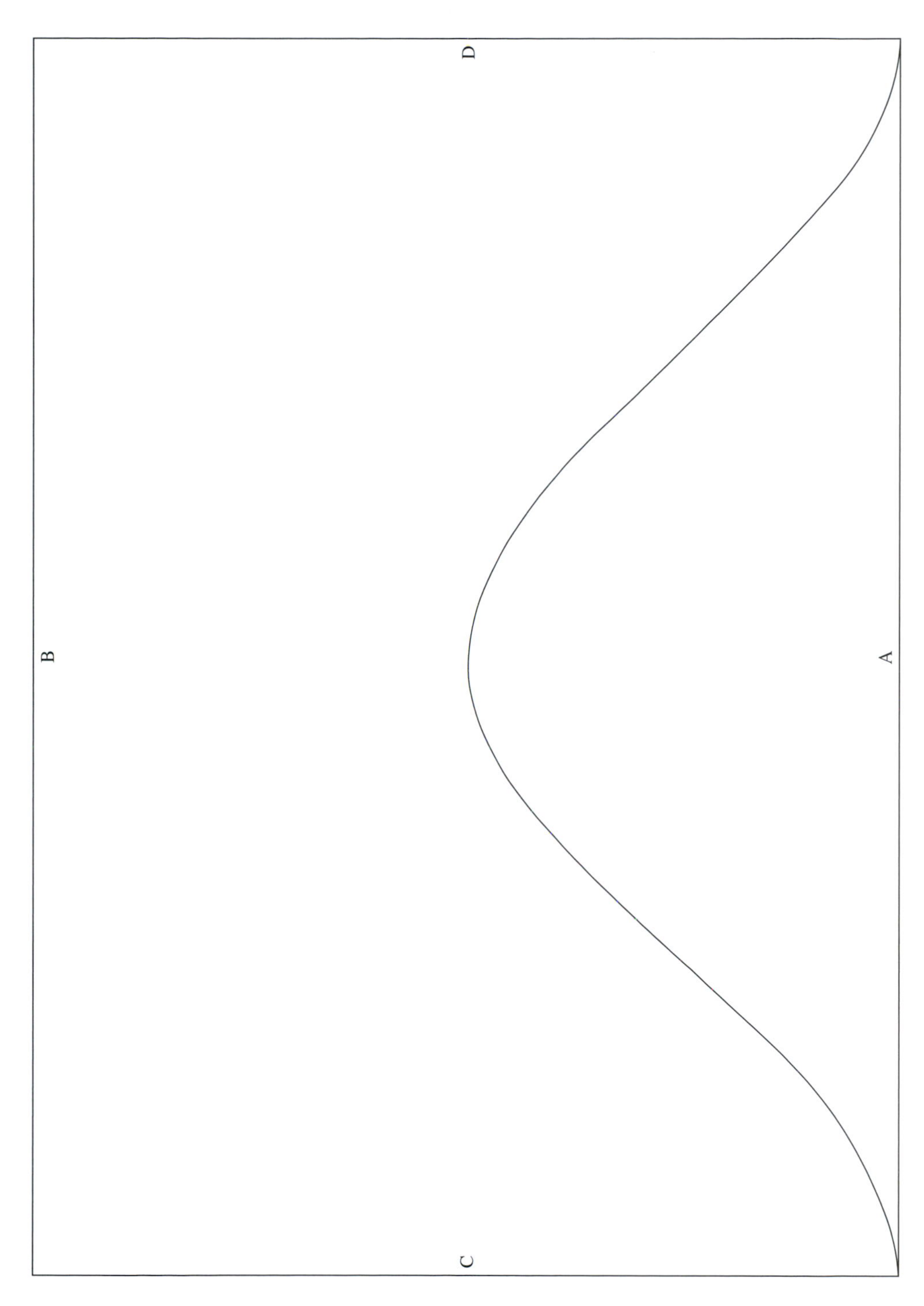

Figure A.21
To Cut Out for Class Activity 11C on Page 373

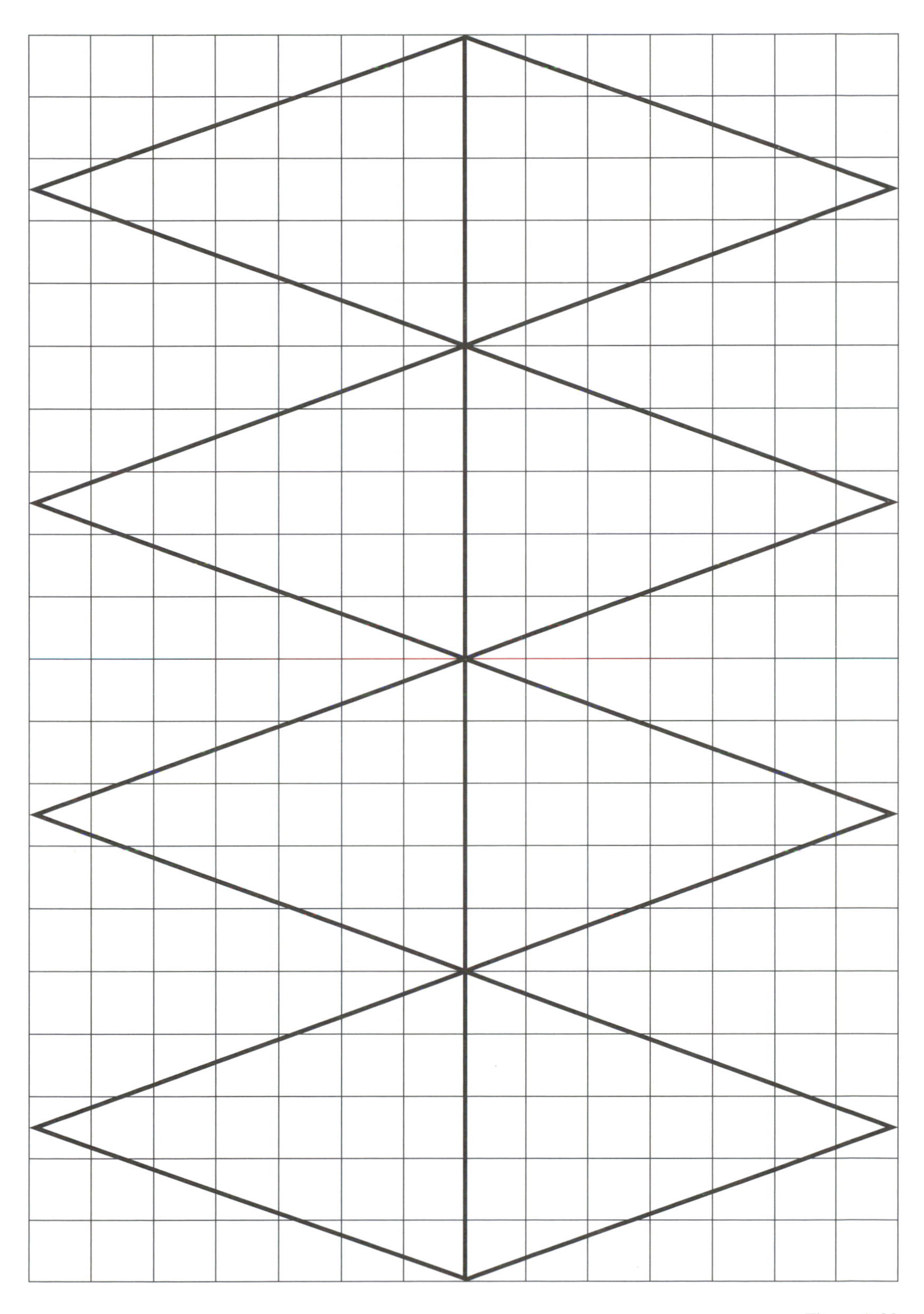

Figure A.22

Triangles for an Octahedron for Class Activity 11C on Page 373

Figure A.23
Rectangles for Class Activity 11A on Page 369

Figure A.24

Rectangles for Class Activity 11A on Page 369

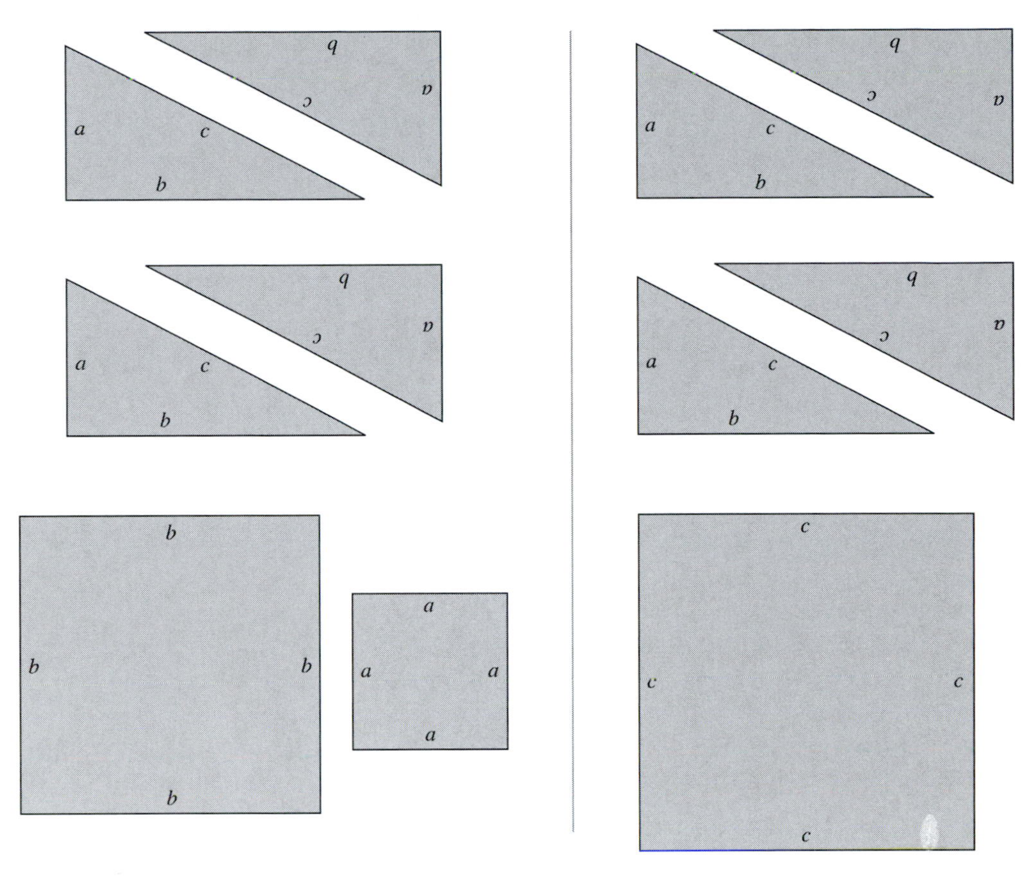

Figure A.25

To Cut Out for Class Activity 11F on Page 376

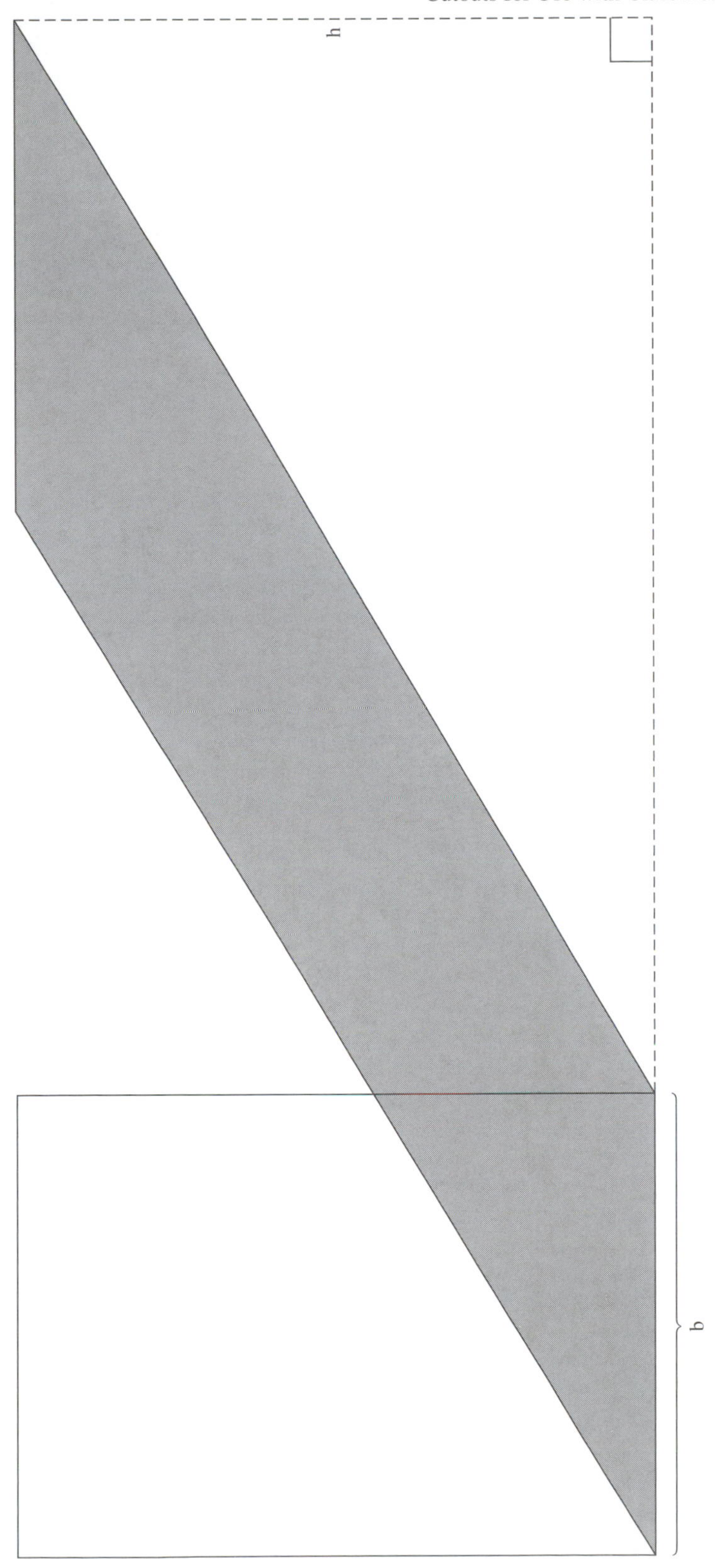

Figure A.26

A Parallelogram to Cut Apart and Rearrange for Class Activity 11K on Page 385

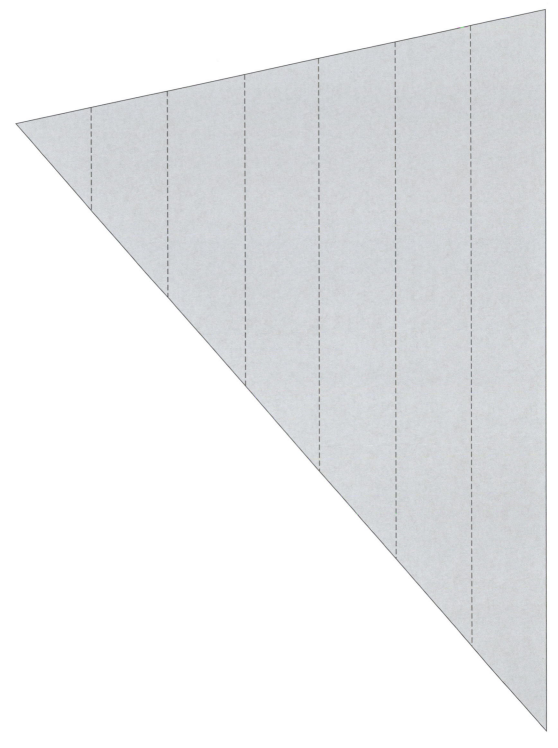

Figure A.27

A Triangle to Shear for Class Activity 11O on Page 389

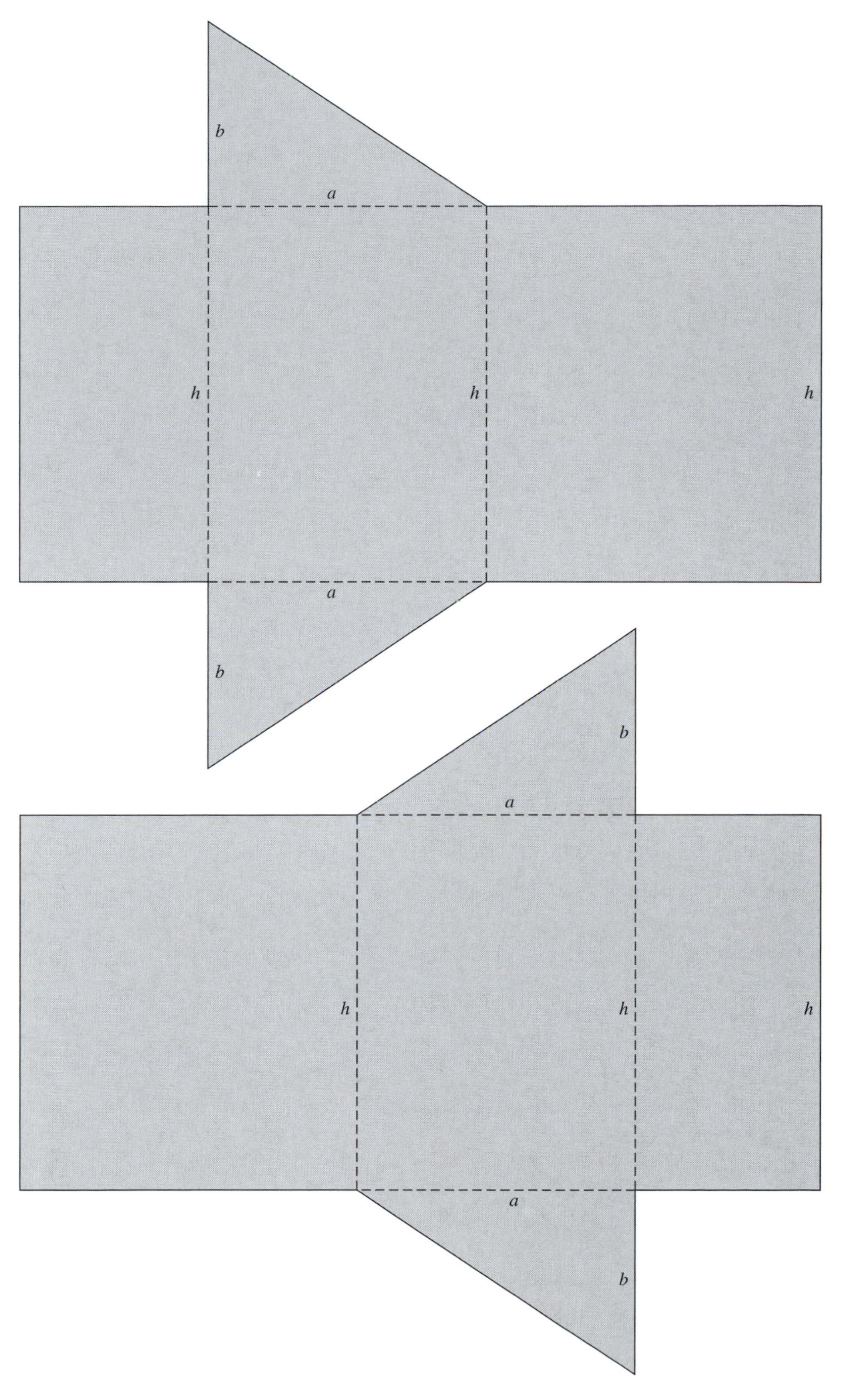

Figure A.28
Patterns for Prisms with Triangle Bases for Class Activity 11Z on Page 409

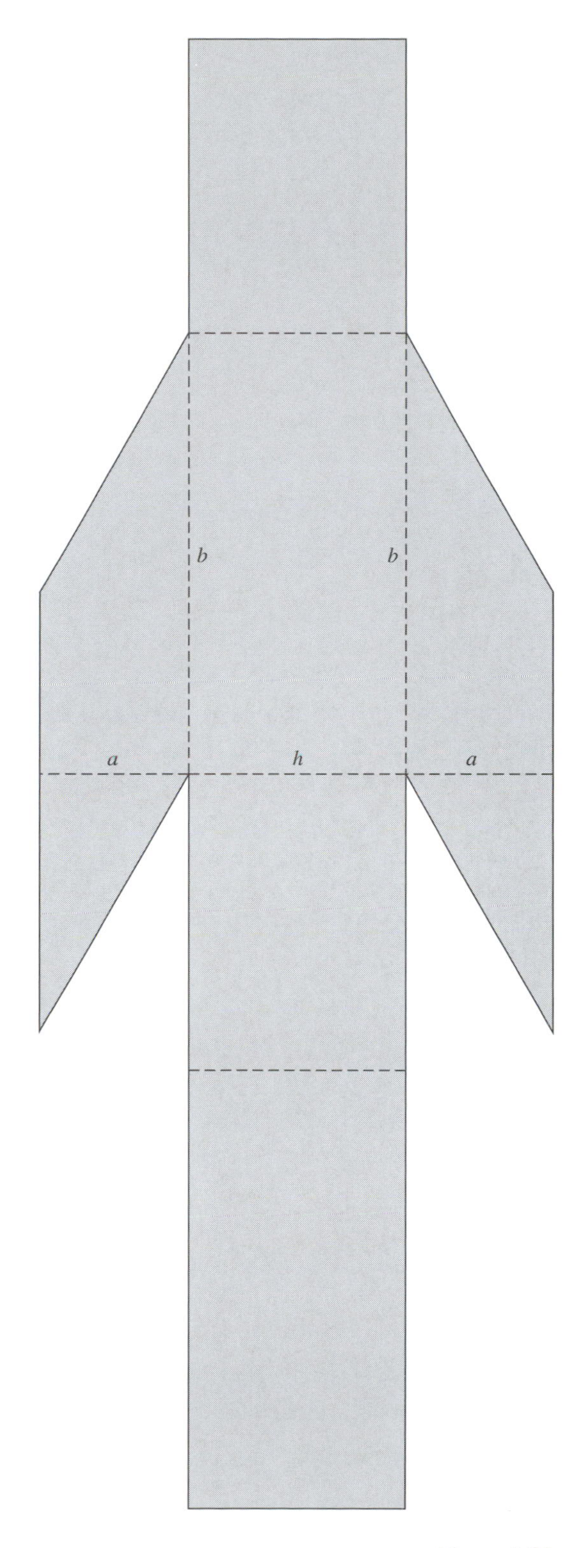

Figure A.29

A Pattern for a Prism with Parallelogram Base for Class Activity 11Z on Page 409

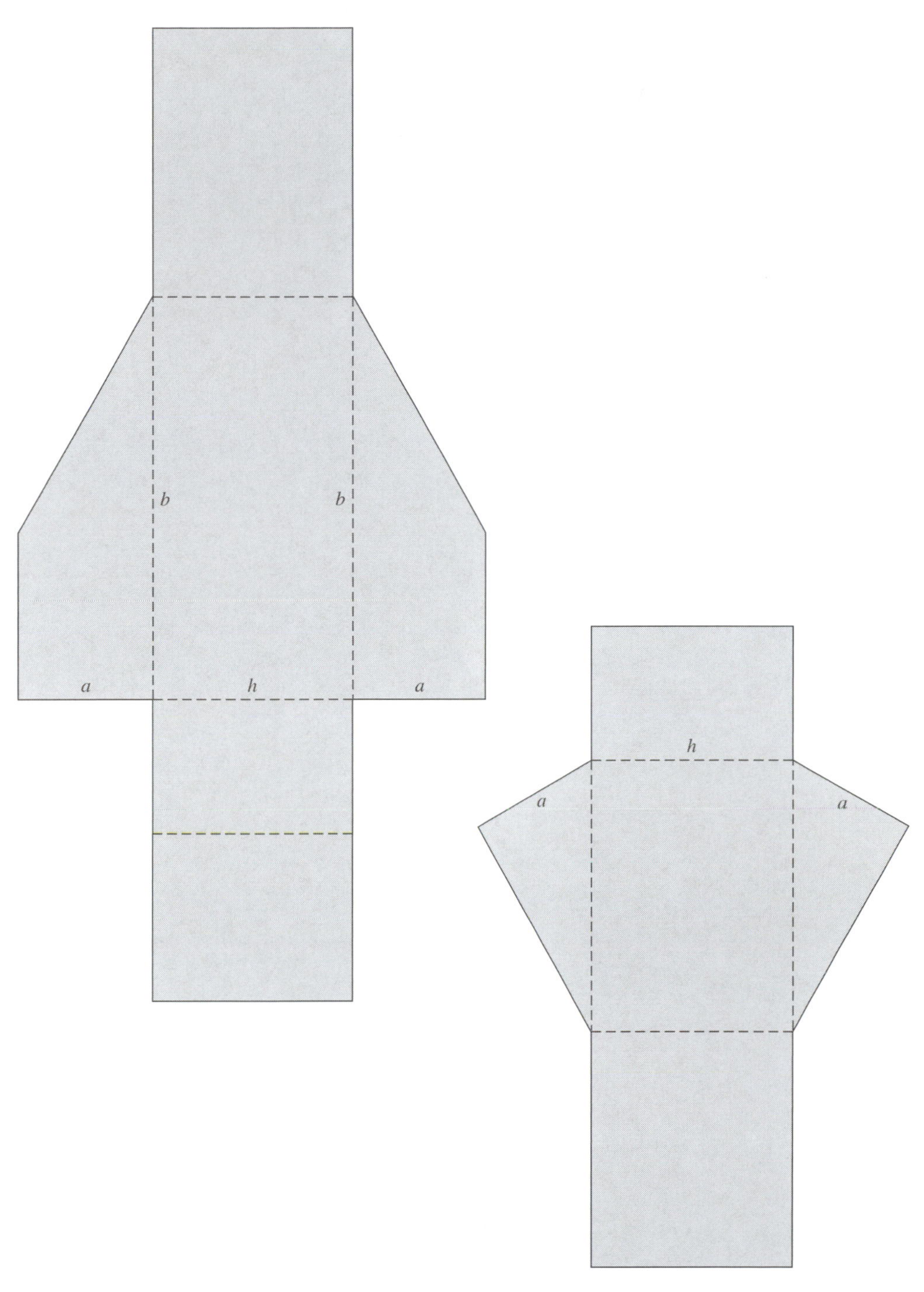

Figure A.30

Patterns for Prisms with Trapezoid and Triagular Bases for Class Activity 11Z on Page 409

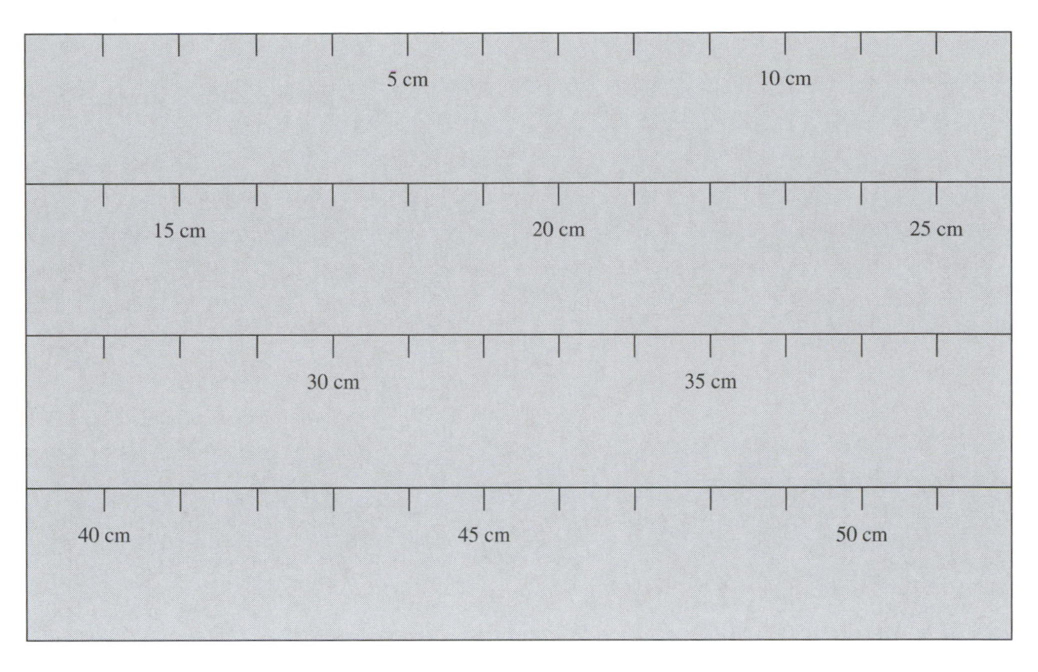

Figure A.31

A Centimeter Tape Measure for Class Activity 11DD on Page 414

Figure A.32

A Pattern for a Rectangular Prism for Class Activity 11BB on Page 412

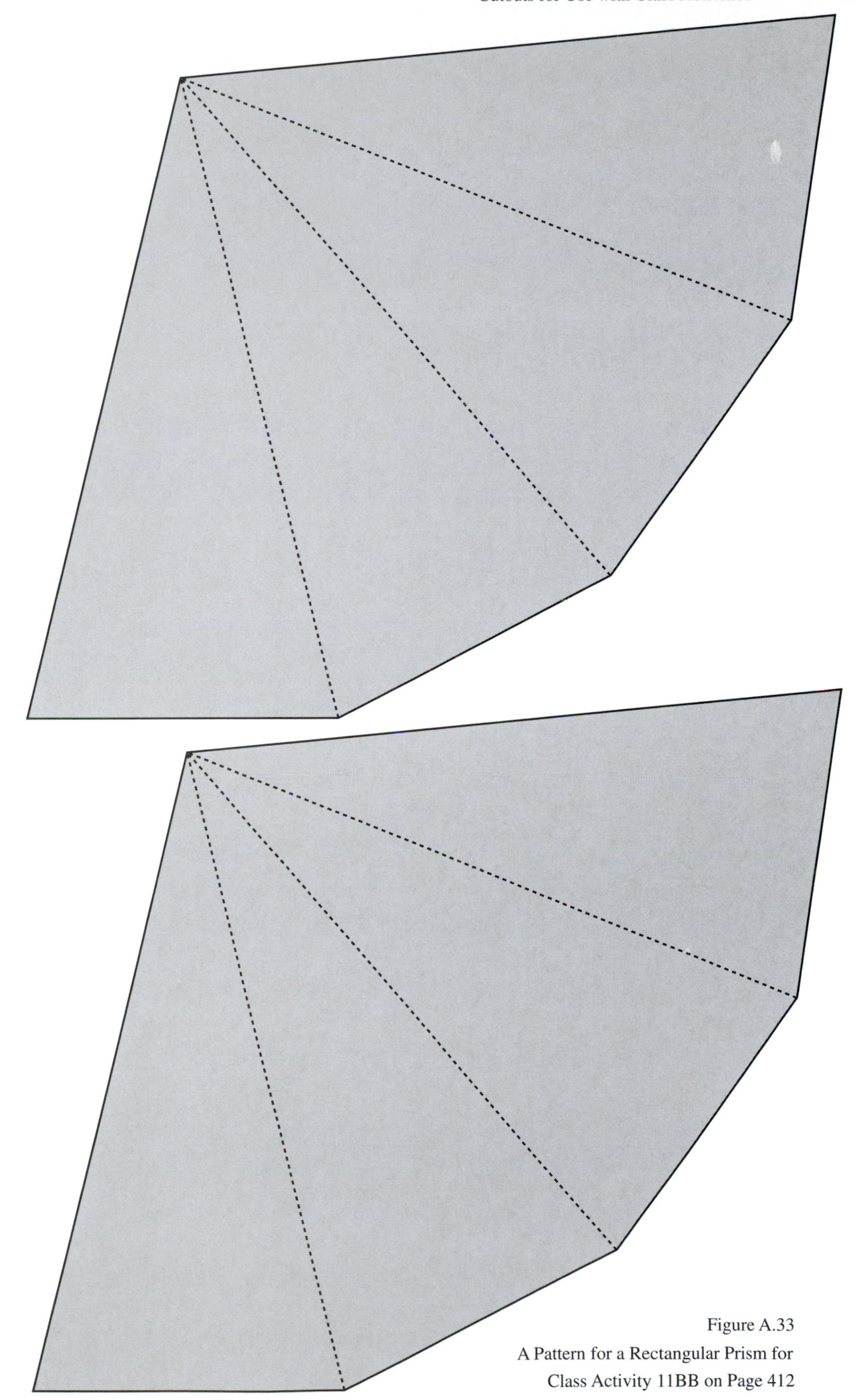

Figure A.33
A Pattern for a Rectangular Prism for
Class Activity 11BB on Page 412

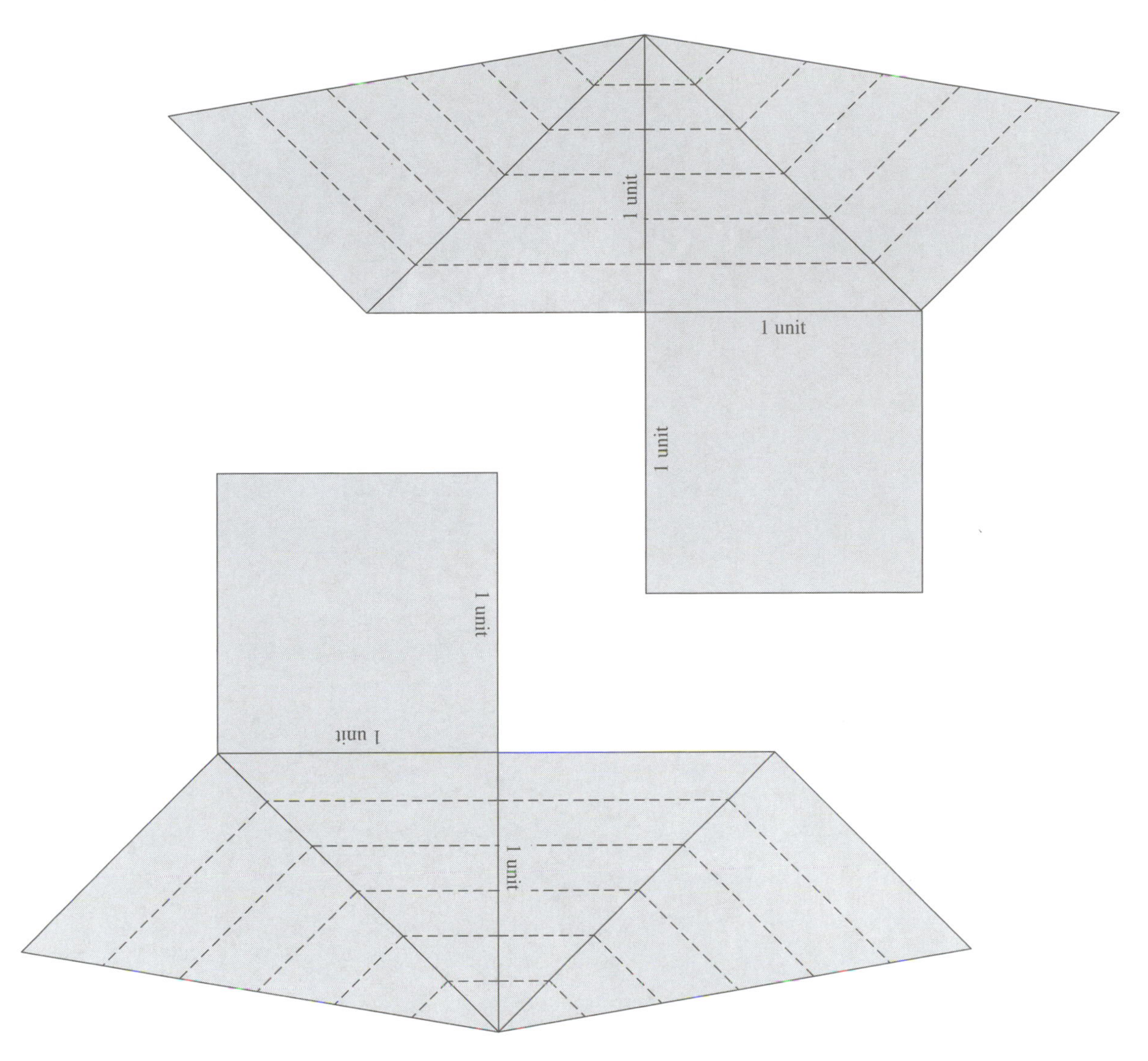

Figure A.34
Patterns for Oblique Pyramids for Class Activity 11CC on Page 413

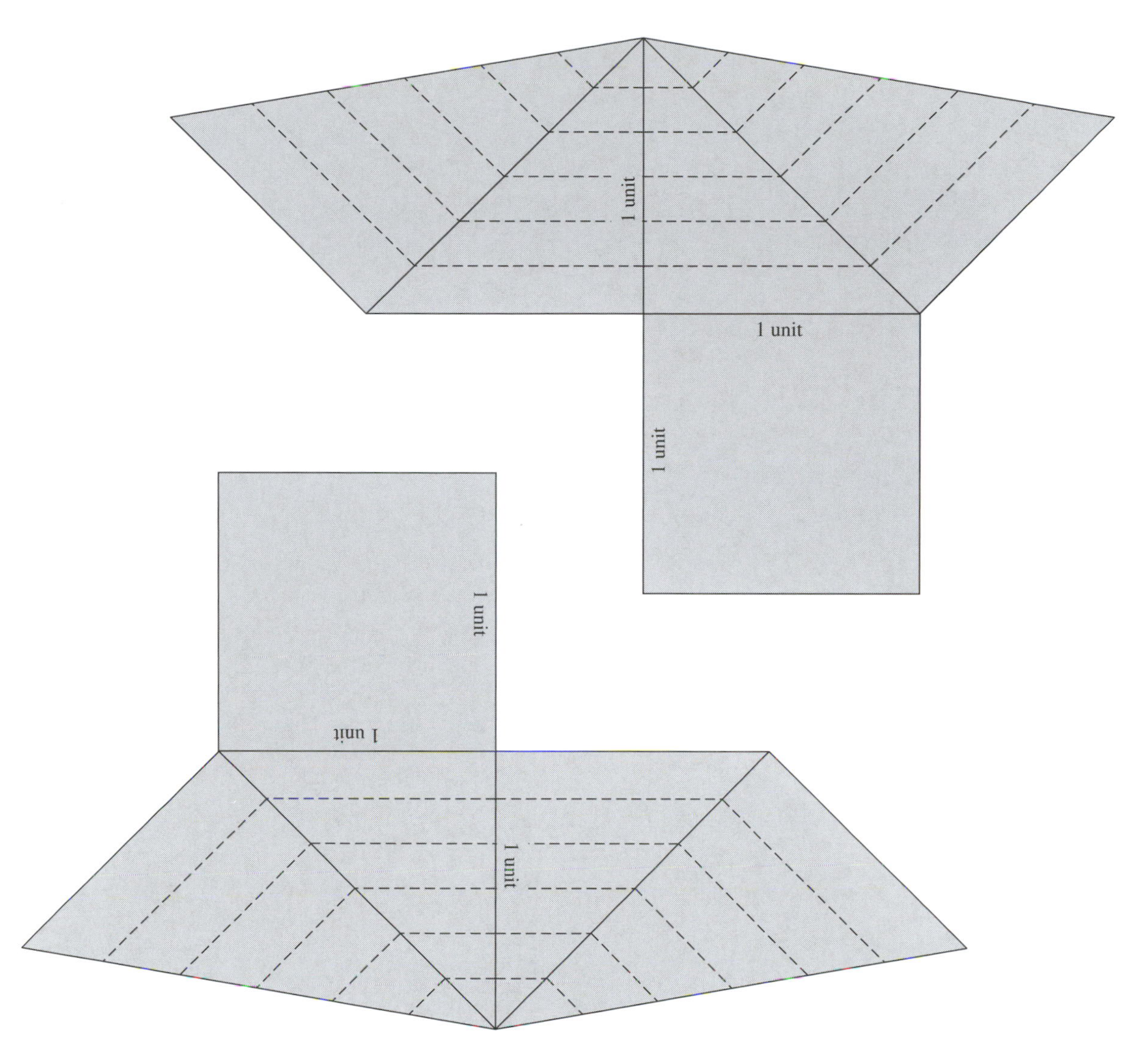

Figure A.35
Patterns for Oblique Pyramids for Class Activity 11CC on Page 413

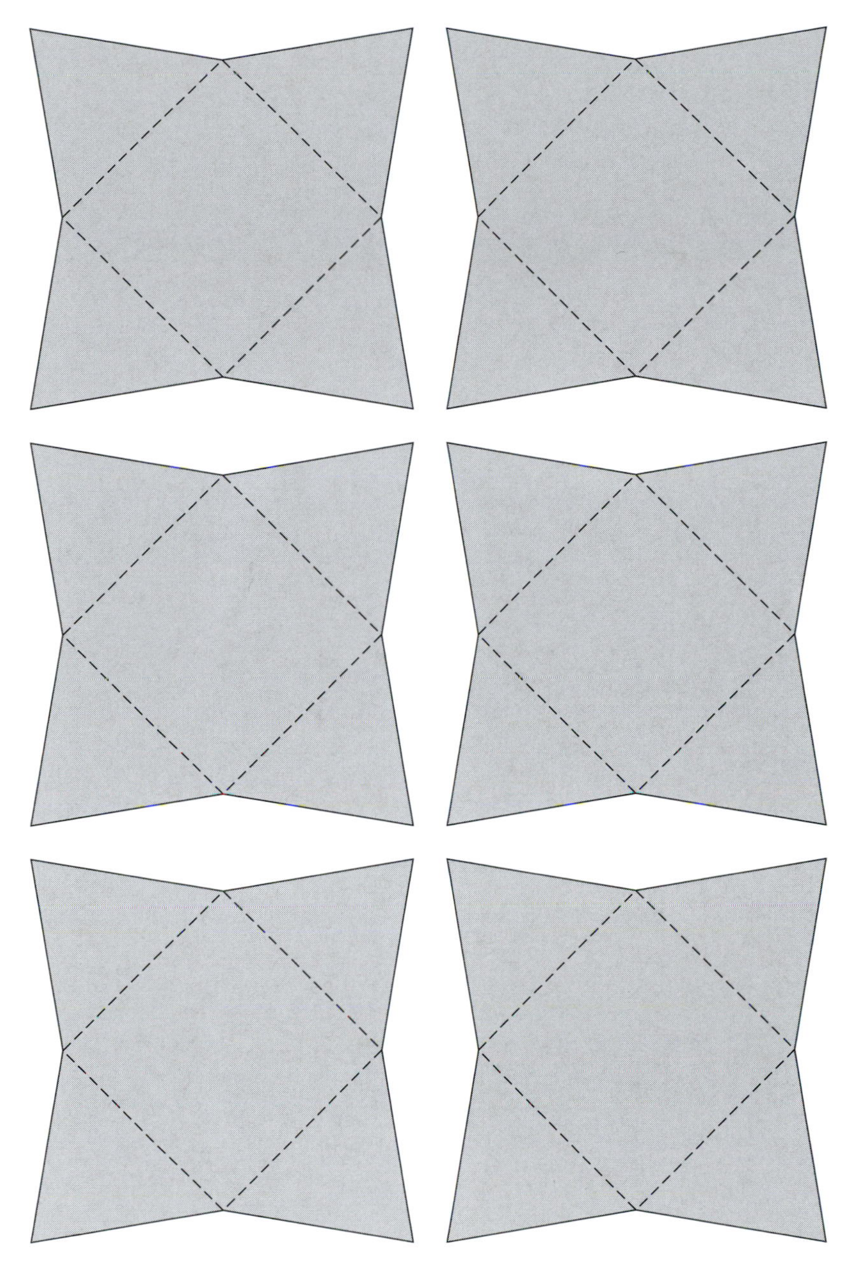

Figure A.36

Patterns for 6 Pyramids for Class Activity 11GG on Page 417

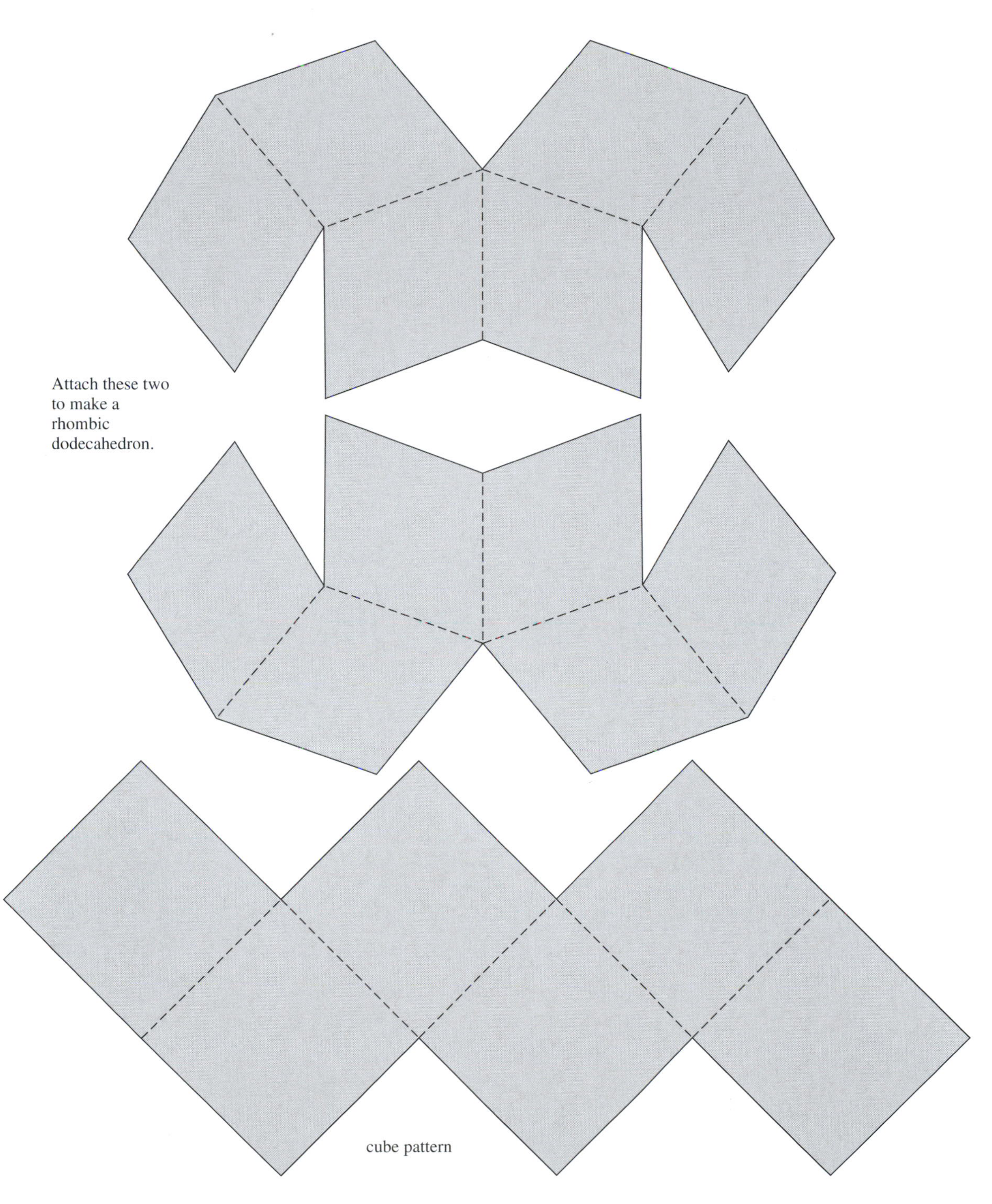
Attach these two
to make a
rhombic
dodecahedron.
cube pattern

Figure A.37
Patterns for a Rhombic Dodecahedron and a Cube for Class Activity 11GG on Page 417

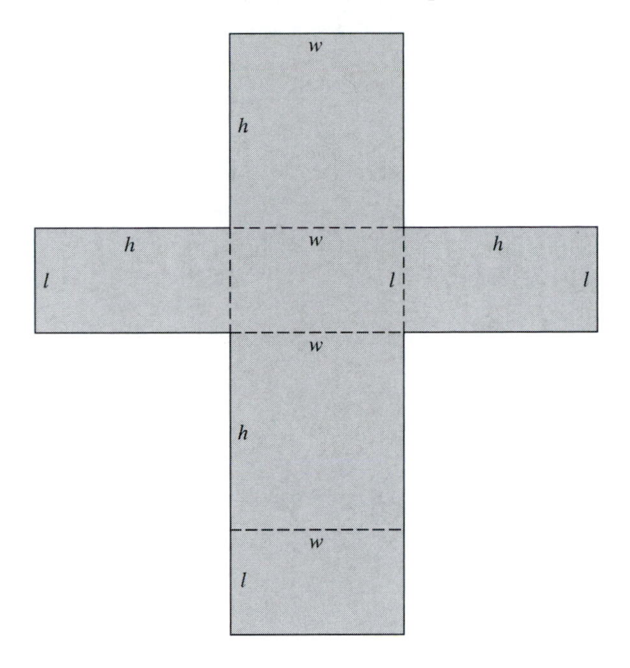

Figure A.38

A Pattern for a Small Box for Class Activity 11HH on Page 418

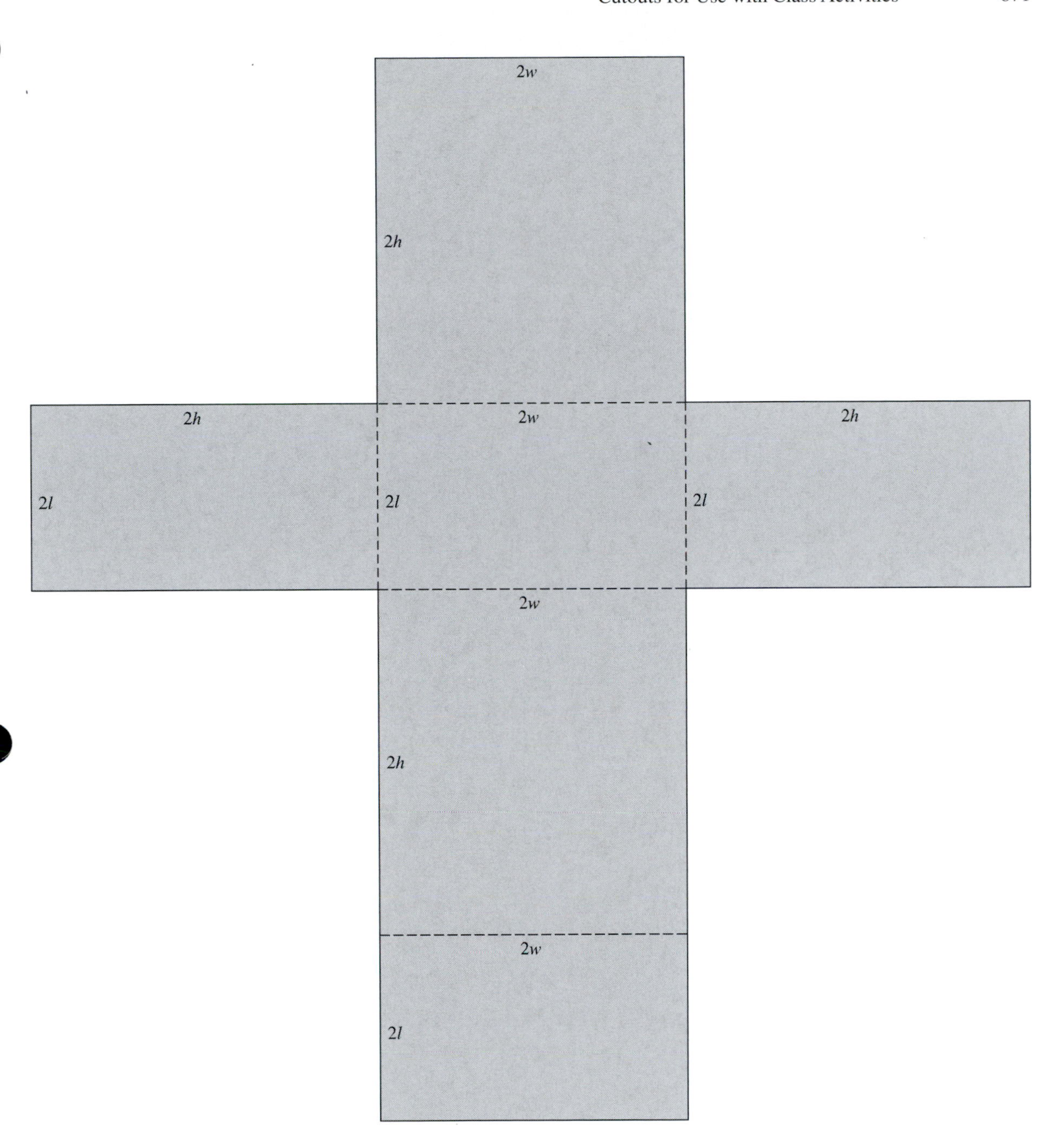

Figure A.39
A Pattern for a Big Box for Class Activity 11HH on Page 418

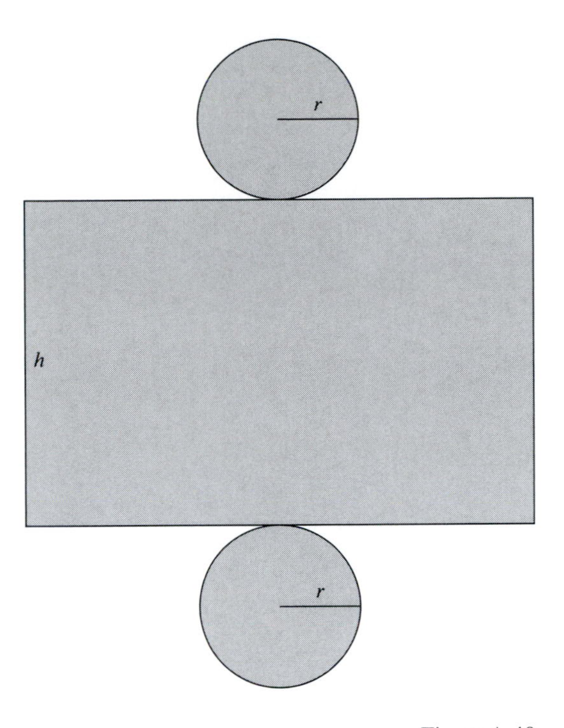

Figure A.40

A Pattern for a Small Cylinder for Class Activity 11II on Page 420

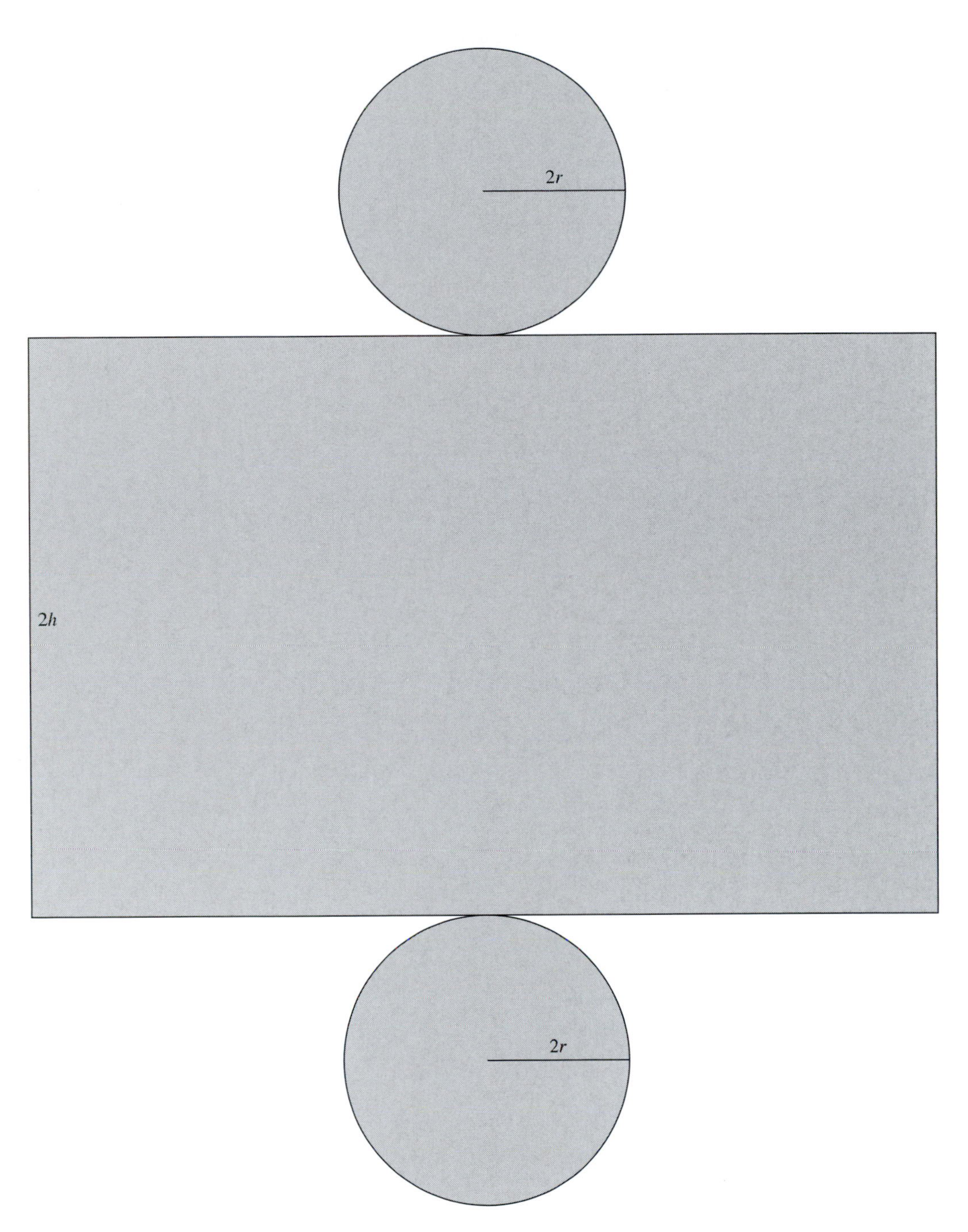

Figure A.41
A Pattern for a Large Cylinder for Class Activity 11II on Page 420

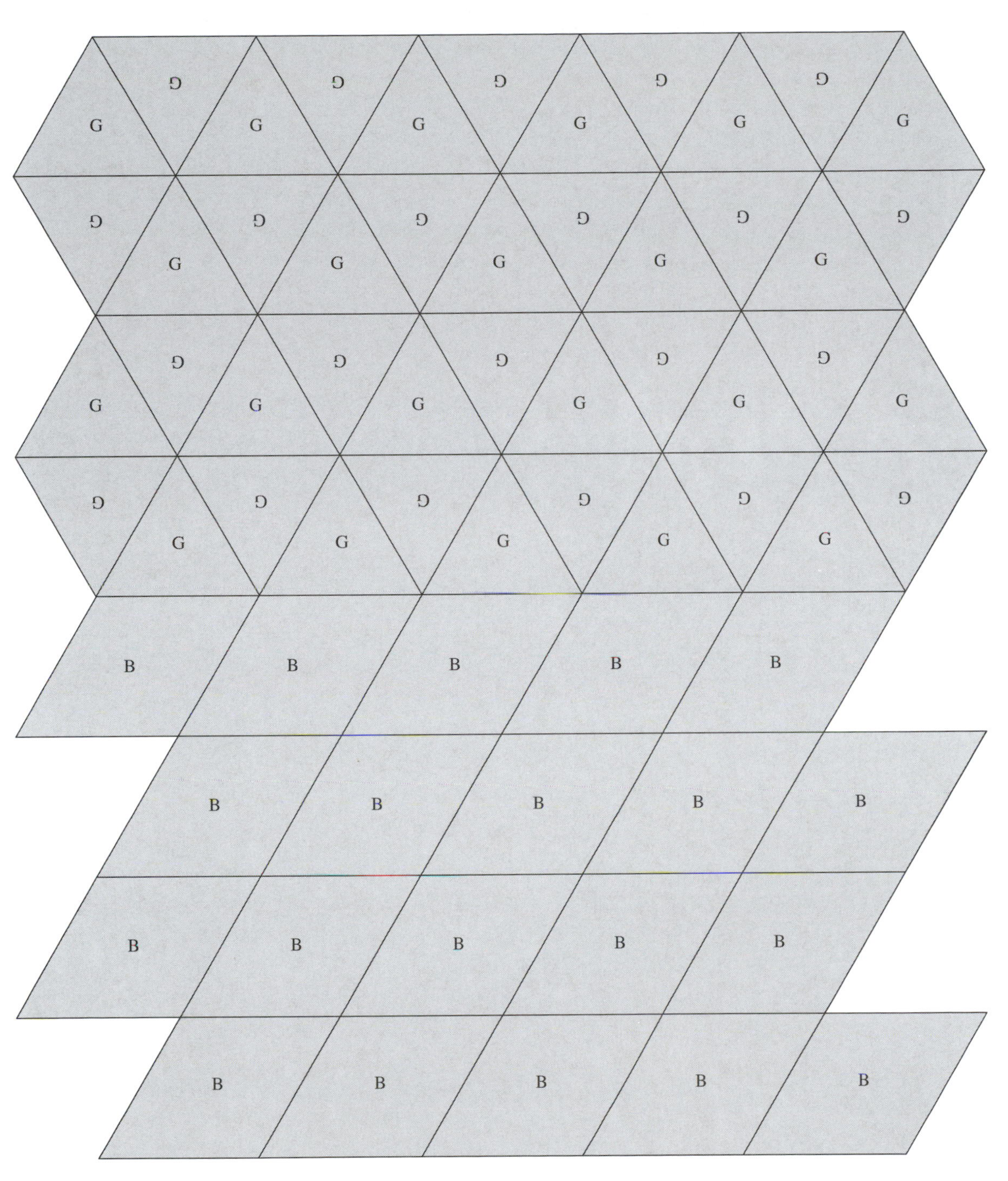

Figure A.42

Pattern Tiles for Class Activity 12U on Page 462

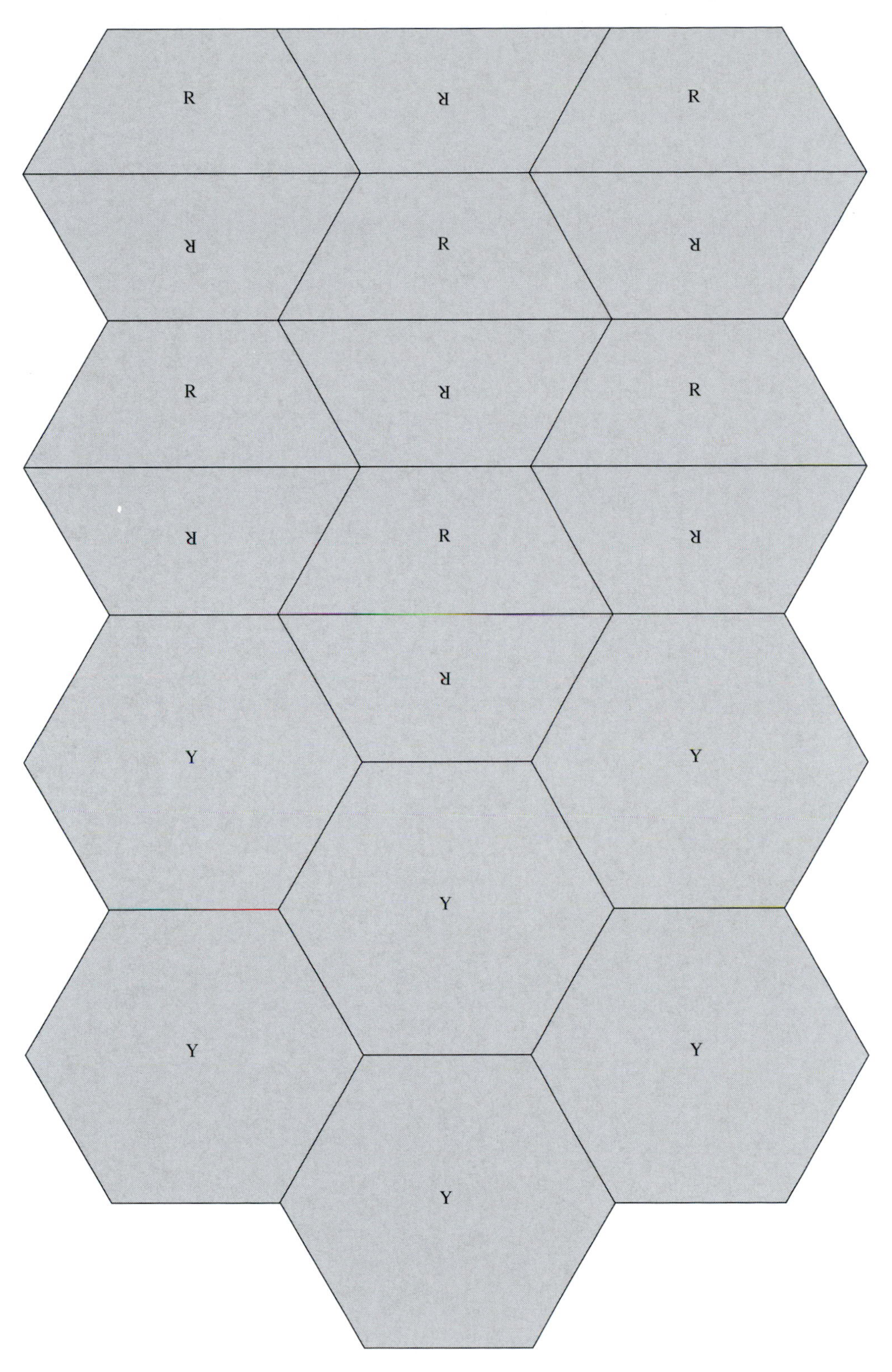

Figure A.43
Pattern Tiles for Class Activity 12U on Page 462

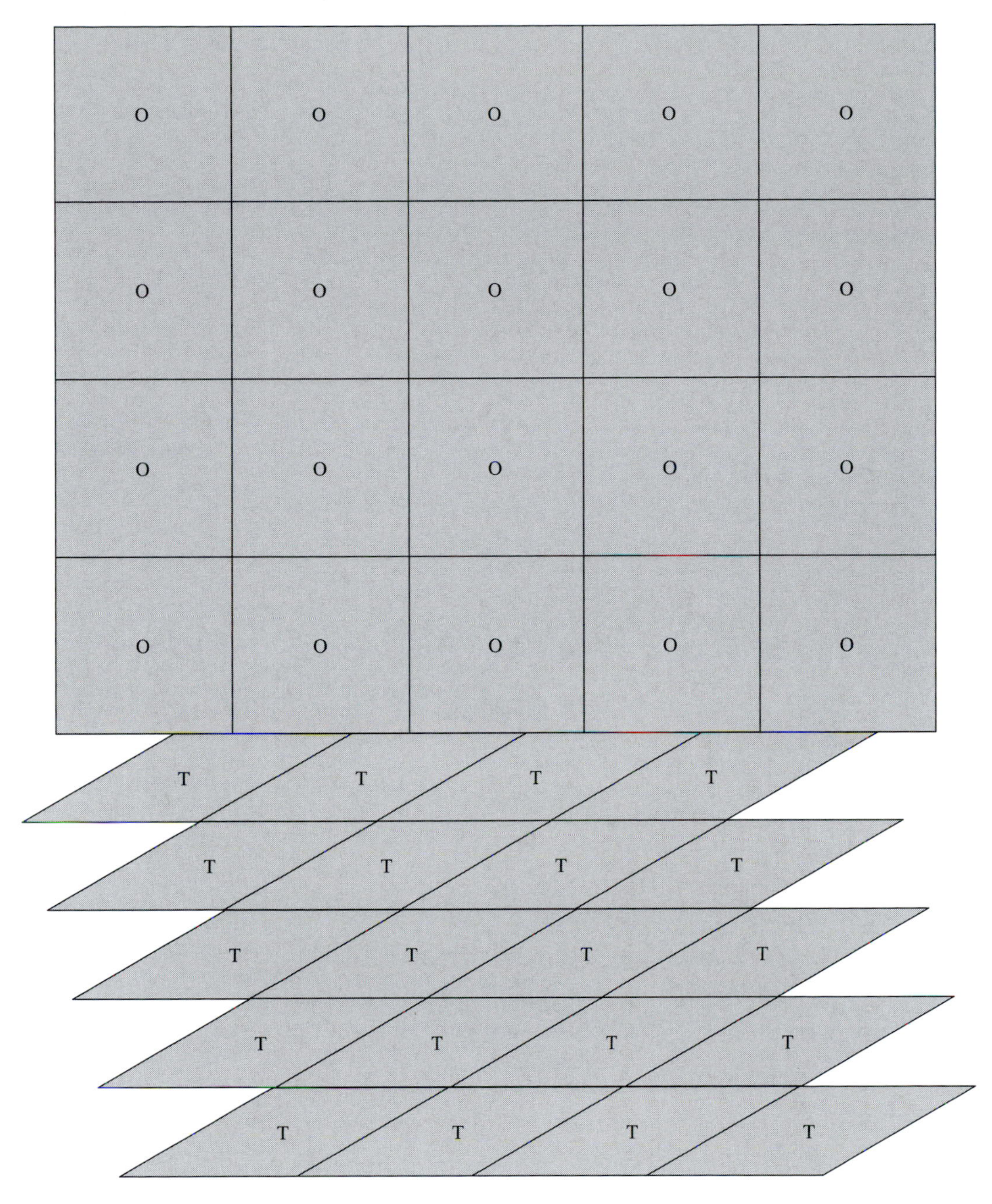

Figure A.44

Pattern Tiles for Class Activity 12U on Page 462